Algorithms for Games

G.M. Adelson-Velsky
V.L. Arlazarov
M.V. Donskoy

Algorithms for Games

With 38 Illustrations

Springer-Verlag
New York Berlin Heidelberg
London Paris Tokyo

G.M. Adelson-Velsky, V.L. Arlazarov,
and M.V. Donskoy
Institute for Systems Studies
9, Prospekt 60-Let Oktyabria
117312 Moscow B-312, USSR

Translator
Arthur Brown
P.O. Box 326
Garrett Park, MD 20766, USA

Library of Congress Cataloging-in-Publication Data
Adelson-Velsky, G.M. (Georgiĭ Maksimovich)
   Algorithms for games.
   Bibliography: p.
   Includes index.
   1. Game theory. 2. Algorithms. I. Arlazarov,
V.L. (Vladimir L'vovich) II. Donskoy, M.V.
III. Title.
QA269.A35 1988   519.3   87-23549

Translation of the original Russian edition
Programmirovanie Igr. © 1978, Nauka, USSR.

ISBN-13: 978-1-4612-8355-3   e-ISBN-13: 978-1-4612-3796-9
DOI: 10.1007/978-1-4612-3796-9

© 1988 by Springer-Verlag New York Inc.
Softcover reprint of the hardcover 1st edition 1988

Text prepared by the translator using the Spellbinder® word processing software
and the STI Math Program.
Typeset by Science Typographers, Inc., Medford, New York.

9 8 7 6 5 4 3 2 1

Contents

Preface

Technicians, economists, industrial managers, and many other specialists often need to find an element belonging to a finite set and having certain given properties (provided, of course, that such an element exists). In principle, this problem can be solved by searching the set, element by element. There is, however, a well-rooted belief that this possibility is purely theoretical because the work involved in the search is enormous. Moreover, any general method (the search method included) may turn out to be either good or bad, depending on the concrete circumstances and the method chosen for implementation.

Suppose, for instance, that we need to find a root of the equation $x^4 + x^3 - 1 = 0$ on the interval $(0,1)$ to within an accuracy of 0.1. Then it is better to evaluate the polynomial on the left side of the equation for the values $x = 0, 0.1, \ldots, 1.0$ than to use the exact Ferrari method, in which we first reduce the equation to the form $y^4 + py^2 + qy + r = 0$ by the substitution $x = y - 1/4$ and then transform this equation to the soluble form

$$\left|\left(y^2 + \frac{p}{2} + \alpha\right)^2 - 2\alpha\left(y - \frac{a}{4\alpha}\right)^2\right| = 0$$

where α is a root of the cubic equation

$$q^2 - 8\alpha\left(\alpha p + \alpha^2 + \frac{p^2}{4} - r\right) = 0,$$

after which we bring it into the form $\beta^3 + p'\beta + q' = 0$ by the substitution $\alpha = \beta - p/3$. Finally we use Cardan's formula

$$\beta = \sqrt[3]{\frac{-q'}{2} + \sqrt{\frac{q'^2}{4} + \frac{p'^3}{27}}}$$
$$+ \sqrt[3]{\frac{-q'}{2} + \sqrt{\frac{q'^2}{4} - \frac{p'^3}{27}}} \, .$$

In essence, a search amounts to the solution of problems arising from a given one when the value of an unknown parameter is fixed in one way or another, and a choice is made among a set of contemplated values that yields the most suitable solution. Often each of the contemplated problems, with the parameter fixed, is solved by a search. Then we speak of a multi-level or hierarchic search. If we adopt the most general definition of a search we can give only trivial recommendations for its implementation (though even these may be useful). Fortunately, the kinds of problems we are dealing with often have features in common aside from their solubility by searches. This permits us to establish, study, and apply general search methods.

This book aims to provide a concrete example of the programming of a two-person game with complete information, and to demonstrate some of the methods of solution; to show the reader that it is profitable not to fear a search, but rather to undertake it in a rational fashion, make a proper estimate of the dimensions of the 'catastrophe', and use all suitable means to keep it down to a reasonable size. The game programming problem would seem to be ideally suited to the study of the search problem, and in general for multi-step solution processes. The clarity and relative simplicity of the rules and the scoring of the results, the availability of suitable experimental methods for comparing various solution algorithms (including experiments on human approaches that arrive at answers produced by informal methods), all act to yield suitable methods for developing and trying out different approaches to the solution of problems by search methods.

A hierarchic search underlies all natural methods for finding a move in a game position. 'If I do this, I reach a position where various possibilities are open to me, but my opponent can answer thus and thus...'—this is a typical basis for the choice of a move. Clearly, it involves a search. As we shall show in the first chapter, however, a so-called *full-width* or *exhaustive* search need not require a complete enumeration of all possible positions that might arise by application of the rules of the game. We have in fact dedicated this book to the study of methods for limiting the extent of a search.

We have written about the programming of games (i.e. we have assumed that the reader would like to write a program that would play a game, serious or not) because we believe this is the best way to bring out the ideas and methods we are attempting to expound. We do not, however, pay attention to the technical problems of programming, essential though these may be. There is no specific technique for programming computers to play games. The fact that many technical paradigms were first conceived in the development of game programs does not contradict this assertion. These paradigms have later found wide development and application in the construction of many programs that are used in many different areas. Their prevalence attests to the fact that many clever programmers have been at work on them, and to the fact that they have stretched the capabilities of their machines to the limit.

We shall study games between two opponents named White and Black. Without loss of generality we may assume that these are zero-sum games. They permit at each stage a finite number of moves, for each of which the permissible replies of the opponent are known. Once the opponent's decision is made, a new position arises and is uniquely defined. Some positions are terminal: no decision is allowed to one of the opponents, the other wins and the one to move loses. If a non-terminal position arises, however, the opponent who has the move must in fact move.

We shall study a narrower class of games–namely games with complete information. In every such game we know which opponent is to move at each turn, and he must move. Thus both opponents know what the resulting position is. The positions in a game with complete information fall into three types: White to move, Black to move, and terminal positions. [We might define the notion of a game with several players and complete information, and carry over or suitably transform some of the results obtained in the book, but we shall not take the time to do so.]

Every game begins with a position which from now on we call the base position. In some games, for example card games, the base positions will vary, perhaps depending on some chance event. In other games, as for example chess, the base position is fixed once and for all. Even in chess, the player's concern is not with this fixed base position but rather with positions that arise either in the course of play or, as in problems and endgame studies, by artifice.

If the base position is prescribed, we may construct a game tree. Its nodes correspond to positions, and every arc leads from one position to another that can arise from it by a legal move. The base position corresponds to the root of the tree. A two-person game with complete information can be formally defined by prescribing the game tree, i.e. specifying the color of non-terminal nodes (designating the person who has the move in the given position) and the score corresponding to the just concluded move. These definitions are given in the first chapter and are widely applied throughout the book. To study the concrete properties of a game, however, we shall deal with positions and moves rather than with the game tree.

In Chapter 1 we develop the branch-and-bound method (the α, β-heuristic). We pay primary attention to the theoretical foundation of the method and to an estimate of the minimum size of the search required for its solution. In some places we relinquish an elegant inductive proof in favor of a more unwieldy one, in an effort to isolate the conditions needed for the existence of an objective score for the base position and for finding the best strategy for each of the opponents.

Chapter 2 is devoted to heuristic (i.e. inexact) methods for choosing a move in a contemplated position. We pay special attention to the problems of establishing such methods, and to a discussion of those properties of specific games that ensure good results when the methods are applied. It is worth noting that we employ a probabilistic approach in establishing our heuristic methods for programming games.

In Chapter 3 we develop the theory of analogical reasoning as a basis for decision-making. This theory is founded on the concept of a move that is independent of the position in which it is made; that is, the same move can be made in many different positions. The sequels to such a move may vary, but in many cases the move leads to roughly identical changes. Suppose that in the position B move Ψ is being studied. We want to know what conditions are sufficient to yield the same score for this move in another position C. We formulate these conditions and prove their sufficiency. They amount to this: the sequence of moves leading from B to C must have no influence on the variation that establishes the score for the move under study. In other words, the sequence must consist of moves that have no relationship to the decisive variation.

In Chapter 4 we take up the probabilistic approach to game programming. This approach has four aspects: a) the methods for formulating the elementary stochastic hypotheses and calculating the probability of correctly scoring a given position and finding the best moves; b) the methods for statistical testing of our hypotheses; c) the construction of more effective methods for computing the score and finding the best moves in a given position, on the basis of an analysis of a stochastic model of the game; and d) the probabilistic approach to the programming of games with complete information.

Since the probabilistic approach to game programming has only recently been applied, many of the results obtained in Chapter 4 must be regarded as preliminary, and some of them have not even been established—e.g. the statistical testing of the stochastic hypotheses and the basis of the probabilistic approach. Nevertheless, the results that have been obtained do in our opinion support the prospects for this new direction in game programming and we have therefore included the chapter.

At the end of the book we present an appendix containing a brief sketch of the work that has been done on algorithms for games and a bibliography which includes some references not cited in the text.

Chapters 3 and 4, which contain new results on two-person zero-sum games with complete information, are mutually independent. Formally they are also independent of Chapter 2, but we nevertheless recommend that the reader interested in probabilistic models of games should read Sections 1 and 2 of Chapter 2.

Since the problems we discuss have an immediate connection with programming, we have felt it worthwhile to introduce some notation taken from programming languages, in particular ALGOL. We use the assignment symbol ': =' and we denote the end of the definition of an operator by the semicolon ';'. Whenever this convention is in conflict with the ordinary rules of syntax we give precedence to the formal language.

We have devoted many years, and still devote our time, to the development of programs for playing chess. Quite naturally, this devotion has influenced our exposition; in particular, the examples we present are often related to chess. Nevertheless, with a few exceptions we have avoided the discussion of problems that arise uniquely in chess. Moreover, we are firmly convinced that one can write strong chess-playing programs without being a good chess player. Therefore, even when the examples presented deal purely with chess, the reader needs only an elementary knowledge of the game.

Translator's Note

The reader should be aware that the Russian text of this book was published in 1978, and that in the years since then, progress in the development of computer chess programs has been rapid. The book is still of interest since it deals with the ideas basic to the problems of search rather than with specific programs and programming techniques. Readers interested in recent developments should consult the publications of the International Computer Chess Association, (*Journal of the ICCA*, edited by H.J. van den Herik, Department of Informatics, University of Technology, Julianalaan 132, 2628 BL, Delft, The Netherlands). Other publications of interest include a recent paper by T. A. Marsland which cites the more important work in this field (*Computer Chess Methods*, *Encyclopedia of Artificial Intelligence*, Wiley, 1987, pp. 159–171), and Hartmut Tanke's very complete *Computer Chess Bibliography* (recently reissued in German). This bibliography can be obtained by writing to: Hartmut Tanke, Kienitzer Str. 104–106, D–1000 Berlin 44, West Germany.

CHAPTER 1

Two-Person Games With Complete Information and the Search of Positions

The Game Tree, Position Score, and Best Move

We recall the recursive definition of a tree. The *elements of a tree* are *nodes* and *arcs*; one of the incident nodes of an arc is called the *beginning* and the other the *end*.

A *tree* is either: a) a single node, or b) a tree with an additional node and an arc beginning at an already existing node and ending at the new node. If the old node is denoted by A and the new one by B, the arc leading from A to B is denoted by (A, B) or $\vec{B}$ *). It is easy to see that every tree is connected. (See Figure 1.)

The *base of a tree* is a node to which no arc leads. A *terminal* node is one from which no arc issues. It follows easily from the definition that every tree has a unique base.

A *subtree of a tree* $\mathfrak{A}$ is a subset of its nodes and arcs that is itself a tree.

Let $\mathfrak{B}$ and $\mathfrak{C}$ be subtrees of a tree $\mathfrak{A}$ with a non-empty intersection $\mathfrak{D}$. Then $\mathfrak{D}$ is a subtree of $\mathfrak{A}$.

A node B is said to be *subordinate* to A if

a) $B \equiv A$ or b) B is the end of an arc issuing from a node subordinate to A. An arc issuing from a node subordinate to A is said to be subordinate to A.

The set of arcs and nodes subordinate to a node A forms a tree with the base A. This tree is called the *A-subtree* of the original tree.

---

* In graph theory, one may assume that the arcs are undirected, i.e. the beginning and end are not distinguished. The object we have just defined is known as a *directed graph*. (The notions of tree and directed graph may also be defined in different ways.) However, since nowhere in the text do we discuss undirected graphs, the term *directed graph* is everywhere replaced by *tree*.

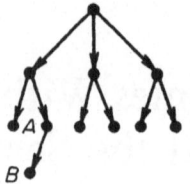

Figure 1

If A is not the base of the original tree, the A-subtree together with the arc $\vec{A}$ is called the *open A-subtree*.

If in a tree $\mathfrak{A}$ we omit the arc leading to a node A and also the A-subtree the remainder $\mathfrak{B}$ is a tree.

The base of a tree will be called a node of rank 0. If the node A is of rank n and there exists an arc (A, B) the node B has rank $n + 1$. The *rank* of each node in a tree is defined in this way.

Let A be a node of a tree and let B be a node of the A-subtree. The *branch* $W(A, B)$ is a sequence of nodes $A_0 = A, A_1, A_2, \ldots, A_k = B$ such that the arcs (A_s, A_{s+1}) exist for all $s = 0, 1, \ldots, k - 1$. If the branch $W(A, B)$ exists it is unique.

Suppose that a color—White or Black—is assigned to every node in a tree, and that real numbers $\mathrm{sc}(F)$, called *scores* are assigned to some of the terminal nodes F. Then we call the tree a *game tree for two opponents*, or a *two-person game tree*.

The nodes of such a tree are called *positions* (the terminal nodes are called *terminal* positions), and the arcs are called *moves*. The moves leading from a position colored White (Black) are called White moves (Black moves). The corresponding positions are called *positions with White (Black) to move*, or *White (Black) positions*. (A game tree may be infinite, and many of our subsequent assertions hold for infinite trees.)

Let $\mathfrak{A}$ be a game tree with White to move in the base position A_0. This means that White must choose a move from among all the possible moves originating at A_0. After this move a new position A_1 arises, at the end of the arc (A_0, A_1) of the game tree. If Black has the move at A_1 he must also choose one move from among those issuing from A_1, and so on. In this way, when White and Black have exercised all their successive choices we arrive at a sequence $A_0, A_1, \ldots, A_n$ in which the last term is a terminal position with the score $\mathrm{sc}(A_n)$. We shall call this sequence a *game* and the score of the terminal position the *outcome of the game*. White chooses his moves with the aim of maximizing the outcome, and Black chooses his with the aim of minimizing it. The *score of the base position* A_0 is the outcome that both White and Black succeed in obtaining. It is not immediately clear, however, that both White and Black should count on obtaining the same outcome.

Let $\mathfrak{A}$ be a game tree. We shall say that $\mathfrak{B}(A)$ is a *W-pruned* tree (relative to $\mathfrak{A}$) if

(1) $\mathfrak{B}(A)$ is obtained from the A-subtree by excluding some number of open B-subtrees, where $\vec{B}$ is a White move.

(2) All terminal positions in $\mathfrak{B}(A)$ are terminal positions in $\mathfrak{A}$, i.e. no new positions arise as a result of the exclusion. A B-pruned subtree is similarly defined for Black.

We use the symbolic inequality

$$\mathrm{sc}(A) > M$$

when there exists a finite W-pruned tree with the base A for which all terminal positions have a score and all scores exceed M. (The inequality symbol may for the moment be regarded as an hieroglyph; we shall show later that in the contemplated game White may count on obtaining a numerical outcome $> M$.)

Similarly, if M exceeds all the terminal scores of some B-pruned tree with the base A we write

$$\mathrm{sc}(A) < M.$$

For the case of weak inequality the change in the definitions is obvious.

Let $\mathfrak{A}$ be a game tree and $\mathfrak{B}$ a W-pruned subtree of it. Then for every Black position $A \in \mathfrak{B}$, $\mathfrak{B}$ contains all the moves leading from A in $\mathfrak{A}$, provided that A is not a terminal Black position. A similar assertion holds with the colors interchanged.

Lemma. *If $\mathfrak{B}_1(A)$ is a W-pruned tree of the game tree $\mathfrak{A}$ and $\mathfrak{B}_2(A)$ is a B-pruned tree, then $\mathfrak{B} = \mathfrak{B}_1(A) \cap \mathfrak{B}_2(A)$ is a tree and all its terminal positions are terminal positions of $\mathfrak{A}$.*

PROOF. $\mathfrak{B}$ is not empty since it contains the base A; it is therefore a tree. Let $B \in \mathfrak{B}$ be a non-terminal node of $\mathfrak{A}$ with White (Black) to move. Then all the moves from B are contained in $\mathfrak{B}_2(A)$ ($\mathfrak{B}_1(A)$), and at least one move is contained in $\mathfrak{B}_1(A)$ ($\mathfrak{B}_2(A)$), i.e. B is not a terminal position in $\mathfrak{B}$. $\square$

Theorem on Consistency. *If $\mathfrak{A}$ is a game tree and $A \in \mathfrak{A}$ the symbolic inequalities*

$$\mathrm{sc}(A) > M,$$

$$\mathrm{sc}(A) \leq M$$

cannot both hold.

PROOF. The symbolic inequality $\mathrm{sc}(A) > M$ implies that there exists a W-pruned tree $\mathfrak{B}_1$ with base position A and scores for all its terminal positions $> M$; the inequality $\mathrm{sc}(A) \leq M$ implies the existence of a B-pruned tree with the same base A and all scores of its terminal positions $\leq M$. The intersection of these two trees contains at least one terminal position F with a score that both exceeds and does not exceed M. $\square$

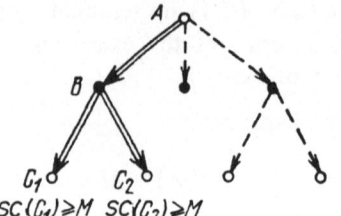

SC(C_1)≥M SC(C_2)≥M Figure 2

Definition. If we have both $\mathrm{sc}(A) \geq M$ and $\mathrm{sc}(A) \leq M$ for some position A in a game tree $\mathfrak{A}$, we say that M is the *score of the position* A.

From here on, all our arguments concerning positions, moves, and pruned trees will hold when colors and directions of inequalities are simultaneously interchanged. As a rule, we will make no specific mention of this fact.

Lemma. *Let A be a position in the game tree $\mathfrak{A}$ with White to move and let $\vec{B}$ be a move issuing from it such that $\mathrm{sc}(B) \geq M$ ($> M$). Then $\mathrm{sc}(A) \geq M$ ($> M$).*

PROOF. Since $\mathrm{sc}(B) \geq M$ there exists a W-pruned tree $\mathfrak{B}$ with base B, for which all terminal positions have scores $\geq M$. Then $\mathfrak{B} + A$ (see Figure 2) is also a W-pruned subtree with the base A and the same terminal positions as $\mathfrak{B}$.   □

Lemma. *Let A be a position in the game tree $\mathfrak{A}$ with White to move and let $\vec{A}_1, \vec{A}_2, \ldots, \vec{A}_n$ be the complete set of moves issuing from A. If the inequality $\mathrm{sc}(A_s) \leq M$ ($< M$) holds for all the positions A_s ($s = 1, 2, \ldots, n$), then $\mathrm{sc}(A) \leq M$ ($< M$).*

PROOF. Each of the inequalities $\mathrm{sc}(A_s) \leq M$ ($< M$) follows from the existence of the B-pruned tree $\mathfrak{B}_s$ with base A_s and all scores of terminal positions $\leq M$ ($< M$). Then $A + \mathfrak{B}_1 + \mathfrak{B}_2 + \cdots + \mathfrak{B}_n$ (see Figure 3) is a W-pruned tree with base A, with all its terminal positions $\leq M$ ($< M$).   □

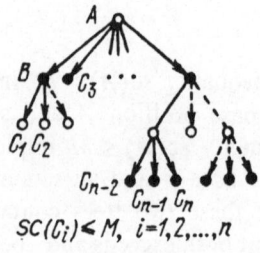

SC(C_i)<M, $i=1,2,\ldots,n$ Figure 3

Theorem on the Transfer of Scores. *Let A be a node of the game tree $\mathfrak{A}$ with White (Black) to move, and let $\vec{A_1}, \vec{A_2}, \ldots, \vec{A_k}$ be all the moves issuing from it. If each node A_s ($s = 1, 2, \ldots, k$) has the score $\mathrm{sc}(A_s)$, the node A has the score $\mathrm{sc}(A) = \max\{\mathrm{sc}(A_s) | s = 1, 2, \ldots, k\}$. If Black has the move at A then $\mathrm{sc}(A) = \min\{\mathrm{sc}(A_s) | s = 1, 2, \ldots, k\}$.*

Theorem on the Existence of Scores. *If $\mathfrak{A}$ is a finite game tree and all its terminal positions have a score, every position A in $\mathfrak{A}$ has a score, which is completely determined by the A-subtree.*

PROOF. All the positions of maximum rank in the A-subtree of the game tree $\mathfrak{A}$ have scores, since they are terminal positions. If all positions A with rank exceeding n have scores, the theorem on the transfer of scores implies that all positions of rank n also have scores.   □

Thus the scores of the positions A in the game tree $\mathfrak{A}$ satisfy the conditions

$$\mathrm{sc}(A) = \max\{\mathrm{sc}(B) | B \in \mathfrak{W}(A)\} \text{ if } A \text{ is a White position,}$$
$$\mathrm{sc}(A) = \min\{\mathrm{sc}(B) | B \in \mathfrak{W}(A)\} \text{ if } A \text{ is a Black position,}$$

where $\mathfrak{W}(A)$ is the set of positions B that are immediate successors of A, i.e. those for which there exists an arc (A, B) issuing from A. We shall call these the Zermelo formulae, since they were established by him in [35] as a definition of the score of a non-terminal position.

John von Neumann [26] gave a different definition of the score. He based it on the notion of a strategy, i.e. an initially formulated choice of the move for each position in a game tree at which the move belongs to one's own color. If White chooses a strategy s from his set of strategies S_w, and Black chooses $s' \in S_b$, the game to be played is uniquely defined. This game normally leads to a terminal position F having the score $\mathrm{sc}(F)$. Then its *result* $r(s, s')$ is defined as $\mathrm{sc}(F)$.

von Neumann proposed that one should choose a strategy such that it yields the best result obtainable against the most damaging strategy that can be chosen by one's opponent. He assigned this result as the score of the base position. In this way he defined two scores for the base position A_0 in the game tree $\mathfrak{A}$:

$$\mathrm{sc}_w(A_0) = \max\{\min(r(s, s') | s' \in S_b) | s \in S_w\},$$
$$\mathrm{sc}_b(A_0) = \min\{\max(r(s, s') | s \in S_w) | s' \in S_b\}.$$

The score for an arbitrary position B is similarly defined.

If the game tree is finite and every terminal position has a score, then for an arbitrary position B we have

$$\mathrm{sc}_w(B) = \mathrm{sc}_b(B) = \mathrm{sc}(B)$$

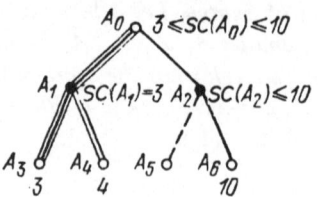

Figure 4

where $sc(B)$ is the score of the position B according to our earlier definition. Thus if one of the players has chosen a strategy s that is optimal in the von Neumann sense, and his opponent has noted his choice, the opponent will know what answer will be made to each of his moves, yet will be unable to improve the result of the game for himself.

When scores are lacking at some of the terminal positions of the game tree $\mathfrak{A}$ with base A_0, there may nevertheless exist W-pruned and B-pruned trees with base A_0 and terminal positions that do have scores (see for example Figure 4). For these trees we will have the symbolic equations

$$sc(A_0) \geq m,$$
$$sc(A_0) \leq M,$$

and the consistency theorem implies that $m \leq M$. It is easy to find a game tree for which the score of the base position exists in our sense, but not in the von Neumann sense (see Figure 5). Since there exist strategies $s' \in S_w$ and $s'' \in S_b$ such that they prescribe the game (A_0, A_2, A_4), which has no result, both $\min\{r(s', s'')|s \in S_b\}$ and $\max\{r(s', s'')|s'' \in S_w\}$ are undefined, and so therefore are $sc_w(A_0)$ and $sc_b(A_0)$ undefined. At the same time there exist W-pruned and B-pruned trees with the base A_0 that define the value of $sc(A_0)$ as 10.

Suppose given the game tree $\mathfrak{A}$, a subset $\mathfrak{F}$ of its terminal positions, and the scores of the positions $C \in \mathfrak{F}$. In some way or other we fix the values of the scores at terminal positions not contained in $\mathfrak{F}$. The resulting game tree $\mathfrak{A}'$ is called an *extension* of the game tree $\mathfrak{A}$ that we began with. Scores exist for the base position, and in general for all positions, of an arbitrary extension $\mathfrak{A}'$. It is therefore meaningful to speak of upper and lower bounds for the score $sc(A_0)$ at the base position A_0 of the game $\mathfrak{A}$, taken with respect to all extensions of $\mathfrak{A}$.

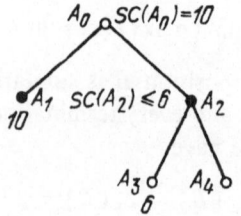

Figure 5

When we seek the score of the position A_0 of some game $\tilde{\mathfrak{A}}$ we may assume that the converse situation obtains. In order to avoid looking at the entire tree $\tilde{\mathfrak{A}}$, we try to find a tree $\mathfrak{A}$ for which our game is an extension, and for which

$$\sup\left\{sc_{\mathfrak{A}'}(A_0)|\mathfrak{A}'\supset\mathfrak{A}\right\} = \inf\left\{sc_{\mathfrak{A}'}(A_0)|\mathfrak{A}'\supset\mathfrak{A}\right\} = sc_{\tilde{\mathfrak{A}}}(A_0).$$

Theorem on the Extensions of a Game. *The inequality*

$$\inf\left\{sc(A_0)|\mathfrak{A}'\supset\mathfrak{A}\right\} > M\ (\geq M)$$

holds if and only if the symbolic inequality $sc(A_0) > M\ (\geq M)$ *holds for the contemplated game* $\mathfrak{A}$, *i.e. there exists a W-pruned subtree* $\mathfrak{A}_w\subset\mathfrak{A}$ *for which all the terminal positions have scores* $> M\ (\geq M)$.

PROOF. A W-pruned subtree $\mathfrak{A}_w$ of the game $\mathfrak{A}$ is a W-pruned subtree of any extension $\mathfrak{A}'$ of $\mathfrak{A}$. Therefore if the scores of all terminal positions of $\mathfrak{A}_w > M\ (\geq M)$ we have for an arbitrary extension $\mathfrak{A}'\supset\mathfrak{A}$

$$sc_{\mathfrak{A}'}(A_0) > M\ (\leq M)$$

and so

$$\inf\left\{sc(A_0)|\mathfrak{A}'\supset\mathfrak{A}\right\} > M\ (\geq M).$$

On the other hand,

$$\inf\left\{sc_{\mathfrak{A}'}(A_0)|\ \mathfrak{A}'\supset\mathfrak{A}\right\} = sc_{\mathfrak{A}^-}(A_0),$$

where the terminal positions with undefined scores in $\mathfrak{A}$ have the score $-\infty$ in $\mathfrak{A}^-$ and then the inequality holds because $sc(A_0)$ is a monotone function of the scores of the terminal positions in the game $\mathfrak{A}'$.

If $\inf\left\{sc(A_0|\mathfrak{A}'\supset\mathfrak{A})\right\} = sc_{\mathfrak{A}^-}(A_0) > M\ (\geq M)$, there exists a W-pruned subtree $\mathfrak{A}_w^-$ of the tree $\mathfrak{A}^-$ for which the scores of terminal positions $> M\ (\geq M)$. In this subtree all the terminal positions belong to the set $\mathfrak{F}$ (the scores of the remaining positions are set equal to $-\infty$) i.e. $\mathfrak{A}_w^-$ is a W-pruned subtree of the game $\mathfrak{A}$. □

The corresponding theorem on the upper bounds of scores of the base position A_0, taken with respect to all extensions of $\mathfrak{A}$, is formulated and proved in a similar fashion. These two theorems imply that the symbolic inequalities $sc(A) > M\ (\geq M,\ < M,\ \leq M)$ may be understood as actual inequalities satisfied by the scores of a position A in an arbitrary extension $\mathfrak{A}'$ of a game $\mathfrak{A}$ (in the game $\mathfrak{A}'$ all terminal positions have scores, and therefore the score of an arbitrary position exists). If the score $sc_{\mathfrak{A}}(A)$ exists, then A will have the same score in an arbitrary extension $\mathfrak{A}'$.

Clearly, there exist game trees for which there are no W-pruned and B-pruned subtrees bounding the score of the base position.

Suppose given the game tree $\mathfrak{A}$ with base position A_0 having the score $sc(A_0)$. Then White's aim is to find the *best move* from the position A_0, i.e a move (A_0, B) leading to a position of rank 1 having the greatest of the possible scores. From what we have shown earlier, if White succeeds then $sc(A_0) = sc(B)$. The best move from a White position A other than the base position is similarly defined, since A is the base of a subtree with White to move.

Now suppose that A is a Black position. From Black's point of view, the best move (A, B) leads to a position having the least of all possible scores. The theorem on the transfer of scores implies that in this case, also, $sc(A) = sc(B)$.

Obviously, there may be more than one best move at any given position.

A branch $W(A, B)$ is said to be *critical* if each of its arcs represents a best move. It is easily seen that the scores of all positions on a critical branch have the same value. This property is diagnostic: if the scores at all positions on a branch $W(A, B)$ coincide, the branch is critical.

Let A be a White position. The move (A, B) will be a best move if and only if there exist a threshold value of M, a W-pruned tree $\mathfrak{A}_w(B)$ with base B for which the scores of all terminal positions are not less than M, and B-pruned trees $\mathfrak{A}_b(B')$ for all other moves (A, B'), in which the scores of all terminal positions do not exceed M. In this case, $sc(B) \geq M$, $sc(B') \leq M$ if $(A, B') \in \mathfrak{A}$ *and* $B' \neq B$. Therefore

$$sc(A) = \max\{sc(B')|(A, B') \in \mathfrak{A}\} = sc(B).$$

Conversely, if (A, B) is a best move, $sc(A) = sc(B)$ and $sc(B') \leq sc(A)$ if $(A, B') \in \mathfrak{A}$. We may take $sc(A)$ as the threshold value M. The conditions for a best move at a Black position are similarly formulated.

When the game tree is not too large and both players are able to investigate it, e.g. to construct W-pruned and B-pruned trees, the definition given above corresponds to the intuitive notion of a best move. Not so when at least one of the players cannot perceive all the possibilities. Let us look at the game depicted in Figure 6, in which White is aware of all the variations to a depth of four plies and Black sees to a depth of two plies only.

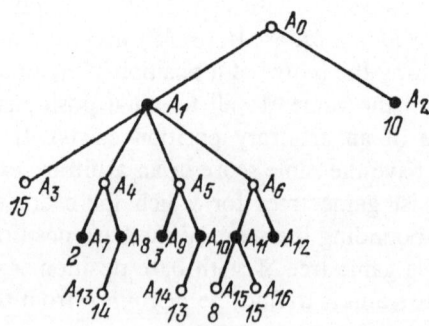

Figure 6

White can make the dismal move (A_0, A_2) leading to the position A_2 with a score of 10. The other move (A_0, A_1) offers a choice for Black: reply with the cautious move (A_1, A_3) and arrive at the position A_3 with a score of 15, or choose one of the three moves (A_1, A_4), (A_1, A_5), (A_1, A_6), for which he cannot see the consequences. His best move (A_1, A_6) leaves White with the score 8, so that the cautious move (A_0, A_2) is indeed best for White. However, we might suppose that Black would act differently. The move (A_1, A_6) is ostensibly no better than two others, which are in fact more enticing since Black perceives the replies (A_4, A_7) and (A_5, A_9) as more favorable for him. In every case where a choice of move must be made repeatedly in a similar situation, an active move such as (A_0, A_1) will lead to better results.

A player who cannot see the whole tree is in the same state as a player in a game with incomplete information. The information he has about the game can be represented by a tree in which some of the terminal positions have unknown scores. One might treat these scores as random variables, estimate in some way or other their stochastic parameters (mean, variance, correlation with scores in other positions) and try to choose a move that will maximize the expected outcome, or maximize the probability of choosing the best move (these objectives are not at all equivalent).

A player who sees to a greater depth than his opponent, and has a representation of the latter's probabilistic model, can build his own probabilistic model, which will enable him to choose risky moves that will increase the probability of winning.

We shall return to the problem of constructing various games that model some given game, but first we must consider some algorithms for finding the best moves in the sense of the formal definition given above.

Searching the Positions in a Game to Define the Score of the Base Position

All, or almost all, of the scores in a game tree $\mathfrak{A}$ can be determined by inspecting the positions in some order $L\{A_1, A_2, \ldots, A_m\}$, perhaps involving repetitions. The choice of successive elements of this sequence is a major part of the work to be done by any algorithm for computing scores. In almost all of the known algorithms the choice is made by traversing the nodes of the tree $\mathfrak{A}$, i.e. the list L satisfies the following condition: if the arc (B, C) belongs to the tree and both its nodes B and C are in L, then the position B is last encountered later than C, i.e. $\max\{i|A_i = B\} > \max\{j|A_j = C\}$.

If all the nodes of $\mathfrak{A}$ appear in L, the latter is called an *exhaustive search*. An exhaustive search allows the determination of the scores of all positions in the game, by application of Zermelo's formula whenever a non-terminal position is last encountered in it.

Many different methods are used in the algorithms for searching the positions in a tree $\mathfrak{A}$, and many of them will be considered in this book. First, however, we must devote some time to the widely used *backtracking* search. An exhaustive backtracking search of a tree $\mathfrak{A}$ is the traversal of a list $L\{A_0, A_1, \ldots, A_0\}$ in which the successive elements satisfy the following conditions:

(1) The first element is the base A_0 of the tree $\mathfrak{A}$.
(2) If the i-th element is the node B, and there exist arcs (B, C) whose ends have not yet been encountered in L, then the $(i+1)$-st element is the end of one of these arcs.
(3) Otherwise the $(i+1)$-st element is the beginning A of a unique arc (A, B) of $\mathfrak{A}$ with its end at the node B (if $B = A_0$ no such arc exists and the i-th element of the list $L(A_0, A_1, \ldots, A_0)$ is the last). In short, we go forward if we can and if we cannot we go backward.

A tree can be so represented that a forward step is always made to the leftmost available node. (See Figure 7.) In this representation nodes of the same rank are taken in order from left to right.

A backtracking search of the nodes of a tree is economical. The terminal nodes are visited only once each. Also, any arc $(B, C) \in \mathfrak{A}$ is traversed twice only—once in the forward direction and once backward—by a *forward step* from B to C and a *backward step* from C to B. Since to any pair A_i, A_{i+1} of neighboring elements in the search there correspond a forward step and a backward step, the number of steps is greater by one than double the number of arcs, or less by one than double the number of nodes (the number of nodes is greater by one than the number of arcs). Thus, on the average each node in the tree is encountered twice.

At every point in an exhaustive backtracking search the subset $\mathfrak{B}_p$ of nodes not yet encountered and the subset $\mathfrak{B}_F$ of nodes that will be encountered again are defined by information about the nodes in the branch $W(A_0, \ldots, A_i)$ connecting the base A_0 to the next element A_i of the sequence. It suffices to know, for each node $A_h \in W$, which of the ends B of the arcs (A_h, B) have already been met in the search. In fact, if the node B is not in W and has already been met in the search, all the nodes in the B-subtree will have been visited and will not be visited again. If B has not yet been visited, the nodes in its subtree will not have been visited either; only the nodes in the branch W have been visited and will be revisited. We have enough information for continuation of the search after such a node is met for the second time. Thus a backtracking search is easily carried out even when the tree is not given explicitly and we have only an algorithm for generating the arcs (A, B) issuing from a given node A, together with their ends B.

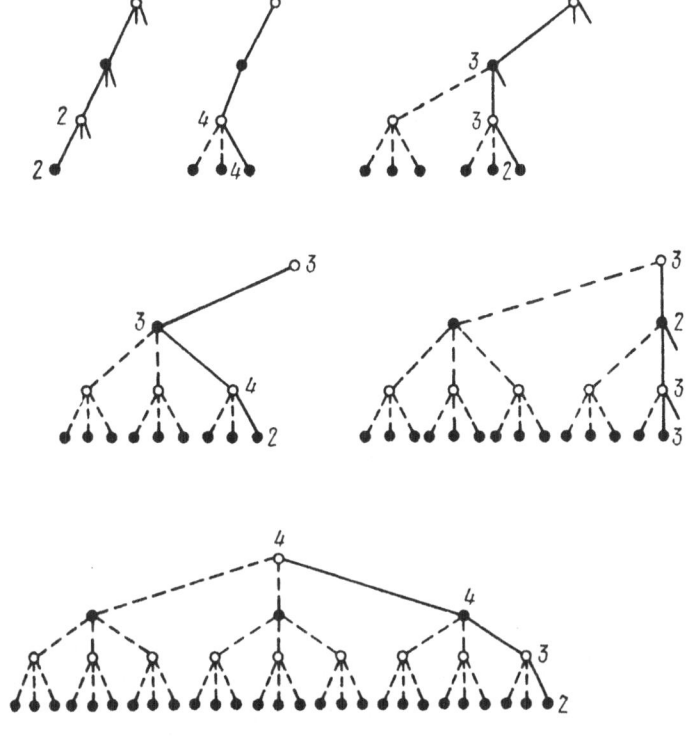

Figure 7

The scores of all the nodes in a game tree can be determined in parallel with the search by recording for all positions $A_h \in W$ the *partial score* $\mathrm{psc}(A_h)$, which has the following values:

$\mathrm{sc}(A_h)$ if A_h is a terminal position;
$\max\{\mathrm{sc}(B)|B \in \mathfrak{B}(A_h)\}$ if A_h is a non-terminal White position;
$\min\{\mathrm{sc}(B)|B \in \mathfrak{B}(A_h)\}$ if A_h is a non-terminal Black position.

Here $\mathfrak{B}(A_h) = \mathfrak{W}(A_h) \cap \mathfrak{B}_p$ is the set of all immediate successors of A_h that have already been investigated. For a non-terminal position with $\mathfrak{B}(A)$ empty we may write $\mathrm{psc}(A) = \pm\infty$ (− for a White position, + for a Black). Thus, if $\mathfrak{B}(A) = \mathfrak{W}(A)$, $\mathrm{psc}(A) = \mathrm{sc}(A)$.

To start the search we set $\mathrm{psc}(A_0) = \pm\infty$; after a forward step to the end of the branch W we add a new node B and define $\mathrm{psc}(B) := \mathrm{sc}(B)$ if B is terminal, $:= -\infty$ if B is a non-terminal White position, $:= +\infty$ if B is a non-terminal Black position. A backward step from a node B occurs when no forward step can be made. This happens whenever all the immediate successors of B have been investigated, and then $\mathrm{psc}(B) = \mathrm{sc}(B)$. We delete B from the branch W, and to its immediate predecessor A we assign the partial score $\mathrm{psc}(A) := \max\{\mathrm{psc}(A), \mathrm{sc}(B)\}$ if A is a White position and $:= \min\{\mathrm{psc}(A), \mathrm{sc}(B)\}$ if A is a Black position.

At any point in the search, the partial scores so computed for positions in the branch W will satisfy the definition of a partial score and will equal the score itself when the position is encountered for the last time.

Let us now see how to find the W-pruned subtree $\mathfrak{A}_w$ and the B-pruned subtree $\mathfrak{A}_b$ that prescribe the score we have found for the base position A_0. To do this we represent the game tree $\mathfrak{A}$ in such a way that at every position the forward steps are chosen in order from left to right, and we *color* the moves as White or Black by the following rules:

(1) If White (Black) has the move in the base position A_0 all moves leaving A_0 are colored Black (White) and the leftmost move (a best move) is also colored White (Black).
(2) If a move of some color leads to a position A with White (Black) to move, the leftmost best move from it is given the same color.
(3) If a move of the opposite color leads to a position A with White (Black) to move, all moves leading from it are also colored in the opposite color.

Thus some W-pruned subtree $\mathfrak{A}_w$ of $\mathfrak{A}$ is colored White, and some B-pruned subtree $\mathfrak{A}_b$ is colored Black; both are minimal with respect to inclusion and they intersect in the leftmost branch W issuing from the base position A_0; the moves in this branch, and only these, are colored in both Black and White. (See Figure 8.)

Lemma on a Branch of the Tree $\mathfrak{A}_w$. *Let $W(A = A_1, \ldots, A_k)$ be an arbitrary branch of the W-pruned tree $\mathfrak{A}_w$. Then* $\mathrm{sc}(A_1) \le \mathrm{sc}(A_2) \le \ldots \le \mathrm{sc}(A_k)$.

In fact if $A_i \in W$ is a White position $(i < k)$ then (A_i, A_{i+1}) is a best move from it and $\mathrm{sc}(A_i) = \mathrm{sc}(A_{i+1})$; if, however, A_i is a Black position, then

$$\mathrm{sc}(A_i) = \min\{\mathrm{sc}(B)|B \in \mathfrak{W}(A_i)\} \le \mathrm{sc}(A_{i+1}).$$

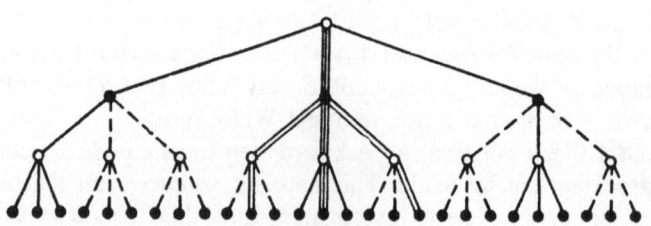

Figure 8

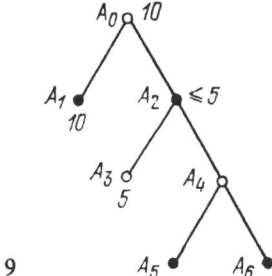

Figure 9

The lemma on branches of the Black-colored tree $\mathfrak{A}_b$ is formulated and proved in a similar way. These lemmas imply that the scores of all positions in the intersection $\mathfrak{A}_w \cap \mathfrak{A}_b$ are equal to $\mathrm{sc}(A_0)$. The move (A_0, A_1), which belongs to the critical branch $W = \mathfrak{A}_w \cap \mathfrak{A}_b$ is a best move; the scores of all terminal positions in the White tree $\mathfrak{A}_w$ are not less than $\mathrm{sc}(A_0)$ and the scores in the Black tree $\mathfrak{A}_b$ are not greater than $\mathrm{sc}(A_0)$. Therefore to determine the score of the position A_0 and find a best move from it we need consider only these subtrees, and any other work on the search is superfluous.

To find a best move from the base position A_0 we do not need to know the scores of all positions in the game tree $\mathfrak{A}$. Suppose for instance that $\mathfrak{A}$ is the game tree depicted in Figure 9, White and Black to move in alternating turns, with White to move in the base position A_0. Assume that we have carried out the portion $L'\{A_0, A_1, A_0, A_2, A_3, A_2\}$ of the search. (The scores of terminal positions and the partial scores of non-terminal positions are shown beside the corresponding nodes in Figure 9.) Since White is to move at A_0, $\mathrm{sc}(A_0) \geq \mathrm{psc}(A_0) = 10$; Black is to move at A_2 and $\mathrm{sc}(A_2) \leq \mathrm{psc}(A_2) = 5$. Moreover

$$\mathrm{sc}(A_0) = \max\{\mathrm{sc}(A_1), \mathrm{sc}(A_2)\} \leq \max\{\mathrm{sc}(A_1), \mathrm{psc}(A_2)\} = 10.$$

Accordingly $\mathrm{sc}(A_0) = 10$, (A_0, A_1) is a best move at the base position, and the search need not be extended. A more complicated example is shown in Figure 10, where White and Black alternate their moves, as in the preceding example. Suppose that we have carried out the portion $L'\{A_0, A_1, A_0, A_2, A_3, A_5, A_7, A_5\}$ of the search and have obtained partial scores for the positions A_0 and A_5. If (A_3, A_5) is a best move, then

$$\mathrm{sc}(A_2) \leq \mathrm{sc}(A_3) = \mathrm{sc}(A_5) \leq \mathrm{psc}(A_5) = 5.$$

Therefore we may skip the positions $A_8, A_9, \ldots, A_{13}$ and go at once to A_6. If it turns out that the score of A_6 is less than 10, we have

$$\mathrm{sc}(A_2) \leq \mathrm{sc}(A_3) = \max\{\mathrm{sc}(A_5), \mathrm{sc}(A_6)\} < 10.$$

i.e (A_0, A_1) is a best move at the base position A_0 and $\mathrm{sc}(A_0) = 10$. If $\mathrm{sc}(A_6) \geq 10$, $\mathrm{sc}(A_2) \geq \mathrm{sc}(A_3) = \mathrm{sc}(A_6) \geq 10$ and (A_0, A_2) is a best move.

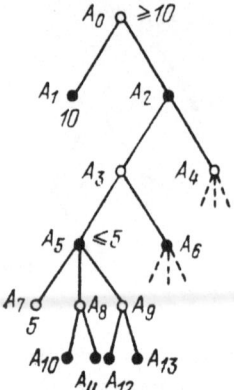

Figure 10

To shorten the search as much as we can we reformulate the second rule stated above to connect it with the partial scores that we are computing for the positions in the branch W:

2′. If the k-th element of the list L is a White (Black) position, and if for all Black (White) positions A in the branch $W(A_0,\ldots,B)$ we have $\mathrm{psc}(A) < \mathrm{psc}(B)$ $(\mathrm{psc}(A) > \mathrm{psc}(B))$, and if there exist arcs (B,C) with ends not yet encountered in the search, then the next element in the search is one of these new immediate successors of the position B.

We shall show later that we can in fact use this method to find a score for the base position A_0 and a best move from it. However, the values we compute, which we have called partial scores, are not really so. Rather, we shall call them *bounds*. The theory of bounds and scores was first considered by A. L. Brudno in his paper [10]. The consequent rule for cutting off a variation during the search, which is essentially equivalent to ours, is often called the α,β-*cutoff* or *branch-and-bound* heuristic. It is useful to reformulate it slightly in order to show more clearly just when the cutoff takes place, i.e. when a whole subtree of the game tree $\mathfrak{A}$ is thrown out of further consideration (pruned). The equivalence of the newly formulated rule with the old one will be obvious.

A *search* is a sequence of steps, to each of which is associated a next position B and either a next move (B,C) leading from B or a next move (A,B) leading to it. (We shall see later that there is one exception.) The next position for the following step is another position, related to the next move, and the next position for the first step is the base A_0. Thus at the beginning of every step we know the related next position. We do not know, however, whether the step is forward or backward (and even less do we know what the next move is). In order to formulate the rules for determining the character of the step we shall need the following notation: $\mathrm{bd}(A)$ is the value of the bound at the position A at the contemplated moment; $\mathrm{Bd}(A)$ is the last value of $\mathrm{bd}(A)$, i.e. the value at the instant of making a step backward from A; W is the branch leading from the base position to the next position; P_w is the set of White positions in W and P_b is the set of Black positions.

The Cutoff (Pruning) Rule: The next step is backward if the next position A is a White position and

$$\text{bd}(B) \geq \min \{\text{bd}(A) | A \in P_b\}$$

or if A is a Black position and

$$\text{bd}(B) \leq \max \{\text{bd}(A) | A \in P_w\}.$$

The Stopping Rule: The next step is backward if B is a terminal position or if all the immediate successors of B have already been examined.

The Rule for Deepening: If neither of the above rules is applicable, the next step is forward.

We now define the next move. If the step is backward it is the move (A, B) from A, the next preceding position but one in the branch W, to the next position B. If the next position is the base position no move is made; this signals the end of the search. If the step is forward the next move is (B, C) from the next position B to one of the positions not yet visited. We shall always suppose that we move to the leftmost of these. Thus after a backward step the branch W is shortened by one position and is lengthened by one position after a forward step.

Finally, we recompute, or determine, the value of the bound at the next position for the following step. If the step is backward, and the next move is (A, B) we have

$$\text{bd}(A) := \max \{\text{bd}(A), \text{Bd}(B)\} \text{ if } A \text{ is a White position,}$$

$$\text{bd}(A) := \min \{\text{bd}(A), \text{Bd}(B)\} \text{ if } A \text{ is a Black position.}$$

If the step is forward and the move is (B, C) we have

$$\text{bd}(C) := \text{sc}(C) \text{ if } C \text{ is terminal,}$$

$$\text{bd}(C) := -\infty \text{ if } C \text{ is a non-terminal White position,}$$

$$\text{bd}(C) := +\infty \text{ if } C \text{ is a non-terminal Black position.}$$

Before the beginning of the search the score of the base position A_0 is defined as if the first step were preceded by a blank move representing a forward step.

Thus at any point in the search the values of the bounds satisfy a condition like the Zermelo formula:

$\text{Bd}(A) = \text{sc}(A)$ if A is a terminal position; $\max \{\text{bd}(B) | B \in \mathfrak{B}(A)\}$ if A is a non-terminal White position; $\min \{\text{bd}(B) | B \in \mathfrak{B}(A)\}$ if A is a non-terminal Black position. Here the maxima and minima are computed over only those immediate successor positions of A that have already been examined.

It is easily seen that the set of all positions examined in the search forms a subtree $\tilde{\mathfrak{A}}$ of the game tree $\mathfrak{A}$ with its root at the base position A_0 and terminal nodes corresponding to those positions of $\mathfrak{A}$ that have assigned scores. It may be looked on as the tree of a model game, with the same next move at each non-terminal position and the same scores at terminal positions as the original tree.

Let us now use the process described above to construct the White and Black subtrees $\tilde{\mathfrak{A}}_w$ and $\tilde{\mathfrak{A}}_b$ and show that these are respectively W-pruned and B-pruned subtrees of the original game tree $\mathfrak{A}$. Then $\mathrm{sc}(A_0) = \mathrm{bd}(A_0)$ and the best move is (A_0, A_1) where $A_1 \in \tilde{\mathfrak{A}}_w \cap \tilde{\mathfrak{A}}_b$. These assertions will be validated by three subsequent lemmas on the changes in the bounds during the search. We formulate and prove them with respect to White positions or positions in the tree $\tilde{\mathfrak{A}}_w$. The corresponding lemmas with an interchange of colors may be formulated and proved in a similar fashion. At any position $A \in \tilde{\mathfrak{A}}$ we write $\mathrm{bd}(A)$ for the next value of its bound, and $\mathrm{Bd}(A)$ for its final value.

Lemma on Monotone Change. *Let $A \in \tilde{\mathfrak{A}}$ be a White position. Then the value of* $\mathrm{bd}(A)$ *does not decrease.*

PROOF. The rules formulated above imply that $\mathrm{bd}(A)$ either takes on the value $\mathrm{sc}(A)$ immediately and remains constant, or takes on the value $-\infty$ and changes subsequently according to the rule

$$\mathrm{bd}(A) := \max\{\mathrm{bd}(A), \mathrm{bd}(B) | (A, B) \in \tilde{\mathfrak{A}}\},$$

i.e. can never decrease.   □

Lemma on Strong Monotonicity. *Let $B \in \tilde{\mathfrak{A}}_w$ be a non-terminal White position. Until we step backward to it from its unique immediate next position in $\tilde{\mathfrak{A}}_w$, we have* $\mathrm{bd}(B) < \mathrm{Bd}(B)$.

PROOF. When we first arrived at B we had

$$\mathrm{bd}(B) := -\infty < \mathrm{Bd}(B).$$

If (B, C) is a move in a forward step from B to a position C in $\tilde{\mathfrak{A}}_w$ it is a best move in $\tilde{\mathfrak{A}}$, and no other forward step is a best move. Therefore, the final values of the bounds for positions to which these other moves lead will be less than $\mathrm{bd}(B) = \mathrm{bd}(C)$ and before a backward step from C to B we have

$$\mathrm{bd}(B) = \max\{\mathrm{Bd}(C') | C' \in \mathfrak{B}(B)\} < \mathrm{Bd}(B).$$ □

Lemma on the Completeness of the Subtree $\tilde{\mathfrak{A}}_w$. *Let $C \in \tilde{\mathfrak{A}}_w$ be a non-terminal White position. Then the cutoff rule is inapplicable at any position immediately preceding C.*

PROOF. Let $W(A_0, \ldots, C)$ be the branch from the base position A_0 to the Black next position C and let $B \in P_b$ be any White position in the branch W. *The final values of* $\mathrm{bd}(B)$ *and* $\mathrm{bd}(C)$ *are equal to the scores of these positions in the game $\tilde{\mathfrak{A}}$. The B-subtree of the W-pruned tree $\tilde{\mathfrak{A}}_w$ is itself a*

W-pruned tree with scores of its terminal positions not less than $\mathrm{bd}(B)$. It contains the position C and so also the C-subtree of $\tilde{\mathfrak{A}}_w$. This C-subtree is also a W-pruned tree. Therefore $\mathrm{Bd}(C) = \mathrm{sc}_{\tilde{\mathfrak{A}}}(C) \geq \min\{\mathrm{sc}(F)|\ F \in \tilde{\mathfrak{A}}_w \cap \mathfrak{F}\} \geq \min\{\mathrm{sc}(F)|\ F \in \tilde{\mathfrak{A}}_w(B) \cap \mathfrak{F}\} = \mathrm{sc}_{\tilde{\mathfrak{A}}}(B) = \mathrm{Bd}(B)$, where $\tilde{\mathfrak{A}}_w(B)$ and $\tilde{\mathfrak{A}}_w(C)$ are the B- and C- subtrees of $\tilde{\mathfrak{A}}_w$ and $\mathfrak{F}$ is the set of all terminal positions in the game $\mathfrak{A}$.

By the inequality proved above, and also by the lemmas on strong monotonicity and on the monotone change in the bound (we need the latter in the formulation for Black positions), the inequalities

$$\mathrm{bd}(B) < \mathrm{Bd}(B) \leq \mathrm{bd}(C) \leq \mathrm{Bd}(C)$$

hold as long as we have not made a backward step to the position B and so whenever we are considering the position C. Accordingly, when C is the next position we have for all $B \in P_w$ the inequality $\mathrm{bd}(B) < \mathrm{bd}(C)$ and the pruning rule cannot be applied.   □

So for every non-terminal Black position $C \in \tilde{\mathfrak{A}}_w$ all the moves from it also belong to $\tilde{\mathfrak{A}}_w$, and this proves the

Theorem on the White Tree of a Search with Pruning. *The White tree $\tilde{\mathfrak{A}}_w$ is a W-pruned subtree of the game tree $\mathfrak{A}$.*

The theorems on the White and Black search trees imply the theorem on the bounds of the base position and the critical branch. After the end of the search $\mathrm{Bd}(A_0) = \mathrm{sc}(A_0)$ and the intersection of the White and Black trees is the critical branch.

Let us now see what information we have about the position B at the instant when we take a backward step from it. Let $W(A_0,\dots,B)$ be a branch leading from the base position A_0 to B; let P_w and P_b be the sets of White and Black positions in this branch, not counting B itself, and write

$$\underline{\lim} := \max\{\mathrm{bd}(A)|A \in P_w\},$$
$$\overline{\lim} := \min\{\mathrm{bd}(A)|A \in P_b\}.$$

where the bounds for the position are evaluated at the instant when a backward step is made from B. They remain unchanged, of course, during the whole time between a forward step to B and the subsequent backward step from B. If $P_w = \varnothing$, $\underline{\lim} = -\infty$; if $P_b = \varnothing$, $\overline{\lim} = +\infty$.

Lemma. $\overline{\lim} > \underline{\lim}$.

PROOF. We may restrict ourselves to the case in which $\overline{\lim} < +\infty$ and $\underline{\lim} > -\infty$. Suppose that

$$\overline{\lim} \leq \underline{\lim}, \qquad \underline{\lim} = \mathrm{bd}(A')|A' \in P_w,$$
$$\overline{\lim} = \mathrm{bd}(A'')|A'' \in P_b.$$

Clearly $A' \neq A''$ since at A' White has the move and at A'' the move is Black's. Suppose for instance that these positions occur in the branch W in the order $W(A_0, \ldots, A', \ldots, A'', \ldots, B)$. Then A'' belongs to the A'-subtree of $\mathfrak{A}$ and B belongs to its A''-subtree, and the search will have been carried out in the following order: First the value of $\mathrm{bd}(A')$ is established as equal to $\underline{\lim}$; then we step forward along the branch W from A', and after a while (perhaps at once) we step forward to A''. After this step the value of $\mathrm{bd}(A'')$ remains fixed at $\overline{\lim}$ and we step forward from A'' along the branch W in the direction of B, which ultimately becomes the next position. However, once we have made the backward step to the position A'' the value of $\mathrm{bd}(A'')$ is equal to $\overline{\lim}$ and we have the equation

$$\mathrm{bd}(A'') = \overline{\lim} \leq \underline{\lim} = \mathrm{bd}(A').$$

i.e., we must apply the pruning rule and therefore step backward from the position A''. But then the position B is never reached in the search, and this contradicts the hypothesis of the lemma.   □

If $\mathrm{Bd}(B) \leq \underline{\lim}$ the position B *is unreachable by White* (when B is a Black position the pruning rule is applied) and if $\mathrm{Bd}(B) \geq \overline{\lim}$, B is *unreachable by Black*. What this means will be explained later.

Theorem on Bounds and Scores. *If for a backward step from a position B we have* $\mathrm{Bd}(B) \geq \overline{\lim}$ *then* $\mathrm{sc}(B) \geq \overline{\lim}$; *if* $\mathrm{Bd}(B) \leq \underline{\lim}$ *then* $\mathrm{sc}(B) \leq \underline{\lim}$; *if however* $\underline{\lim} < \mathrm{Bd}(B) < \overline{\lim}$, *then* $\mathrm{Bd}(B) = \mathrm{sc}(B)$.

PROOF. We consider an auxiliary game $\tilde{\mathfrak{A}}(B)$ for which the tree consists of the positions $\tilde{A}_0, \tilde{A}_1, \tilde{A}_2, \tilde{A}_3$, and the moves $(\tilde{A}_0, \tilde{A}_1)$, $(\tilde{A}_0, \tilde{A}_2)$, $(\tilde{A}_2, \tilde{A}_3)$, $(\tilde{A}_2, B)$, plus the B-subtree of the game $\mathfrak{A}$ (see Figure 11). The move at $\tilde{A}_0$ belongs to the side that has the move at B, and belongs to the opposite side at $\tilde{A}_2$ (we assume that B is a White position). The positions $\tilde{A}_1$ and $\tilde{A}_3$ are terminal and have the scores $\mathrm{sc}(\tilde{A}_1) = \underline{\lim}$, $\mathrm{sc}(\tilde{A}_3) = \overline{\lim}$ (if B is a Black position the converse holds, i.e. $\mathrm{sc}(\tilde{A}_1) = \overline{\lim}$, $\mathrm{sc}(\tilde{A}_3) = \underline{\lim}$). The turn to move at non-terminal positions in the B-subtree and the scores of terminal positions are the same as in the game $\mathfrak{A}$. Then that portion of the search of the tree $\mathfrak{A}$ that begins after the step forward from B and ends with a step backward from it is identical with the search of the game $\tilde{\mathfrak{A}}(B)$ after the initial steps with the next positions $\tilde{A}_0, \tilde{A}_1, \tilde{A}_0, \tilde{A}_2, \tilde{A}_3, \tilde{A}_2, B$, and after completion of this segment the search of the tree $\tilde{\mathfrak{A}}(B)$ ends in two backward steps with the next positions $\tilde{A}_2$ and $\tilde{A}_0$.

The theorem on the bounds of the base position implies that $\mathrm{Bd}(\tilde{A}_0) = \mathrm{sc}(\tilde{A}_0)$—remembering that $\mathrm{Bd}(\tilde{A}_0)$ is the value of $\mathrm{bd}(A_0)$ at the end of the search algorithm. Moreover,

$$\mathrm{sc}(\tilde{A}_0) = \max\left\{\underline{\lim}, \mathrm{sc}(\tilde{A}_2)\right\},$$

$$\mathrm{Bd}(\tilde{A}_0) = \max\left\{\underline{\lim}, \mathrm{Bd}(\tilde{A}_2)\right\}.$$

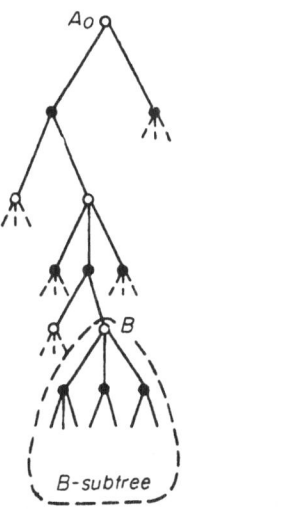
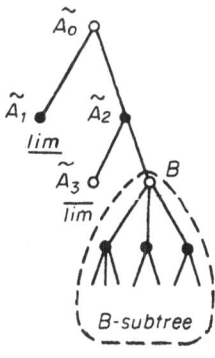

Figure 11

So, either $\mathrm{sc}(\tilde{A}_0) = \underline{\lim}$ and then

$$\mathrm{sc}(\tilde{A}_2) \leq \underline{\lim}, \qquad \mathrm{Bd}(\tilde{A}_2) \leq \underline{\lim}, \qquad \text{or}$$

$$\mathrm{sc}(\tilde{A}_0) = \mathrm{sc}(\tilde{A}_2) = \mathrm{Bd}(\tilde{A}_2) > \underline{\lim}.$$

In the first case $(\tilde{A}_2, B)$ is a best move from the position $\tilde{A}_2$, and another move leads to a position $\tilde{A}_3$ with the score $\overline{\lim} > \underline{\lim} \geq \mathrm{sc}(\tilde{A}_2)$. Accordingly

$$\underline{\lim} \geq \mathrm{sc}(\tilde{A}_2) = \mathrm{sc}(B),$$

$$\underline{\lim} \geq \mathrm{Bd}(\tilde{A}_2) = \min\{\overline{\lim}, \mathrm{Bd}(B)\} = \mathrm{Bd}(B).$$

In the second case either

$$\mathrm{sc}(\tilde{A}_2) = \min\{\overline{\lim}, \mathrm{sc}(B)\} = \overline{\lim} \leq \mathrm{sc}(B),$$

$$\overline{\lim} = \mathrm{sc}(\tilde{A}_2) = \mathrm{Bd}(\tilde{A}_2) = \min\{\overline{\lim}, \mathrm{Bd}(B)\} \leq \mathrm{Bd}(B).$$

or

$$\overline{\lim} > \mathrm{sc}(B) = \mathrm{sc}(\tilde{A}_2) = \mathrm{Bd}(\tilde{A}_2) = \min\{\overline{\lim}, \mathrm{Bd}(B)\} = \mathrm{Bd}(B).$$

Zermelo's formula implies that the position B has the same score in both of the games $\mathfrak{A}$ and $\tilde{\mathfrak{A}}(B)$. This concludes the proof of the theorem. A similar proof holds when B is a Black position.   □

Thus a position B is unreachable by White if its score does not exceed $\underline{\lim}$, and is unreachable by Black if its score is not less than $\overline{\lim}$.

Theorem on Candidate Positions for the Critical Branch. *Unless a position is reachable by both White and Black it cannot lie on the critical branch defined by the search process.*

PROOF. Suppose for instance that

$$\mathrm{Bd}(B) \geq \overline{\lim} = \mathrm{Bd}(A) < +\infty, \ A \in P_b,$$

i.e. B is unreachable by Black. If it belonged to the critical branch $W(A_0,\ldots, A,\ldots, B)$, the latest values of $\mathrm{bd}(A)$ and $\mathrm{bd}(B)$ would be equal. But after a step backward from B the value of $\mathrm{bd}(B)$ does not change and the value of $\mathrm{bd}(A)$ cannot increase, since A is a Black position. Therefore $\mathrm{bd}(A)$ is already equal to $\mathrm{Bd}(B)$ and cannot change in the future. Meanwhile

$$\mathrm{bd}(A) = \min\{\mathrm{Bd}(B')|B' \in \mathfrak{B}(A)\} = \{\mathrm{Bd}(B'')|B'' \in \mathfrak{B}(A)\}.$$

Therefore the branch $W(A_0,\ldots, A, B'',\ldots)$, consisting of the positions examined during the search is also the critical branch and lies to the left of the branch W; that is, the latter is not constructed during the search, which is what we were to prove.

We treat the case $\mathrm{bd}(B) \leq \underline{\lim}$ in the same way.   □

Suppose we are interested only in knowing whether the condition $\mathrm{sc}(A) > m$ is satisfied at some position A in the game $\mathfrak{A}$, while we do not need an exact value of the score and are not interested in knowing the best move from A. Then we may search the A-subtree of the game, after setting the initial value of the bounds at non-terminal White positions to m rather than $-\infty$. We carry out the search as though at each White position B that we contemplate there existed a move (B, B') to a terminal position with $\mathrm{sc}(B) = m$, and this move is to be examined first. Then if $\mathrm{Bd}(A) > m$ it coincides with $\mathrm{sc}(A)$ and we can determine the best move. If, however, $\mathrm{Bd}(A) = m$ we have $\mathrm{sc}(A) \leq m$ and the best move is unknown. In the same way, when we wish to know whether $m < \mathrm{sc}(A) < M$ we write, at each forward step to a position B, $\mathrm{Bd}(B) := \mathrm{sc}(B)$ for a terminal position, $\mathrm{Bd}(B) := m$ for a non-terminal White position, and $\mathrm{Bd}(B) := M$ for a non-terminal Black position.

On the Number of Positions Examined in Determining the Score of a Position and Finding the Best Move

A game $\mathfrak{A}$ need not be defined by specifying its game tree and the scores of terminal positions. For instance, a chess position is defined by (a) the distribution of the pieces on the board, (b) the assignment of the turn to move, and (c) a small number of supplementary data (the right to castle on either side, capture en passant, drawn game after the third repetition of a position or by application of the fifty-move rule; these relate to the branch W connecting the given position to the base chess position). This information determines uniquely whether there is a succeeding terminal position and if so what its score is, or if there is no such position then what moves may be made. These specific details allow us, in many positions, to determine the score or the best move without pursuing the subordinate subtree all the way to its terminal positions.

If however the game tree is specified in some suitable (possibly indirect) way while the scores of the terminal positions are given explicitly and perhaps arbitrarily, the theorems on extensions of games for the definition of scores and best moves force us to consider all the terminal positions of certain W-pruned and B-pruned subtrees $\mathfrak{A}_w(A)$ and $\mathfrak{A}_b(A)$. On the other hand, the construction of these trees, and of the scores of terminal positions that satisfy the condition

$$\min\left\{\operatorname{sc}(C)|C \in \mathfrak{F}_w(A)\right\} = \max\left\{\operatorname{sc}(C)|C \in \mathfrak{F}_b(A)\right\}$$

suffices to define the score at A and the best move from A. Here $\mathfrak{F}_w(A)$ and $\mathfrak{F}_b(A)$ are respectively the sets of terminal positions of the subtrees $\mathfrak{A}_w(A)$ and $\mathfrak{A}_b(A)$.

If, as is usually the case, the W-pruned and B-pruned subtrees of the game $\mathfrak{A}$ are strongly branched because as a rule we can choose more than one move from any one of their non-terminal positions, the total number of positions in these subtrees is of the same order of magnitude as the number of their terminal positions. Therefore if we are looking for the most economical algorithm (with respect to number of positions examined) for finding position scores and best moves, we may look for it among those algorithms that examine the entire branch $W(A,\ldots,C)$ leading from the position of interest A to a terminal position C whenever C is under consideration in the process of finding the score and best move at A. We may suppose that A is the base position A_0, else we examine the game $\mathfrak{A}(A)$ whose tree is the A-subtree of the original game $\mathfrak{A}$, with the initial move belonging to the side that originally had it.

The branches $W(A_0,\ldots,C)$ define a subtree $\mathfrak{A}' \subset \mathfrak{A}$ which, by what we have said above, contains the W-pruned and B-pruned subtrees $\mathfrak{A}_w$ and $\mathfrak{A}_b$ with, respectively, the lower and upper bounds of the scores at the terminal positions. Since the game tree $\mathfrak{A}$ is finite these subtrees contain minimally inclusive subtrees $\mathfrak{A}'_w$ and $\mathfrak{A}'_b$ with the same property. We naturally expect the algorithm that examines the least number of position to be found among those that consider only the positions in the smallest W-pruned and B-pruned subtrees defining the score at the base position A_0.

Lemma on Minimality. *The W-pruned subtree $\mathfrak{A}_w$ of the game $\mathfrak{A}$ which is minimal inclusive contains precisely one move from every non-terminal White position B.*

In fact if $(B, C) \in \mathfrak{A}_w$ and $(B, C') \in \mathfrak{A}_w$, the second move and the C'-subtree may be excluded while the tree is still W-pruned.   □

The lemma obtained by changing color also holds.

Lemma on the Critical Branch. *The minimal-inclusive W-pruned and B-pruned subtrees of a game tree $\mathfrak{A}$ intersect in a branch $W(A_0, \ldots, C)$ containing the base position A_0 and some terminal position C. If these subtrees determine the score at A_0, W is the critical branch.*

PROOF. Suppose that the non-terminal position $B \in \mathfrak{A}_w \cap \mathfrak{A}_b$, where $\mathfrak{A}_w$ and $\mathfrak{A}_b$ are the smallest W-pruned and B-pruned subtrees respectively. If B is a White (Black) position, then $\mathfrak{A}_w$ ($\mathfrak{A}_b$) contains exactly one move (B, C) leading from it and $\mathfrak{A}_w$ ($\mathfrak{A}_b$) contains all such moves. Thus $\mathfrak{A}_w \cap \mathfrak{A}_b$ contains a single move (A_0, A_1) leading from the base position A_0 to some first-rank position A_1, a single move from A_1 to a second-rank position A_2, and so on, until we arrive at the terminal position C. The intersection of two subtrees of the game tree $\mathfrak{A}$ is a tree, i.e. it is connected, and therefore the intersection $\mathfrak{A}_w \cap \mathfrak{A}_b$ contains no positions nor moves other than those making up the branch $W(A_0, \ldots, C)$. If the subtrees $\mathfrak{A}_w$ and $\mathfrak{A}_b$ determine the score at A_0, then for any position $B \in \mathfrak{A}_w$ we have

$$\mathrm{sc}(B) \geq \mathrm{sc}(A_0)$$

and for any position $B \in \mathfrak{A}_b$ we have

$$\mathrm{sc}(B) \leq \mathrm{sc}(A_0).$$

So, all the positions B in the branch $W(A_0, \ldots, C)$ have scores equal to $\mathrm{sc}(A_0)$, i.e. W is the critical branch, which is what we were to prove.   □

Let $\mathfrak{A}_w$ and $\mathfrak{A}_b$ be W-pruned and B-pruned subtrees determining the score of the base position A_0 in the game $\mathfrak{A}$. Then there exist ways of choosing the next move in a search algorithm for the tree $\mathfrak{A}$ such that the algorithm leads only to nodes in the union $\mathfrak{A}_w \cup \mathfrak{A}_b$ of those subtrees. If $\mathfrak{A}_w$ and $\mathfrak{A}_b$ are W-pruned and B-pruned subtrees with the smallest number of nodes in their union, the contemplated variant of our algorithm is minimal with respect to all algorithms determining the score at the base position or the best move from it (without making use of any specific information about the scores at terminal positions). In choosing the next move, however, we assume that the trees $\mathfrak{A}_w$ and $\mathfrak{A}_b$ are known. Nevertheless, the proof we shall give below shows that if we are successful enough in guessing the best move when we are choosing the next moves in our forward steps, the number of positions examined in our search is more or less minimal.

The rule for choosing the next move is conveniently expressed, in the representation of the game tree $\mathfrak{A}$ on a plane surface, as the requirement that we choose the next move as the leftmost of all those possible. To

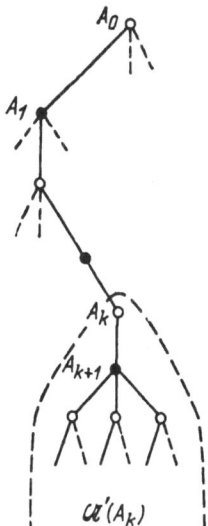

Figure 12

confine the search to the nodes in the union $\mathfrak{A}_w \cup \mathfrak{A}_b$ we must make the leftmost move $(B,C) \in \mathfrak{A}_w$ at every White node $B \in \mathfrak{A}_w$, and make the move $(B,C) \in \mathfrak{A}_b$ at every Black node $B \in \mathfrak{A}_b$. The rule for choosing the move at other nodes in the game tree $\mathfrak{A}$ is arbitrary. So when B lies on the critical branch $W = \mathfrak{A}_w \cap \mathfrak{A}_b$, we first attend to a move on this branch. Therefore at the beginning of the search we move through the nodes $A_0, A_1, \ldots, A_m$ of the branch W and we obtain the scores

$$\mathrm{Bd}(A_m) = \mathrm{sc}(A_m) = \mathrm{sc}(A_n),$$

and then from time to time we come back along W.

Lemma on the Bounds in the Critical Branch. *After a backward step to the position* $A_k \in W$, *we have* $\mathrm{Bd}(A_k) = \mathrm{sc}(A_k) = \mathrm{sc}(A_0)$.

PROOF. In fact the portion of the search that begins with the first forward step from the position A_k and ends with the first backward step to it may be taken as a search for the purpose of determining the score at the base position A_k of the game $\mathfrak{A}'(A_k)$, whose tree is the open A_{k+1}-subtree of $\mathfrak{A}$, and where the move $(A_k, A_{k+1}) \in W$. (See Figure 12.) Accordingly, after the first backward step to the position A_k, the theorem on bounds implies the equations

$$\mathrm{bd}(A_k) = \mathrm{sc}_{\mathfrak{A}'(A_k)}(A_k) = \mathrm{sc}_{\mathfrak{A}'(A_k)}(A_{k+1})$$

$$= \mathrm{sc}(A_{k+1}) = \mathrm{sc}(A_0) = \mathrm{sc}(A_k),$$

which is what we were to prove.   □

Theorem on the Optimality of the Search. *Let $A_k \in W$ be a White position. Then the set of positions and moves that will be examined in the search, from the first step backward to A_k until the step backward from it, belongs to the subtree $\mathfrak{A}_b$.*

PROOF. It suffices to show that the set of contemplated positions and moves that are subordinate to the immediate successor nodes B of A_k and that do not lie on the branch W do lie in the subtree $\mathfrak{A}_b$. Since $A_k \in \mathfrak{A}_b$ and is a White position, $B \in \mathfrak{A}_b$. Let us consider the portion of the search that begins after a forward step with the next move (A_k, B), and see when we next arrive at a position not lying in $\mathfrak{A}_b$. If we are at some position $D \in \mathfrak{A}_b$ and step backward from it, we land on an immediate predecessor position C that also belongs to $\mathfrak{A}_b$, since (C, D) is the only move in $\mathfrak{A}$ leading to D. If we step forward from $D \in \mathfrak{A}_b$ we again arrive at a position $E \in \mathfrak{A}_b$ since the move (D, E) is the leftmost. Finally, if we take any step forward from a White position we arrive at a node $E \in \mathfrak{A}_b$ since all moves from D lead to a position in $\mathfrak{A}_b$.

There remains the consideration of a backward step from a node $D \in \mathfrak{A}_b$ subordinate to B (and perhaps coinciding with it). We set the value of the bound at D as equal to the score at one of the terminal positions to be inspected after the first forward step from D. If the first forward step from the node B does not bring us to the end of the subtree $\mathfrak{A}_b$, none of the scores of the contemplated positions exceed $\mathrm{sc}(A_0)$. Thus $\mathrm{bd}(D) \le \mathrm{sc}(A_0)$. But, by the lemma on the bounds in the critical branch, $\mathrm{bd}(A_k) = \mathrm{sc}(A_0)$, and we must step backward on account of the pruning rule, since A_k is a position in the branch $W'(A_0, \ldots, D)$ leading from the base position A_0 to the next position D.

Thus until we return to the position A_k and even more, until we take a backward step from A_k—the search moves only within the minimally inclusive B-pruned subtree $\mathfrak{A}_b$ of the game tree $\mathfrak{A}$. This is what we were to prove.   □

There is a similar proof that after a first backward step to a Black position in the critical branch W and before any backward step from this position we contemplate only positions lying in the minimally inclusive W-pruned subtree $\mathfrak{A}_w$. Thus until we are forced to make a backward step from the base position A_0 we do not go outside the union $\mathfrak{A}_w \cup \mathfrak{A}_b$.

Let us now see how many terminal positions we need to inspect for the determination of the score at the base position when the moves alternate between the colors, i.e. when a White move (B, C) necessarily leads to a Black position, and vice versa. Suppose for instance that the base position, the only one of rank 0, is a White position. Then all positions of rank 1, and in general all positions of odd rank, are Black positions. The score at the base position can be obtained by a search employing the minimum amount of information about the scores at terminal positions.

We shall make a few remarks about the process. The backward steps and their next moves may be divided into three classes. If the value of $\mathrm{bd}(B)$ changes after a backward step with the next move (B, C) {in the way desired by the side that has the move at B} and begins or continues to satisfy the inequalities

$$\underline{\lim} < \mathrm{bd}(B) < \overline{\lim},$$

such a step and its next move (B, C) are said to be an *improving* step and move. After an improving step the pruning rule does not apply.

If $\mathrm{bd}(B)$ changes so that the pruning rule must be applied on the next step, the given step and its next move (B, C) are called a *refutation* step and move (one of the opponent's preferred moves in the branch). After a refutation step the values of $\mathrm{Bd}(B)$ and $\mathrm{Bd}(C)$ satisfy the conditions $\mathrm{Bd}(B) = \mathrm{Bd}(C) \geq \overline{\lim}$ if B is a White position, $\mathrm{Bd}(B) = \mathrm{Bd}(C) \leq \underline{\lim}$ if B is a Black position.

In the remaining cases the backward step and its next move are said to be *bad*, although in fact one of the moves in the branch preferred by the side of the same color may be bad. After a step backward from the position C we have $\mathrm{Bd}(C) \leq \mathrm{bd}(B) < \overline{\lim}$ if B is a White position, and $\mathrm{Bd}(C) \geq \mathrm{Bd}(B) > \underline{\lim}$ if B is a Black position. As in an improving step, so here the pruning rule does not apply to the following step. We will present some further discussion of White and Black positions, noting that in what we say we may simultaneously interchange colors, $\underline{\lim}$ and $\overline{\lim}$, and the inequality signs $<$ and $>$, etc. Let (B, C) be the next move for some step in the search. The branch $W'(A_0, \ldots, B, C)$ leading from the base position A_0 to the position C is an extension of the branch $W(A_0, \ldots, B)$ from A_0 to B. Then the sets of positions in these branches with moves of one color that determine the values of $\underline{\lim}(B)$, $\overline{\lim}(B)$, $\underline{\lim}(C)$, $\overline{\lim}(C)$ satisfy the conditions

$$P_w(C) = P_w(B) \cup B,$$
$$P_b(C) = P_b(B).$$

If (B, C) is an improving move the inequality $\mathrm{Bd}(C) > \mathrm{bd}(B)$ holds before a backward step from C and the inequalities $\underline{\lim}(B) < \mathrm{bd}(B) = \mathrm{Bd}(C) < \overline{\lim}(B)$ hold after it. Thus, before such a step we have

$$
\begin{aligned}
\overline{\lim}(C) &= \min\{\mathrm{bd}(A) | A \in P_b(C) = P_b(B)\} \\
&= \overline{\lim}(B) > \mathrm{Bd}(C) > \max\{\mathrm{bd}(B), \underline{\lim}(B)\} \\
&= \max\{\mathrm{bd}(B), \max\{\mathrm{bd}(A) | A \in P_w(B)\}\} \\
&= \max\{\mathrm{bd}(A) | A \in P_w(B) \cup B = P_w(C)\} = \underline{\lim}(C).
\end{aligned}
$$

The theorem on bounds of positions implies that after a backward step from C we have $\mathrm{Bd}(C) = \mathrm{sc}(C)$. Thus improving moves lead to positions with scores that are determined by the search. We call these *pseudocritical* positions.

If (B, C) is a refutation move, we have after a backward step from C

$$\mathrm{Bd}(C) = \mathrm{Bd}(B) \geq \overline{\lim}(B) = \overline{\lim}(C).$$

Accordingly, after a backward step from C we have, by the theorem on bounds of positions,

$$\mathrm{sc}(B) \geq \overline{\lim}(B) \text{ and } \mathrm{sc}(C) \geq \overline{\lim}(B) = \overline{\lim}(C),$$

i.e. the position B cannot be reached by the opponent, and the (Black) position C cannot be reached by its own side.

If (B, C) is a bad move we have before a backward step from C either $\mathrm{Bd}(C) \leq \mathrm{Bd}(B)$ or $\mathrm{Bd}(C) \leq \underline{\lim}(B)$. Therefore

$$\mathrm{Bd}(C) \leq \max\{\mathrm{Bd}(B), \underline{\lim}(B)\} = \underline{\lim}(C),$$

and by the theorem on bounds of positions, $\mathrm{sc}(C) \leq \underline{\lim}(C)$, i.e. the position C cannot be reached by the opponent.

The sets E, F, and G consisting respectively of pseudocritical positions, positions unreachable by the opponent, and positions unreachable by their own side, may be resolved into subsets of positions all of the same rank: $E = \bigcup E_i$, $F = \bigcup F_i$, $G = \bigcup G_i$.

We use the lower case characters e_i, f_i, and g_i to denote the numbers of elements in the corresponding subsets E_i, F_i, G_i. The base position A_0 belongs to the critical branch. Accordingly,

$$E_0 = \{A_0\}, \qquad F_0 = G_0 = \varnothing, \qquad e_0 = 1, \qquad f_0 = g_0 = 0.$$

Theorem on the Next Moves. *All moves from a position in the set G_i lead to a position in the set F_{i+1}; moves from positions in the set E_i lead to positions in the set $E_{i+1} \cup G_{i+1}$ and so all moves from positions in the set $E_i \cup G_i$ lead to positions in the set $E_{i+1} \cup F_{i+1}$. At least one move leads from any non-terminal position $B \in E_i$ to a position in the set E_{i+1}, and from any non-terminal position $B \in F_i$ exactly one refutation move leads into the set G_{i+1} (see Figure 13).*

PROOF. If $B \in E_i \cup G_i$ is a non-terminal position and (B, C) is the next move backward at B, and $\mathrm{bd}(B)$ is the bound at B after this backward move and $\mathrm{Bd}(B)$ is its last value, then

$$\mathrm{Bd}(C) \leq \mathrm{bd}(B) \leq \mathrm{Bd}(B) < \overline{\lim}(B) = \overline{\lim}(C).$$

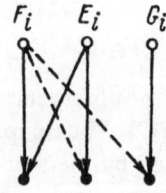

G_{i+1} E_{i+1} F_{i+1} Figure 13

The position C belongs to $E_{i+1} \cup G_{i+1}$ since it is a Black position and $\text{Bd}(C) < \overline{\lim}(C)$. Moreover, on no return to B can the pruning rule be applied, so that all immediate successors of B belong to this set as well. If $B \in G_i$, even stronger conditions are satisfied. After a forward step at B, $\text{bd}(B)$ takes on the value $-\infty < \underline{\lim}$. If before some backward step to B we have $\text{bd}(B) \leq \underline{\lim}(B)$, then

$$\text{Bd}(C) \leq \text{Bd}(B) \leq \underline{\lim}(B) = \max\{\text{bd}(B), \underline{\lim}(B)\} = \underline{\lim}(C),$$

i.e. $C \in F_{i+1}$. If, finally, $B \in F_i$ and (B, C) is the move determining the last value of $\text{Bd}(B)$, then

$$\text{Bd}(C) = \text{Bd}(B) \geq \overline{\lim}(B) = \overline{\lim}(C),$$

i.e. $C \in G_{i+1}$. For any preceding step backward to B from some position C' we have

$$\text{Bd}(C') < \text{Bd}(B) = \overline{\lim}(C').$$

So $C \notin G_{i+1}$ and exactly one move leads from B into the set G_{i+1}. □

Now suppose that the search is optimal, i.e. the leftmost move from any position in the set E is a best move and a move from any position in the set F is a refutation. Then after the first step to the position $B \in E_i$ with the next move (B, C) we have

$$\text{bd}(B) = \max\{-\infty, \text{Bd}(C)\} = \max\{\text{Bd}(C')|(B, C') \in \mathfrak{A}\} = \text{Bd}(B),$$

and after succeeding steps with the next moves (B, C') we have

$$\text{Bd}(C') \leq \text{Bd}(B) \leq \max\{\text{bd}(A)|A \in P_b(C')\} = \underline{\lim}(C),$$

i.e. the moves (B, C) are bad. After the first step backward to a position $B \in F_i$ with next move (B, C) we have $\text{bd}(B) \geq \overline{\lim}(B)$, i.e. the pruning rule must be applied at the next step. Then we must to some extent correct the graph of relationships given in Figure 13 for the sets E_i, F_i, G_i and $E_{i+1}, F_{i+1}, G_{i+1}$. See Figure 14, in which the unique moves available at non-terminal positions are depicted in boldface.

Let us consider three successive ranks, of order $i, i+1, i+2$. Figure 15 shows what moves can be made from positions in the sets E_i, F_i, G_i, E_{i+1}, F_{i+1}, and G_{i+1}. (The dotted lines show the non-considered moves from positions in the set F.) All moves from the set $E_i \cup G_i$ lead into the set $E_{i+1} \cup F_{i+1}$; the moves from the set F_i either lead into G_{i+1} or are ignored.

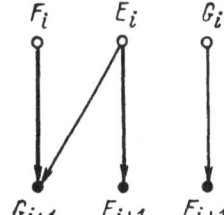

Figure 14 G_{i+1} E_{i+1} F_{i+1}

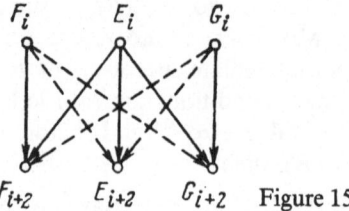

F_i E_i G_i

F_{i+2} E_{i+2} G_{i+2} Figure 15

Accordingly the average number of moves from positions in the set $E_i \cup G_i$ is equal to $\mu_i = (e_{i+1} + f_{i+1})/(e_i + f_i)$.

From every non-terminal position in the set $E_{i+1} \cup F_{i+1}$ one move leads into the set $E_{i+2} \cup G_{i+2}$ (from E_{i+1} it leads into E_{i+2}, and from F_{i+1} into G_{i+2}), and positions in the latter set cannot arise in any other way. Therefore the fraction of non-terminal positions in the set $E_{i+1} \cup G_{i+1}$ is equal to

$$\varepsilon_{i+1} = \frac{e_{i+2} + g_{i+2}}{e_{i+1} + f_{i+1}} \leq 1.$$

Thus

$$e_{i+2} + g_{i+2} = \mu_i \varepsilon_{i+1}(e_i + g_i),$$

whence it follows that

$$e_i + g_i = \prod_{j=1}^{\lceil i/2 \rceil} \varepsilon_{i-2j+1} \mu_{i-2j},$$

and

$$e_i + f_i = \mu_{i-1} \prod_{i=1}^{\lceil i-1/2 \rceil} \varepsilon_{i-2j} \mu_{i-2j-1}.$$

In the optimal search we are now considering, the successor positions are only those in the minimally inclusive W-pruned and B-pruned subtrees $\mathfrak{A}_w$ and $\mathfrak{A}_b$, which intersect in the critical branch W. Only the positions in this branch belong to the set E. Therefore $e_i = 1$ for $i \leq l$, where l is the rank of a terminal position in W, and $e_i = 0$ for $i > l$. If the rules for selecting the next move in the contemplated step are not specifically directed toward the minimization of the μ_i, the values obtained will usually be close to the mean numbers ν_i of all moves from positions in the i-th rank. Moreover, in many interesting cases there are almost no terminal positions of non-maximal rank (we shall see later that this property is shared by many model games played by programs). Then $\varepsilon_i \approx 1$, $\ln \varepsilon_i \approx 0$; we again assume that the values of the $\ln \nu_i$ are approximately equal as between odd and even ranks, and that there is an essential branching in the trees $\mathfrak{A}$, $\mathfrak{A}_w$, and $\mathfrak{A}_b$ for which the number of positions is approximately equal to the total number of terminal positions. Then we can show that the number of positions considered in an optimal search is of the same order of magnitude as the square root of the total number of all positions in the game tree $\mathfrak{A}$. In fact, let h_i be the number of positions of rank i in the tree, let $\overline{\ln}\mu$ and $\overline{\ln}\nu$ denote the means of the $\ln\mu_i$ and $\ln\nu_i$ averaged over all ranks, and let k be the maximum rank in the game tree $\mathfrak{A}$. Then

$$\ln(e_k + f_k) = \ln\mu_{k-1} + \sum_{i=1}^{\lceil(k-1)/2\rceil} \left(\ln\varepsilon_{i-2j} + \ln\mu_{i-2j-1}\right)$$

$$\approx \frac{k}{2}\overline{\ln}\mu + O(k) \approx \frac{k}{2}\overline{\ln}\nu + O(k),$$

$$\ln(e_k + g_k) = \sum_{i=1}^{\lceil k/2\rceil} \left(\ln\varepsilon_{i-2j+1} + \ln\mu_{i-2j}\right)$$

$$\approx \frac{k}{2}\overline{\ln}\mu + O(k) \approx \frac{k}{2}\overline{\ln}\nu + O(k),$$

$$\ln(e_k + f_k + g_k) \leq \ln(2\max(e_k + f_k, e_k + g_k))$$

$$\approx \frac{k}{2}\overline{\ln}\mu + O(k) \approx \frac{k}{2}\overline{\ln}\nu + O(k),$$

$$\ln h_k = \sum_{i=0}^{k-1} \ln\nu_j = k\overline{\ln}\nu,$$

where $\lceil x\rceil$ is the nearest integer not exceeding x (Knuth's notation [19]), and everywhere $O(k) < \lambda k$ for $\lambda \ll 1$. Then

$$h_k = e^{k\overline{\ln}\nu},$$

$$e_k + f_k + g_k \sim e^{\frac{k}{2}\overline{\ln}\nu + O(k)} = \left(\sqrt{n_k}(1+\lambda)\right)^k,$$

where $\lambda \ll 1$.

This conclusion may be sharpened for a completely uniform game $\mathfrak{A}_{m,k}$, in which all positions in the ranks of order $0, 1, \ldots, k-1$ are non-terminal and have the same number m of moves, while the positions of order k are all terminal. For such games

$$e_i = 1,$$

$$f_i = \mu_{i-1} \prod_{j=1}^{\lceil(i-1)/2\rceil} \varepsilon_{i-2j}\mu_{i-2j-1} - 1 = m^{\lceil(i+1)/2\rceil} - 1,$$

$$g_i = \prod_{j=1}^{\lceil i/2\rceil} \varepsilon_{i-2j+1}\mu_{i-2j} - 1 = m^{\lceil i/2\rceil} - 1, \qquad i = 0, 1, \ldots, k.$$

Thus the number of terminal positions that define the score of the base position, is equal to $\lceil(k+1)/2\rceil + \lceil k/2\rceil - 1$, and the number of all positions in the union of the minimal W-pruned and B-pruned subtrees is

$$\frac{m^{\lceil(k+3)/2\rceil} + 2m^{\lceil(k+2)/2\rceil} + m^{\lceil(k+1)/2\rceil}}{m-1} - \left(k + 2 - \frac{2}{m-1}\right).$$

In all, there are $M = (m^{k+1} - 1)/(m - 1)$ positions in the tree $\mathfrak{A}$. Of course, the minimum number of positions examined in a search is of the order of $\sqrt{M}$ for k even, and $\sqrt{mM}$ for k odd.

If the best and refutation moves do not immediately come to mind, a larger number of positions must be inspected. It can be shown that there are $(c\bar{m})^{\bar{k}/2}$ of these, where $\bar{k}$ is the mean depth of the tree, $\bar{m}$ is the mean number of moves at a generic position and $c > 1$ is of the order of the sum of the mean numbers of moves at positions in the set E and the non-refutation moves at positions in the set F. We content ourselves with a definition of the mathematical expectation of the number of positions considered during a search of a completely uniform game $\mathfrak{A}_{m,k}$ when the numbers of improving moves at all positions $B \in E$ are independent random variables with mean γ (clearly $\gamma \geq 1$), and the numbers of improving moves and bad moves at positions $B \in F$ are also independent random variables with means δ and ε. The independence of these random variables implies that the means of the numbers of positions in the trees are summed over the means of the numbers of positions in the subtrees in precisely the same way as individual values of these numbers in any concrete instance.

Let Ω_i be the mathematical expectation of the number of terminal positions in the B-subtree of the search tree, for positions $B \in E_i$ and let Φ_i and Ψ_i be the corresponding quantities for the sets F_i and G_i. Then

$$\Omega_0 = \Phi_0 = \Psi_0 = 1,$$
$$\Omega_{i-1} = \gamma\Omega_i + (m - \gamma)\Phi_i,$$
$$\Phi_{i-1} = \delta\Omega_i + \varepsilon\Phi_i + \Psi_i,$$
$$\Psi_{i-1} = m\Phi_i, \qquad i = 1, 2, \ldots, k.$$

The system of $3k$ equations represented by the last three lines above is linear and homogeneous in $3k + 3$ unknowns, and has rank $3k$. Accordingly it suffices to find three linearly independent solutions of the system. These we seek in the form

$$\Omega_i = \Omega t^{k-i}, \qquad \Phi_i = \Phi t^{k-i}, \qquad \Psi_i = \Psi t^{k-i}.$$

Substituting these values yields the three equations

$$\Omega t = \gamma\Omega + (m - \gamma)\Phi,$$
$$\Phi t = \delta\Omega + \varepsilon\Phi + \Psi,$$
$$\Psi t = m\Phi.$$

For these equations to have a non-zero solution the determinant

$$\begin{vmatrix} t - \gamma & -(m - \gamma) & 0 \\ -\delta & t - \varepsilon & -1 \\ 0 & -m & t \end{vmatrix}$$
$$= t^3 - (\gamma + \varepsilon)t^2 - [m(1 + \delta) - \gamma(\delta + \varepsilon)]t + \gamma m$$

must vanish.

Since $1 \le \gamma \le m$ one of the roots, say t_1, of the equation

$$t^3 - (\gamma + \varepsilon)t^2 - [m(1 + \delta) - \gamma(\delta + \varepsilon)]t + \gamma m = 0$$

must lie in the half-open interval $(0, \gamma]$. We write

$$q\delta = \frac{m - \gamma}{m + \varepsilon t_1 - t_1^2}, \quad \delta = \frac{\gamma - t_1}{t_1}.$$

In the cases that interest us, γ, δ, and ε are much smaller than m. Then t_1 and q depend on δ as shown in Figure 16, and $q \approx 1$. The value of t_1 is conveniently expressed in terms of q:

$$t_1 = \frac{\gamma}{1 + q\delta}.$$

The second and third roots are solutions of the quadratic equation

$$t^2 - (\gamma + \varepsilon - t_1)t - \frac{\gamma m}{t_1} = 0$$

and are equal to

$$t_{2,3} = \frac{\gamma q\delta + \varepsilon(1 + q\delta)}{2(1 + q\delta)} \pm \sqrt{m(1 + q\delta) + \left(\frac{\gamma q\delta + \varepsilon(1 + q\delta)}{2(1 + q\delta)}\right)^2}.$$

Thus the general solution of our system of equations has the form

$$\Omega_i = \Omega^{(1)}t_1^{k-i} + \Omega^{(2)}t_2^{k-i} + \Omega^{(3)}t_3^{k-i},$$

$$\Phi_i = \Phi^{(1)}t_1^{k-i} + \Phi^{(2)}t_2^{k-i} + \Phi^{(3)}t_3^{k-i},$$

$$\Psi_i = \Psi^{(1)}t_1^{k-i} + \Psi^{(2)}t_2^{k-i} + \Psi^{(3)}t_3^{k-i},$$

where

$$\Phi^{(1,2,3)} = \frac{t_{1,2,3} - \gamma}{m - \gamma}\Omega^{(1,2,3)},$$

$$\Psi^{(1,2,3)} = \frac{m}{m - \gamma}\frac{t_{1,2,3} - \gamma}{t_{1,2,3}}\Omega^{(1,2,3)}.$$

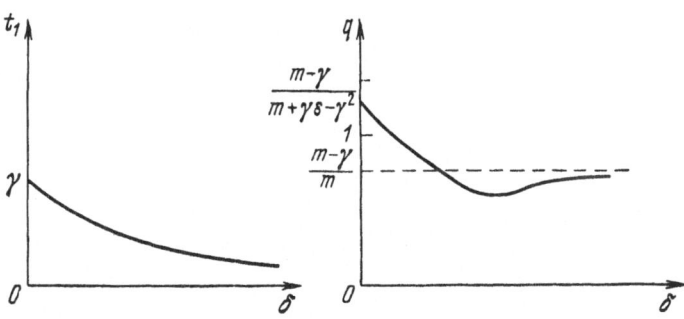

Figure 16

The values $\Omega^{(1,2,3)}$ may be found from the equations $\Omega_k = \Phi_k = \Psi_k = 1$.

Thus the mathematical expectation of the number of terminal positions considered in the tree $\mathfrak{A}_{m,k}$ is equal to

$$\Omega_0 = \frac{(m-t_2)(m-t_3)}{m(t_1-t_2)(t_1-t_3)} t_1^{k+1} + \frac{(m-t_1)(m-t_3)}{m(t_2-t_1)(t_2-t_3)} t_2^{k+1}$$
$$+ \frac{(m-t_1)(m-t_2)}{m(t_3-t_1)(t_3-t_2)} t_3^{k+1}.$$

CHAPTER 2

Heuristic Methods

Control of the Tree Size and Evaluation Functions

The goal of a game-playing program is to recommend a move in every position presented to it. The recommended move need not be the best, for instance in the sense defined in the preceding chapter, but it should have the highest quality possible, as seen by the professional player, and must be chosen within a preselected time limit. An arbitrary algorithm for choosing a move in the positions of a given game may be looked on as an algorithm for choosing the best move in the base position of the game tree of another game, which we shall call a *model* game. The tree for the model game is a subtree of the original game tree.

In constructing the model game we may adopt the goal of finding, so far as we can, the best move for the original game, or we may be guided by other considerations mentioned earlier. In practically all research on game-playing programs it has been assumed that one is to find the best move. Either an exhaustive search or a pruned search of the game tree may be used. In the course of such an algorithm, or its analogues, one constructs *W*-pruned and *B*-pruned subtrees of the contemplated game tree with their roots at the base position. Thus their processing time is of an order of magnitude not less than the number of positions in these subtrees. We are interested in games with strongly branching trees, such that even with maximal pruning the search cannot be completed within the available time. There are two ways to save time: a) build as small a model game as possible (with as few positions as possible in the subtrees to be studied), or b) perfect the method of choosing the best move for a fixed game tree.

We shall discuss methods based on the use of meaningful properties of the contemplated games. These methods allow us to choose moves that are

more or less satisfactory. If an algorithm for an arbitrary game is to be
significantly more effective than a pruned search, it must not construct
W-pruned and B-pruned subtrees, except perhaps in a preliminary and
implicit way. For such an algorithm we must define precisely what we mean
by calling a chosen move *fairly good*. If the algorithm is deterministic, then
by using the theorem on extensions of game trees we can find a game such
that in its winning positions one will always win the game against errorless
play by the opponent (if the game includes random factors, substitute
almost always for always). Of course, there exist concrete games to which
the notion of a *fairly good move* cannot be applied.

We shall call a method *exact* when it can be shown to solve the assigned
task correctly, and *heuristic* when no such proof has yet been found. (The
term *heuristic* is used in various senses; we have chosen one that suits us.)
The exact method constructs a model equivalent to the original game, i.e.
the best moves in the base positions of its trees are best moves in the
corresponding original game. A method for accelerating the choice of a best
move in the base position of a fixed game tree may be called exact if a proof
exists that its application does in fact shorten the search, i.e. decrease the
number of positions examined.

We begin with the heuristic methods that up to now have been the most
thoroughly elaborated. The first of these was described by Shannon [39]. In
a game tree with its root at a given position, he proposed to consider only
positions in the low ranks; that is, Shannon's algorithm has a parameter n,
the depth of the search, which is the maximum rank of the positions to be
examined. All positions in the n-th rank are considered terminal. We must
specify their scores; these of course need not coincide with the true scores,
which we do not know. We select an ensemble of features to be evaluated,
from among those occurring in the various theories of the given game.
Shannon's paper discusses a chess program and such features as the
material balance, weak and strong squares, attacks by the pieces on them,
elements of the pawn structure, etc.

The score for a terminal position in the model game must be equal to the
value of some function, that is reasonably easy to compute, of the features
characterizing the position. (Of course, an actual terminal position in the
original game is terminal in the model also; the won and lost positions have
model scores denoting won and lost games; a drawn position has an
intermediate score.) The contemplated function can be computed at any
position in the game; we call it an *evaluation function*. Let us now compare
models having varying depths of search but the same scoring function.
Since the game tree in chess is finite, Shannon's game tree with sufficiently
large n will coincide with it. In that case the model will choose the actual
best move.

On the other hand, if we could define an evaluation function that would
yield the true score in every position, the model need have only the depth
$n = 1$. This is often referred to as looking ahead one ply (taking account of
the fact that a whole move consists of the union of a move by White and a

move by Black). What significance can we attach to a search with depth $n > 1$ that does not reach the end of the game? Wins and losses that occur in very few moves, the so-called catastrophes, will of course be examined. But this is wholly insufficient for what would be called decent play. If the evaluation function can only distinguish between a position after a catastrophe has occurred and a prior position, it is of no use for anything else. However, for certain postulates on the character of the game and on the properties of the evaluation function one can prove that even in the absence of catastrophes the quality of play increases monotonely with the depth of the search.

Using Shannon's model, we now describe a somewhat idealized game and develop a naive probabilistic basis for an effective choice of best move in its positions. Let $\mathfrak{A}$ be a completely uniform game, i.e. White and Black move alternately, at every position the number m of available moves is the same, and all terminal positions have the same (sufficiently large) rank N. Suppose, moreover, that the number of winning moves at every position where they exist is the same, and is equal to 3.Let the terminal positions have scores 0 or 1. We postulate that the evaluation function $f(A)$ is a random variable such that in a White (Black) position won for White (Black) we have $f(A) = 1$ (0) with probability p, and $f(A) = 0$ (1) with probability $(1 - p)$; in lost positions the corresponding probabilities are $q < p$ and $(1 - q)$. At different positions A the values of $f(A)$ are independent random variables. It is easy to see that in this game there are no catastrophes.

As usual, we assume that White is to move in the base position A_0. A game-playing program using Shannon's model will make an errorless move whenever the position B arising in the game $\mathfrak{A}$ after a winning move has the score 1 and all positions B' to which losing moves (A_0, B') lead have score 0. We must determine the probabilities of the scores 1 and 0 for winning and losing moves of rank 1 in the original game $\mathfrak{A}$, using the Shannon model of depth n.

Let us begin with a Shannon model of depth 1 (the search omits some of the positions B of rank 1 whose scores are compared in searching for the best move). If for a position B of rank 1

$$\mathrm{msc}(B) = \min\{f(C)|(B,C) \subset \mathfrak{A}\} = 1,$$

the values of the evaluation function $f(C)$ are equal to 1 for all immediate successors C of B. Because these random variables are independently distributed for different positions C (as shown in Figure 17 the $C_1, C_2, \ldots, C_m$ are immediate successors of B), the probability of this event is equal to the product of the probabilities for every event $f(C_i) = 1$:

$$\mathbf{P}(\min\{f(C)|(B,C) \in \mathfrak{A}\} = 1)$$
$$= \mathbf{P}(f(C_1 = 1)) \cdot \mathbf{P}(f(C_2) = 1) \cdots \mathbf{P}(f(C_m) = 1).$$

We note that all the $C_1, C_2, \ldots, C_m$ are Black positions.

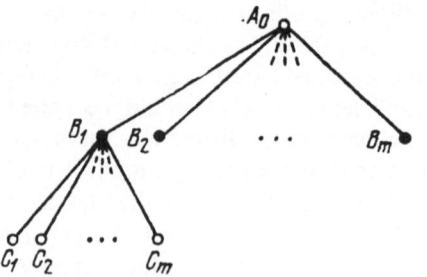

Figure 17

When B is a won position for White in the original game we shall denote the corresponding probability by $1 - Q_1$, otherwise by $1 - P_1$ (it is natural to assume that $P_0 = p$, $Q_0 = q$). In the first case all the positions $C_1, C_2, \ldots, C_m$ are also won for White. Accordingly,

$$1 - Q_1 = p^m.$$

In the second case, with s positions won for Black and the rest for White, we have

$$1 - P_1 = p^{m-s} q^s.$$

Let P_n be the probability of obtaining a score 1 for the base position A_0 in Shannon's model game when A_0 (a White position, as always) is won for White in the original game, and let Q_n be the probability of obtaining the score 1 if it is lost for White. Then the probability that the positions C_i ($1 \le i \le m$) in Figure 17 will obtain the score 1 in a Shannon model with depth $n + 1$ is P_n if C_i is in fact a won position and Q_n if it is a lost position. As in the model of depth 2,

$$1 - Q_{n+1} = P_n^m,$$
$$1 - P_{m+1} = P_n^{m-s} Q_n^s.$$

If P_n is near enough to 1 and Q_n to 0, P_{n+1} will be even larger and Q_{n+1} even smaller. Thus the probability of obtaining a correct score for the positions, and so finding the best move, increases with the depth n of the search and tends to 1 so that it is practically equal to 1 for $n \ll N$. Table 1 displays the values of P_n and Q_n as functions of n for $m = 10$, $s = 2$, $p = 0.99$, $q = 0.1$. However, if p is only slightly smaller than 0.99 or q is somewhat larger than 0.1, both P_n and Q_n fail to converge. After some number of oscillations both tend to 1 for one parity of n and to 0 for the other parity.

This result—the increase in the reliability of the Shannon model decision as the depth of the search increases—follows from an important property of the games in question, namely that in a won position there are more than several winning moves and that we can construct a relatively simple evaluation function that correlates well with the true score. Seemingly, both these properties are necessary if we are to count on a good result from

Table 1. Probability of Correct Model Score for Shannon Model Games

Notation

   m—number of moves at non-terminal positions
    s—number of winning moves at winning non-terminal positions
   n—depth of the model
   P_n—probability of a win in the model game for a won position in the original game
   Q_n—probability of a win in the model game for a lost position in the original game

	$m = 10$,	$s = 2$				
n	P_n	Q_n	P_n	Q_n	P_n	Q_n
0	0.9900	0.1000	0.9900	0.1200	0.9880	0.1000
1	0.9908	0.0956	0.9867	0.0956	0.9909	0.1137
2	0.9915	0.0885	0.9918	0.1252	0.9880	0.0872
3	0.9927	0.0817	0.9853	0.0792	0.9931	0.1139
4	0.9937	0.0709	0.9944	0.1374	0.9877	0.0669
5	0.9952	0.0612	0.9819	0.0543	0.9959	0.1162
6	0.9964	0.0467	0.9974	0.1666	0.9869	0.0398
7	0.9979	0.0355	0.9728	0.0252	0.9986	0.1234
8	0.9988	0.0210	0.9995	0.2410	0.9850	0.0142
9	0.9996	0.0123	0.9421	0.0051	0.9998	0.1407
10	0.9998	0.0044	0.9998	0.4490	0.9802	0.0018

applying Shannon's model, or others like it that we shall describe in Section 3 below. It may turn out that for some games the fulfillment of these conditions would be strictly proved or disproved. As of now such proof is lacking and we must resort to statistical tests of their fulfillment.

We should not be discouraged by the stringent requirements placed on the parameters p and q that define the correlation between the scores and the evaluation function. These requirements stem from the simplifications we have adopted, in particular from the postulate that the evaluation functions evaluated before and after a move are independently distributed, and from the non-differentiability of their values (0 and 1 only). Moreover, we must keep in mind that one purely qualitative achievement in the construction of the evaluation function, namely its positive correlation with the true score, may fail to yield any advantage (also, the required depth of the search depends on the level of this correlation and on circumstances that are not random). We shall return to this question in Chapter 4.

The evaluation function may depend on factors other than the true scores of the corresponding positions. If at some position a win can be achieved in 10 moves, and at another position in 4 moves, the scores of the two positions are identical. Yet for the players the two positions are not equivalent; it is always best to choose a move leading to a desired result in the shortest possible time. Therefore we may define the evaluation function in such a way that conditions near the end of the game can increase the score (when we expect a win). It is especially important to have such values for the evaluation function at positions where the theory of the game specifies a win (this does not mean that we can give a strict proof of a win).

For instance, in easily won chess endgames, where a lone King is to be mated, one may avoid a failure to win, so long as the 50 move rule and the rule of triple repetition of a position do not apply. However, even when the evaluation function agrees with the true score, it is impossible to guarantee a mate when the search depth $n \leq 10$. At the same time, if the evaluation function yields a better value for the attacking side when the lone King is near the edge of the board and the corners (when the mate is to be given by Bishop and Knight the corner must be of the same color as the Bishop), and the attacking King is near the opposing King, then a program with depth 4 will win, and the fourth ply is needed only to test whether a mate or a stalemate occurs on the third ply. For a mate by two minor pieces a specially selected score is needed for a few standard positions with the opposing King in or near a corner of the board.

As we have said, an evaluation function should be relatively easy to compute. The simplest would be linear. We shall define a collection of features of a position—predicates $p_i(A)$ which have the value 1 if A possesses the i-th feature and the value 0 otherwise. If q_i is the weight of the i-th feature ($i = 1, 2, \ldots, N$) the evaluation function is defined by the formula

$$\Phi(A) = \sum_{i=1}^{N} q_i p_i(A),$$

where the weights may be either positive or negative. It can be shown that when the number of features to be taken into account is large, then even for a very simple game there exists no linear evaluation function that is monotonely related to the true scores of the positions [22]. Therefore the problem of complicated methods for computing the evaluation function is unavoidable. However, the determination of model scores of positions by Zermelo's formula, starting from the value of an evaluation function in pseudoterminal positions, is also a method for computing evaluation functions and we may confine our considerations to processes that require significantly less work.

The use of threshold logic provides an example of a moderate complication of the methods for computing an evaluation function. Aside from the initial set of features—the system of predicates $\{p_i(A)\}_1^n$—we consider a set of evaluation functions

$$\left\{ \Phi_j(A) := \sum_{i=1}^{N} q_i^j p_i(A) \right\}_1^M$$

and thresholds R_j for them. The evaluation function $\Phi(A)$ is defined by the formula

$$\Phi(A) = \sum_{j=1}^{M} Q_j \Pi_j(A),$$

where $\{Q_j\}_1^M$ is a set of weights and

$$\Pi_j(A) := \begin{cases} 1, & \text{if } \Phi_j(A) = \sum_{i=1}^{M} q_i^j p_i(A) \geq R_j, \\ 0, & \text{if } \Phi_j(A) < R_j, \qquad j = 1, 2, \ldots, M. \end{cases}$$

The ability of threshold logic to represent essential properties of certain objects of study was investigated by Minsky and Papert [25], who elucidated the constraints on the nature of the properties, the structure of the logical scheme, and the number of elements. However, the properties they studied that are to be reflected by threshold logic are quite different from those properties of a position that are to be represented by an evaluation function.

In the existing game programs the approach to the use of threshold logic to compute the evaluation functions $\Phi(A)$ is somewhat different. It is connected with the concept of the *position type* borrowed from chess theory. An elementary feature of a type is computed by means of the value of a linear function

$$\Psi_j(A) := \sum_{i=1}^{N} \psi J_i^j p_i(A)$$

of the elementary features $p_i(A)$ of a position A and the thresholds R_i ($i = 1, 2, \ldots, M-1$). The position type is defined by logical functions of these features. For example, the predicates $P_k(A)$ indicating that A is of type k may be defined by the formulae

$$P_1(A) := \{\Psi_1(A) \geq R_1\},$$

$$P_k(A) := \underset{j=1}{\overset{k-1}{\&}} \{\Psi_j(A) < R_j\} \& \{\Psi_k(A) \geq R_k\},$$

$$k = 2, \ldots, M-1,$$

$$P_M(A) := \underset{j=1}{\overset{M-1}{\&}} \{\Psi_j(A) < R_j\} \& \{\Psi_M(A) \geq R\}.$$

For every type a linear evaluation function

$$\Phi_j(A) := \sum_{i=1}^{N} \phi_i^j P_i(A), \quad j = 1, 2, \ldots, M,$$

is defined; the functions ϕ_i^j of the features $p_i(A)$ with non-zero coefficients in this evaluation function are usually not the same as those with non-zero coefficients in the functions Ψ_j that define the type. Thus the value of the evaluation function for an arbitrary position A is given by

$$\Phi(A) = \sum_{j=1}^{M} \Phi_j(A) P_j(A).$$

Another way to develop a complex evaluation function, still within the framework of linear functions, is to use not only elementary features given *a priori* but also logical functions of them. An arbitrary function of the elementary predicates $P_i(A)$ ($i = 1, 2, \ldots, N$) defined above can be defined in the same way. In fact, let $\Phi_{\varepsilon_1 \varepsilon_2 \ldots \varepsilon_N}$ be the value of the contemplated function for the arguments $p_i(A) = \varepsilon_i$, where ε takes on the values 0 or 1; then

$$\Phi(A) = \sum_{\varepsilon_1 = 0}^{1} \sum_{\varepsilon_2 = 0}^{1} \cdots \sum_{\varepsilon_N = 0}^{1} \Phi_{\varepsilon_1 \varepsilon_2 \ldots \varepsilon_N} \{ p_1(A)$$

$$= \varepsilon_1 \} \& \cdots \& \{ p_N(A) = \varepsilon_N \}.$$

In such a representation the number of terms is very large (equal to 2^N), and only relatively short formulae may be easily computed. In practical applications the evaluation functions usually do have non-elementary terms, but only a few of them.

So, in actual game programs the evaluation functions either depend linearly on features of a position that are sufficiently simple in their definition or they are superpositions of linear and logical functions. In constructing a model game we must decide what features to take into account, and what weights to assign to them. Shannon proposed deriving a collection of elementary features from the theory of any specific game. In the theories of chess, checkers (draughts), bridge, etc. there are a series of such features that monotonely influence the outcome of the game. Almost all of them, however, lack a precise formal definition (material balance is an exception). We need to construct formal features that coincide at least approximately with those we are interested in. It is important to note that essential features have been found which are unknown in the theories of actual games. Also, there are interesting features for which no algorithmically expressible equivalents (even approximate) have been found.

Many papers have been devoted to the problem of weights to be assigned to features, and two basic concepts have been elaborated. The first says that in every position there is one dominant feature, and the weights should be so chosen that the presence of an arbitrary positive feature would be suppressed only by the presence of a more important feature, and not by a combination of less important features. If an arbitrary logical function of the elementary features may be thought of as a new feature that affects the value of the evaluation function, the first concept is meaningless. Moreover, when the amount of computing is limited and therefore there are not too many terms in the linear expression, the concept in its pure form conflicts with the empirically based method of computing the material balance (the values of the pieces in chess or bridge). Attempts to apply it even to positional features have not succeeded (the strong game of Samuel's program [33] for American checkers would appear to be unconnected with the fact that he used this concept in assigning weights to features).

The second concept approaches the assignment of weights empirically. They are introduced as parameters in the program and after various games have been played the values chosen are those for which the program plays most strongly. It has also been proposed that an automatic process be set up, in which the values of the parameters change in the course of a single game. Such programs are sometimes referred to as *self-learning*, but the appropriateness of this term is disputable. The authors have made a statistical study of the way in which various values of the weights affect the strength of a chess program. It turns out that random variation of the weights, within wide limits around the values established *a priori* on the basis of certain chess-theoretical considerations, has very little influence on the strength of the play (of course, we are speaking of positional features). When, however, the weights are so chosen that some positional features have practically no influence on the score, the quality of play worsens significantly. It is also important to note that no positional factor will dominate a group containing a substantial number of other features.

Some studies, e.g. [12], have taken up the problem of automatically choosing the logical functions (of the elementary features) from which the evaluation function is built up. The procedure is to enumerate the logical functions, compute the correlation of these with scores made by experts for various test positions, and choose the most informative of the logical functions. The work done by such a program may be more rightfully called learning than can changing the weights of features in a given program. The *teacher* — the expert who specifies the collection of test positions and their scores—plays a fundamental role in this process. Such methods have achieved definite results in studies on pattern recognition (cf. [6]). However, the learning of programs for the diagnosis of position scores in a game has not yet yielded any substantial results.

The first functioning chess programs implemented Shannon's ideas literally and played very weakly. The search was comparatively shallow, many technical programming problems were solved in a far from satisfactory fashion, and the pruning method (the α, β heuristic) was not used. A much more salient shortcoming of the programs was their use of a fixed search depth and their failure to vary it depending on the character of the position. For instance, an exchange might be broken off at an arbitrary instant when the material balance on the board failed to reflect the true balance. The shortcomings of the first models were removed in various ways, some of which will be discussed in the next section. No really strong chess program has yet been developed, although very strong programs exist for simpler games. A program with rather simple ideas plays Russian draughts [36–38] at a good level, and Samuel's program for American checkers (mentioned earlier) plays at a level that need not yield to professionals. There also exist programs that correctly solve the problem of mate in a few moves or play correctly in simple endgames such as King and Pawns against King.

On the Order in which Positions are Searched in the Game Tree

Let us begin with algorithms for a search with pruning (the α, β heuristic). As we proved in Chapter 1, the minimum number of positions to be examined in the game tree $\mathfrak{A}$ in order to establish the score of the base position A_0 and the best initial move, is of the order of $m^{k/2}$, where m is the mean number of moves at the positions to be examined and k is the depth of $\mathfrak{A}$. To attain this minimum we must examine the best move first when in the critical path W, and we must examine one of the refutation moves first in positions that are unreachable by the opponent. If, on the other hand, we examine the moves at any given position in increasing order of quality, i.e. each move examined is better than its predecessor, we will be making a practically exhaustive search and will look at $O(m^k)$ positions. It is therefore essential to determine a sequence for looking at positions that is as close as possible to the optimal.

However, in order to guarantee optimality of the search sequence, we need to know which is the best move at most positions in the game tree $\mathfrak{A}$, i.e. we must be able to solve at these positions the same problem we needed to solve only at the base position A_0. We are interested in the case in which we do not know how to solve this problem and can only more or less rapidly guess the necessary best or refutation move. The search process itself may be looked on as a test of the correctness of our guesses. The methods for guessing may be formulated as a sharpening of the rule for deepening a search with pruning, which tells us how to choose the next move in a step forward.

Since the search sequence to be constructed cannot be guaranteed to be optimal, there may be errors in it. Therefore it is useful to know how the number of positions in the search depends on errors of different kinds and what errors we should above all try to avoid, even at the cost of an increase in the number of errors of other kinds. In Section 3 of Chapter 1 the quality of a search sequence was described by three parameters:

γ—the mean number of improving moves at positions with determined scores (candidates for the critical branch);

δ—the mean number of improving moves at positions from which later refutation moves can be found (unreachable by the opponent);

and

ε — the mean number of bad moves at the same positions.

In a completely uniform game $\mathfrak{A}_{m,n}$ of depth n with m moves at each non-terminal position, when the numbers of improving and bad moves are equal to the mean values corresponding to the position type, the number of terminal positions in the search tree is defined by the formula

$$\Omega_{m,n} = \Omega^{(1)}t_1^n + \Omega^{(2)}t_2^n + \Omega^{(3)}t_3^n,$$

where $t_1, t_2, t_3, \Omega^{(1)}, \Omega^{(2)}, \Omega^{(3)}$ are determined by the parameters $\gamma, \delta, \varepsilon$, and m. Thus the way in which the growth of the number of positions depends on the depth n of the tree is determined by the largest of the numbers t_1, t_2, t_3; these are the roots of the cubic equation

$$t^3 - (\gamma + \varepsilon)t^2 - (m(1+\delta) - \gamma(\delta + \varepsilon))t + \gamma m = 0.$$

If $\gamma, \delta, \varepsilon$ are much less than m, the roots t_1, t_2, t_3 can be expressed in terms of a parameter q which is near to 1:

$$t_1 = \frac{\gamma}{1 + q\delta},$$

$$t_{2,3} = \pm \sqrt{ m(1+q\delta) + \frac{\left(\dfrac{q\gamma\delta}{1+q\delta} + \varepsilon\right)^2}{4} } + \frac{\dfrac{q\gamma\delta}{1+q\delta} + \varepsilon}{2}.$$

$$0 < t_1 \le \gamma,$$

$$|t_2| = \sqrt{m(1+q\delta)} + O(\gamma + \varepsilon),$$

$$|t_3| = \sqrt{m(1+q\delta)} - O(\gamma + \varepsilon).$$

The number of positions to be examined increases with increasing values of $\gamma, \delta, \varepsilon$ but these parameters have differing effects on the rate of increase. Roughly speaking, γ and ε form additive terms, whereas δ determines the multiplier $\sqrt{1+q\delta}$ in the formula for t_2, which specifies the order of the growth rate in the number of terminal positions as it depends on the depth n of the tree $\mathfrak{A}_{m,n}$. The values of $\Omega_{m,n}$ are shown in Table 2 for various values of γ, δ, and ε.

Now let us ask how individual errors in determining the search sequence in the search tree $\tilde{\mathfrak{A}}$ influence the number of positions. Let A be a next position of rank $l < n$ which is unreachable by the opponent. If instead of examining the refutation move (A, B) we choose the improving move (A, B'), we include the B'-subtree in the tree $\tilde{\mathfrak{A}}$ to define the score at B'. This is equivalent to the search tree of the game $\mathfrak{A}_{m,n-l-1}$ and, as we found in §3 of Chapter 1, in the absence of other errors this tree contains

$$\frac{\sqrt{m}^{\,n-l-2}}{2}\left((\sqrt{m}+1)^2 + (-1)^{n-l}(\sqrt{m}-1)^2\right) - 1$$

$$= m^{\lceil (n-l)/2 \rceil} + m^{\lceil (n-l-1)/2 \rceil} - 1$$

Table 2. Number of Positions in the Search Tree as a Function of the Number of Moves at Non-Terminal Positions and the Move-Inspection Sequence

Notation

m—number of moves at non-terminal positions,

n—depth of the search tree,

γ—mean number of improving moves at candidates for the critical branch,

δ—mean number of improving moves at positions unreachable by the opponent,

ε—mean number of bad moves at positions unreachable by the opponent,

A_n—number of terminal positions in the tree $\mathfrak{A}_n$ that defines the score of the base position,

B_n—number of terminal positions in the tree $\mathfrak{B}_n$, which determines that the base position is unreachable by the opponent,

C_n—number of terminal positions in the tree $\mathfrak{C}_n$, which determines that the base position is unreachable by oneself.

$$\gamma = 1, \qquad \delta = 0, \qquad \varepsilon = 0$$

	$m = 2$			$m = 10$			$m = 40$		
n	A_n	B_n	C_n	A_n	B_n	C_n	A_n	B_n	C_n
0	1	1	1	1	1	1	1	1	1
1	2	1	2	10	1	10	40	1	40
2	3	2	2	19	10	10	79	40	40
3	5	2	4	109	10	100	1639	40	1600
4	7	4	4	199	100	100	3199	1600	1600
5	11	4	8	1099	100	1000	65599	1600	64000
6	15	8	8	1999	1000	1000	127999	64000	64000
7	23	8	16	10999	1000	10000	2623999	64000	2560000

$$\gamma = 2, \qquad \delta = 0, \qquad \varepsilon = 0$$

	$m = 2$			$m = 10$			$m = 40$		
n	A_n	B_n	C_n	A_n	B_n	C_n	A_n	B_n	C_n
0	1	1	1	1	1	1	1	1	1
1	2	1	2	10	1	10	40	1	40
2	4	2	2	28	10	10	118	40	40
3	8	2	4	136	10	100	1756	40	1600
4	16	4	4	352	100	100	5032	1600	1600
5	32	4	8	1504	100	1000	70864	1600	64000
6	64	8	8	3808	1000	1000	202528	64000	64000
7	128	8	16	15616	1000	10000	2837056	64000	2560000

Table 2 (*Continued*)

$$\gamma = 1, \qquad \delta = 1, \qquad \varepsilon = 0$$

	m = 2			m = 10			m = 40		
n	A_n	B_n	C_n	A_n	B_n	C_n	A_n	B_n	C_n
0	1	1	1	1	1	1	1	1	1
1	2	2	2	10	2	10	40	2	40
2	4	4	4	28	20	20	118	80	80
3	8	8	8	208	48	200	3238	198	3200
4	16	16	16	640	408	480	10960	6438	7920
5	32	32	32	4312	1120	4080	262042	18880	257520
6	64	64	64	14392	8392	11200	998362	519562	755200
7	128	128	128	89920	25592	83920	21261264	1753562	20782480

$$\gamma = 1, \qquad \delta = 0, \qquad \varepsilon = 1$$

	m = 2			m = 10			m = 40		
n	A_n	B_n	C_n	A_n	B_n	C_n	A_n	B_n	C_n
0	1	1	1	1	1	1	1	1	1
1	2	2	2	10	2	10	40	2	40
2	4	4	4	28	12	20	118	42	80
3	8	8	8	136	32	120	1756	122	1680
4	16	16	16	424	152	320	6514	1802	4880
5	32	32	32	1792	472	1520	76792	6682	72080
6	64	64	64	6040	1992	4720	337390	78762	267280
7	128	128	128	23968	6712	19920	3409108	346042	3150480

$$m = 10$$

$$\gamma = 1.5, \qquad \delta = 0.5$$

	ε = 0.5			ε = 1.0		
n	A_n	B_n	C_n	A_n	B_n	C_n
0	1	1	1	1	1	1
1	10	2	1	10	2	10
2	32	16	20	36	18	25
3	184	44	160	203	61	175
4	650	274	440	820	337	606
5	3304	902	2740	4096	1353	3372
6	12623	4843	9020	17648	6773	13534
7	60100	17753	48430	84046	29132	67734

Table 2   (*Continued*)

$$\gamma = 1.5, \qquad \delta = 1$$

	$\varepsilon = 0.5$			$\varepsilon = 1.0$		
n	A_n	B_n	C_n	A_n	B_n	C_n
0	1	1	1	1	1	1
1	10	2	10	10	3	10
2	36	21	25	40	23	30
3	235	72	212	256	94	230
4	963	483	719	1179	580	935
5	5554	1924	4834	6697	2694	5798
6	24685	11351	19239	32943	15188	26939
7	133508	49599	113507	178512	75069	151879

$$\gamma = 2, \qquad \delta = 0.5$$

	$\varepsilon = 0.5$			$\varepsilon = 1.0$		
n	A_n	B_n	C_n	A_n	B_n	C_n
0	1	1	1	1	1	1
1	10	2	10	10	2	10
2	36	16	20	40	18	25
3	200	46	160	220	62	175
4	768	283	460	940	348	625
5	3800	986	2830	4660	1442	3475
6	15484	5223	9855	20860	7248	14425
7	72750	20208	52228	99700	32102	72475

$$\gamma = 2, \qquad \delta = 1$$

	$\varepsilon = 0.5$			$\varepsilon = 1.0$		
n	A_n	B_n	C_n	A_n	B_n	C_n
0	1	1	1	1	1	1
1	10	2	10	10	3	10
2	40	21	25	44	23	30
3	250	76	212	272	97	230
4	1105	500	756	1320	599	970
5	6212	2111	5003	7432	2889	5990
6	29316	12271	21114	37976	16311	28890
7	156803	56566	122713	206440	83177	163110

terminal positions. If we first examine the bad move (A, B'') instead of the move (A, B), then in the absence of other errors we must also examine the minimal W- and B-subtrees that bound the score of the position B'' from one side. These contain

$$\frac{\sqrt{m}^{n-l-2}}{2}\left((\sqrt{m}+1)+(-1)^{n-l}(\sqrt{m}-1)\right) = m^{\lfloor(n-l-1)/2\rfloor}$$

terminal positions, i.e. a lesser number.

Now suppose that as a result of the search the next position A turns out to be a candidate for the critical branch. After each improving move (A, B) we must determine the score of the position B, i.e. in the absence of errors in the search of the B-subtree of the game $\mathfrak{A}_{m,n}$ we must consider

$$m^{\lceil(n-1)/2\rceil} + m^{\lceil(n-l-1)/2\rceil} - 1$$

terminal positions; after a bad move (A, B) we must consider, in the best case, $m^{\lceil(n-l-1)/2\rceil}$ such positions (l is the rank of the position A). If a bad move (A, B') is examined by mistake before an improving move, it may appear to be improving. In that case, in the absence of other errors, we must consider $m^{\lceil(n-l)/2\rceil}$ superfluous terminal positions, i.e. fewer than if we had preferred an improving move to a refutation move for examination among the positions of the same rank. Moreover, when the number of errors in the search sequence is not very large, the number of candidates for the critical branch is much less than the number of positions unreachable by the opponent, and when the probabilities of error in positions of the first type are identical, the number will still be less.

Thus, the lower the rank of the position in which an error in defining the search sequence occurs, the more serious the consequences of the error. Therefore some extra work is worth doing in sharpening the search sequence in next positions of low rank. On the average, first place should be given to attempts to find refutation moves quickly, even if the result is to increase the probability of examining bad moves and decrease the probability of examining the best move. In the base position A_0, however, which necessarily lies on the critical branch, it is especially important to find the best move as soon as possible. If the tree $\mathfrak{A}$ of the contemplated game is far from uniform, an additional requirement is to give precedence to moves leading to *simple* positions, i.e. bases of subtrees of comparatively small volume.

Thus we may formulate the following requirements on the order in which positions are examined:

(1) First choose moves that have a good chance of being a best move.
(2) First choose moves that have a good chance of being a refutation, even if these have high chances of turning out to be bad.
(3) In approximately equal circumstances, prefer moves leading to subtrees of smaller dimension; for instance those in which the opponent has fewer responses.

(4) The determination of the search sequence at positions of high rank l, near the middle depth of the game tree $\mathfrak{A}$, must not be too laborious.

The first and second of these requirements conflict to a certain extent (as far as the authors know, the second is omitted in all existing game programs). At the base position A_0, and possibly in low-rank positions, the first requirement is most important. In fact, the first move (A_0, B) chosen at the base position A_0 is an improving move, even if it is rather bad. After a move satisfying the first requirement has been selected, examined, and has yielded a score, we should examine 'sharp' moves, which may turn out to be a refutation or bad. If there is reason to suppose that the opponent has already made a bad move on the branch leading to the next position A from the base position A_0, we may be guided by this requirement and choose an improving move rather than a refutation move.

Let us now take up two questions: What features of positions and moves may have a bearing on the search sequence? and how do we formulate the rules for choosing the next move in a forward step. We may use the same features that we applied in computing the value of the evaluation function $f(A)$. Suppose that the differences $f(B_i) - sc(B_i)$ between the evaluation function values and the true scores, at the positions B_i reached by the moves (A, B_i) from the next position A, are independently distributed random variables with identical distribution functions. Then these moves should be considered in the order of decreasing (increasing) values of the evaluation function $f(B_i)$ if White (Black) is to move at A. In fact, in an arbitrary subset of such moves from a White position, the move to the position B_i with the highest value of the evaluation function has the highest probability of being a best or refutation move, independently of what other moves may have been examined; when A is a Black position, the minimum value of $f(B_i)$ yields the highest probability.

The assumptions justifying such an order of search are generally not satisfied for the games in which the authors (and the reader) are interested. The assumption that all the random variables $f(B_i) - sc(B_i)$ are identically distributed is extremely dubious; to the contrary, there is reason to assume that one can find easily computed position features with high expected absolute values of these differences. The above method for choosing the next move suffers from two fundamental shortcomings:

(1) If the chosen move $\Psi_i = (A, B_i)$ is among the first and turns out to be a refutation move, then it is a waste of time to process the position A through all the positions $B_{i'}$ that result from the moves $\Psi_{i'} = (A, B_{i'}) \in \mathfrak{A}$ and are excluded from the search after evaluation of the function $f(B_{i'})$.

(2) A move (A, B_i) with a value of $f(B_i)$ that favors one's own side has a good chance of being an improving move rather than a refutation.

For the first reason, if A is a position with high rank, it is inexpedient to order the moves Ψ_i $(i = 1, 2, \ldots, m)$ leading from it in an order of decreasing or increasing value of $f(B_i)$ at their destinations B_i. Instead, we may order the moves by means of an instant evaluation function $\phi(\Psi_i)$ which can be computed without knowing the positions B_i. For example, consider the game of noughts and crosses, where the aim is to place five of one's own symbols (a nought or a cross) in an unbroken line (vertical, horizontal, or diagonal) of squares on a board of large horizontal and vertical dimensions (number of squares). A move after which there arises either an *open triplet*, consisting of three of one's own pieces in a line unblocked by the opponent, or a *half-open quadruplet* consisting of four of one's own pieces in line and blocked by the opponent at one end only, is said to be *dangerous* since it may lead to a win. It is possible to decide that a proposed move has this character without actually making it. (Such a move satisfies the third requirement: the opponent must reply either by blocking the contemplated configuration or by posing an even greater threat. Obviously the number of replies that need to be considered is much smaller than in positions without threats.)

To order the moves we may use the notion of identical moves in different positions; this concept is applicable to many games. Intuitively, two moves may be considered identical if the change in the positions of the pieces is the same, without regard to the disposition of the non-moving pieces. With this concept, the number of different moves is far less than the total number of different positions. In chess there are about 10 000 different moves, compared to not less than 10^{60} different positions. Some of them, for instance the incursion of a White Rook into an uncontested square in the 7th or 8th rank, or a Black Rook into the 1st or 2nd, or the capture of a piece having a value higher than the attacker's, may be taken as having great weight, independently of the position in which they originate. Another —castling, a standard developmental move—takes on great weight in the presence of easily determined features of the base position, e.g. the presence of many pieces on their original squares.

These methods may be regarded as static in that they are based on studying a move and the positions before and after it. The study of the dynamic circumstances, connected with the search process itself, is much more important. We can gain much more useful information as a by-product of the search than we can by an over-laborious analysis of a move and the positions before and after it. To be sure, the dynamic method relates to other positions; however, neighboring positions in a search tree differ only in the locations of a few pieces, and the properties of the positions are relatively stable under such changes.

In fact, the values of the evaluation function depend linearly, or almost linearly, on features defined by the location of only a few pieces; there are many features and each piece influences a number of them. But, in positions close to each other in the game tree $\mathfrak{A}$ most of the pieces are in identical locations, and the positions all have or do not have the corresponding features. Also, the rules of the game allow many identical moves to be made in these positions, and the features change in similar ways after these moves (all the more so when the features depend on the moves and not on the positions). Finally, many features depend on the locations of slowly moving pieces—such as Pawns and Kings in chess—which if they do move do not move very far. The same property is exhibited by features that do not play the same role in ordering the search sequence as they play with respect to the value of the evaluation function, e.g. when they determine the probable accuracy of the function.

The above arguments imply that moves which are best or improving in one position have a good chance of being best or improving in neighboring positions, and that moves leading to the root of a small subtree will probably preserve this property. In the search process, therefore, it is worth while to collect statistical information on the quality of the moves and use it for choosing the next move. The portion of the program that attends to such information and to the choice of move on the basis of it, is called the *best move table* (ordinarily we do not distinguish between best and improving moves).

This routine may use an ample statistical data base, requiring a large memory, capable of holding all the desired information for say 10 000 moves in chess. The principal difficulty lies in the need to search this massive memory for the information required about each admissible move at each next position, or conversely, for many moves having relevant information stored in the memory, to determine whether they are or are not admissible in a given position. This takes a great deal of machine time and is undesirable, at least for positions of high rank in the search tree.

Normally the best move table preserves information only about moves that have been explored in the search process, producing a rough list of best and improving moves. One method is the use of the so-called substitution scheme. Suppose the program has calculated and preserved l best moves for each side. Then we have two arrays $\Xi_w[i]$ and $\Xi_b[i]$, each holding l elements. When we want to apply the deepening rule during the search and we must choose the next move from the White (Black) position A, we first look for the move $\Xi_w[i]$ ($\Xi_b[i]$) with the lowest index k, $1 \le k \le l$, that is admissible at A within the rules of the model game, and not yet examined. If there is no such move, we select moves in an order determined by a static process.

The arrays $\Xi_w[i]$ and $\Xi_b[i]$ are updated when we step backward. Suppose such a step is made from a White (Black) next position B. If this position is unreachable by the opponent, we do not update. Otherwise we know the best or improving move $H = (B, C)$ for Black (White) from the position C. This move may or may not be in the corresponding array $\Xi_w[i]$ ($\Xi_b[i]$). If it is, the move H is pushed upward in the array, i.e. if $H = \Xi_{w(b)}[i]$ ($i = 1, 2, \ldots, l$) we make the substitutions

$$R := \Xi_{w(b)}[i];$$
$$\Xi_{w(b)}[i-1] := \Xi_{w(b)}[i];$$
$$\Xi_{w(b)}[i] := R;$$

where R is a temporary entry used to exchange elements in the array. Obviously if $H = \Xi_w[1]$ $(\Xi_b[1])$ it is already at the top of the stack and cannot be pushed upward.

If H is a new move (perhaps a best move not preserved in memory) it is inserted in the k-th place in the array; elements with index from k to $l-1$ are pushed down and the l-th element is erased:

$$\Xi_{w(b)}[i] := \Xi_{w(b)}[i-1] \ (i = k+1, \dots, l);$$

$$\Xi_{w(b)} := H.$$

The substitution scheme is parameterized by k and l. Given maximum confidence in the data base, $k = l$, i.e. a new move is placed at the last position; given minimal confidence, it is placed first, i.e. $k = 1$.

Maximum confidence is founded on the assumption that information in the data base about best moves is mostly reliable, i.e. consists of frequently occurring best moves, especially in positions approximating the one under examination; the appearance of a new move often turns out to be a random nuance. Minimum confidence is founded on the assumption that the set of *candidates for best move* changes rather quickly, and the appearance of a new move most often signals such a change. Both of these assumptions are extremes, and the value of k is usually set somewhere between 1 and l.

Let us now consider a somewhat more complex best move service routine, in which we take account (even though roughly) of three factors: the frequency with which the moves turn out to be best or refutations; the size of the corresponding subtrees; and the nearness of the positions in which these moves occurred to the one under examination. We are to choose the next move $H = (A, B)$, from a position A during a step forward to a position B among those accessible from A and not yet investigated. We compute the value of a *priority function* $\Psi(H)$ depending on the above parameters and on static features, after which we choose the move H having the highest priority value. (To save time, we may use a simpler method of choosing a move for positions of high rank.)

As in the simpler variant, the arrays in the tables are updated during a backward step, when the best or refutation move $H(B, C)$ from the next position B is known. We now describe the updating of the elements of the array for the corresponding color. We have the following data bases:

$\Xi[i]$—moves appearing in the table;

$\quad \nu_w[i]$, $\nu_b[i]$—the number of cases in which the corresponding moves were best moves and refutations;

$\quad \gamma_w[i]$, $\gamma_b[i]$—the mean number of positions, normalized with respect to the standard depth, in the corresponding subtrees of the search;

$\quad \rho_w[i, j]$, $\rho_b[i, j]$—the shortest distances, measured in the search tree $\tilde{\mathfrak{A}}'$, between the positions A_j in the branch $W(A_0, A_1, \dots, A_h)$ that connects the base position A_0 with the next position A_h, and the positions in which the move $\Xi[i]$ appears as a best or refutation move, respectively.

On a backward step from A_h, when the best (or refutation) move $H = (A_h, B)$ is known, the move is inserted in the array $\Xi[i]$. If it is a new move it displaces some move $\Xi[i]$ and takes its place, while we define the initial values of the parameters as:

$$\nu_w[i] := \nu_b[i] := \gamma_w[i] := \gamma_b[i] := 0;$$

$$\rho_w[i, j] := \rho_b[i_0, j] := \infty \qquad (j = h, h - 2 \ldots).$$

The symbol ∞ is to be interpreted as a number larger than any possible distance between nodes of the search tree $\tilde{\mathfrak{A}}'$. The displaced move must be rarely encountered and far away from the positions A_j.

Depending on whether the move $H = \Xi[i]$ is a best or refutation move we next modify the parameters with subscript b or r:

$$\gamma_{w(b)}[i] := \frac{\gamma_{w(b)}[i] \times \nu_{w(b)}[i] + \mu / M_{w(b)}(k)}{\nu_{w(b)}[i] + 1};$$

$$\nu_{w(b)}[i] := \nu_{w(b)}[i] + 1;$$

$$\rho_{w(b)}[i, j] := \min\left(\rho_{w(b)}[i, j], h - j\right) \qquad (j = 0, 1, \ldots, h);$$

Here μ is the size of the B-subtree of the search tree $\mathfrak{A}$ and $M_{w(b)}(k)$ is the mean size of the corresponding subtree at a k-th rank node among the candidates for the critical branch $(M_w(k))$ and the positions unreachable by the opponent $(M_b(k))$. These values are determined experimentally or calculated theoretically on the basis of various assumptions about the nature of the search and the tree of the model being used.

On a forward step from a k-th rank position A_k the distances $\rho_w[i, k]$ and $\rho_b[i, k]$ must be recalculated for all moves in the table $\Xi[1]$, $\Xi[2], \ldots, \Xi[l]$:

$$\rho_{w(b)}[i, k] := \rho_{w(b)}[i, k - 1] + 1.$$

Strictly speaking, these distances are calculated for both sides, White and Black. But in games where White and Black move alternately, the distances between positions need to be saved for one color only. Then we introduce a fictitious rank of order -1, and the distances for backward and forward steps are calculated by the formulae:

$$\rho_{w(b)}[i, j] := \min\left\{\rho_{w(b)}[i, j], h - j | j = h, h - 2, \ldots, 0 \text{ or } -1\right\};$$

$$\rho_{w(b)}[i, k] := \rho_{w(b)}[i, k - 2] + 2 | i = 1, 2, \ldots, l.$$

In the games we have studied we have found moves after which the standard best moves are unsuitable, and specifically tailored moves are required. Arguments like those given above show that the property of being a specific reply Θ to a move H must also be relatively stable. A supplementary routine for generating best replies was therefore proposed, in addition to the best move generator; this would maintain pairs of moves (H, Θ) if it turned out that the best or refutation reply to a move H from some position

A was not derived from the best move generator. There are far fewer specific best replies than best moves.

At the outset of our work, the tables of best moves (and replies) were either empty or were filled with the results of the search for the preceding move in the play (they could have been filled with some standard entries before beginning a game or, more exactly, before the first move out of the opening book, but this was not done). Since the data base was still small, these tables were badly fouled by random moves. The best move routine quickly generated moves of the required quality. A proper development of the best reply was possible only after this was done, since earlier the best and refutation moves were rarely found in the array, and moves were often random. However, when the best move generator began to work well, moves to be inserted in the best reply array were encountered only very rarely, and their updating and correction was slow. This slowness, and the rather large output of the best move generator, may explain why the best reply generator produced no significant effect.

Let us take up the question of what preliminary work can be done in low-rank positions to sharpen the order in which positions are searched. We might, for example, apply a more complex logic in the best move generator; in high-order positions we might limit ourselves to a simple displacement scheme. But principally we should look at the possibility of applying a supplementary search in low-order positions. One of the methods to be applied is the so-called *iterative search*. Rather than using a single model $\mathfrak{A}'$ of the original game tree $\mathfrak{A}$, the algorithm uses several models (usually two), which increase in volume. Normally the models are of a single type, with increasing depth of search n.

At the outset the search is carried out in a small game tree $\mathfrak{A}'$, while the search tree $\tilde{\mathfrak{A}}'$ and all the final values of the pseudoscores $\mathrm{bd}(A)$ for positions $A \in \tilde{\mathfrak{A}}'$ are stored in memory. This allows us to determine which move among the candidates for the critical branch is best, and which was the refutation move at positions unreachable by the opponent. In the search of the larger game tree $\mathfrak{A}''$ in the positions $A \in \tilde{\mathfrak{A}}'$ we first choose the moves that appear as best or refutations in the game $\mathfrak{A}'$. These will often carry over their properties to the larger model. However. if we must consider a low-ranked position $A \notin \tilde{\mathfrak{A}}'$ we apply the methods described above for choosing the move. The number of positions in the tree grows at least like $m^{n/2}$, where m is the mean number of moves (*fanout*) at the positions in the tree and n is the mean depth of the search; therefore the preliminary search of the reduced tree $\mathfrak{A}'$ has little influence on the overall search time.

But this means that it is reasonable to use more information in the preliminary search than the α, β-heuristic can provide. In particular we could find scores in the search at small depths for all positions B_i after the moves (A_0, B_i) $(i = 1, 2, \ldots, \mu)$ from the base position A_0. The α, β-heuristic needs to be applied only for positions of rank higher than 1. However, a different method seems better for extending the search tree and enlarging the information base at low-rank positions. We can weaken the pruning rule so that it applies only when the pseudoscores for the next position, as defined by the positions already examined, significantly exceed the attainable bounds:

$$\mathrm{bd}\left(A_k\right) \geq \overline{\lim} + \delta = \max\left\{\mathrm{Bd}\left(A_i\right) + \delta | A_i \in \Pi_b\right\} \text{ if } A_k \text{ is a White position,}$$

$$\mathrm{bd}\left(A_k\right) \leq \underline{\lim} - \delta = \min\left\{\mathrm{Bd}\left(A_i\right) - \delta | A_i \in \Pi_w\right\} \text{ if } A_k \text{ is a Black position,}$$

where A_k is the next position, of rank k, and Π_w, Π_b are respectively the sets of those positions in the branch $(A_0, A_1, \ldots, A_k)$, leading from the base position A_0 to the next position A_k, for which the next moves are White and Black; $\delta > 0$ is chosen so that we investigate moves of sufficiently high quality.

Finally, we try to extract from the search of the model tree $\mathfrak{A}'$ enough information to prescribe the next approximation to the original game—the model $\mathfrak{A}''$. But this is not related to the determination of the order in which the search is conducted. We note only that if we follow this path we had best develop our model game in stepwise fashion by deciding at each step which position is to be added to our existing model tree.

In many games we find equivalent positions, which arise at different places in the game tree. For instance, in games where pieces move around on a board, equivalent positions occur with the same collocations of the pieces and the same side having the move (in chess, we must also know whether the position has arisen earlier, and in some positions whether the right to castle in one direction or the other has been lost, and whether capture *en passant* is permissible). A study of the game of *odnomastka* (the name means 'single-suit') will be found in [11]; this is played by algorithms rather than by people (it is a simplified version of such human games as Boston or bridge). A stronger definition of equivalence of positions can be given for this game.

In odnomastka there are $2s$ pieces, ordered by strength, which are dealt equally by some process to two players. The play consists of a sequence of cycles; in each cycle one player selects one of his pieces, and the second player does the same thing; the selected pieces are compared, the player with the stronger piece scores one point, and the two pieces are discarded. White chooses first on the first cycle; afterward, the scoring player chooses first. The goal of the game is to get the highest point score. After each cycle the number of pieces in the hand of each player is reduced by 1. A position may be *compressed* by discarding from the ordered sequences belonging to both players those pieces that will not be played in the future. Clearly, positions with the same configurations of pieces after such compression, and with the same turn to move, are equivalent.

To shorten the search it may be useful to know whether a position equivalent to the next one has been met before, and if so, what can be said about its score. If we merely preserve all the previously encountered positions and information about their scores, in the form of a single table, or as a memory of the structure of the tree, we will spend as much time on the average in finding an equivalent position as we would spend in searching the subtree of the given position. Some special table structures have been developed that allow a given element to be found quickly, without searching the entire table. For these, the table is said to be *dynamic*. Its makeup varies in the course of its use. Several papers have been devoted to the development of effective structures for dynamic tables, e.g. [3,17,24].

For the study of games with large but searchable numbers of non-equivalent positions an effective search can be made upward from the terminal positions rather than downward from the base position. This is the way in which chess endgames with five to ten pieces in play have been investigated: King, Queen, and Pawn against King and Queen; or King, Rook, Pawn against King and Rook (see [14], [21]). In such endgames, and those that can be derived from them by promotion of Pawn and captures of pieces, there are more than 10^9 non-equivalent positions, and in searching the corresponding subtrees a substantially greater number of nodes must be examined.

For simplicity we shall assume that in the games we now contemplate White and Black will move alternately and that there are only two outcomes, e.g. a win by the stronger side, and the impossibility of such a win. As in problems and studies, we shall assume that White is the stronger side. Positions won for White may be classified by rank, and we shall denote by $R[i]|(i=0,1\ldots)$ the sets of positions won in i moves. The set $R[0]$ can be determined by a single inspection of all Black positions to see whether they are terminal and what their score is. In fact, $R[0]$ contains all the positions at which we can determine either immediately or after a short search that they are won for White; for instance, positions in which White can promote a Pawn to Queen or Rook and Black has no stalemate nor can check or immediately capture a White piece.

The basic tool for the investigation is a *backward step* to a position from which a legal move can be made to the given position. Suppose that we have already constructed certain sets $R[0], R[1], \ldots, R[k]$ of positions with rank not exceeding k and $M[i]|i=1,2,\ldots,k$ of positions not yet ordered and belonging to the corresponding color:

$$M(2i) = M_b \setminus \bigcup_{j=0}^{i} R[2j],$$

$$M[2i+1] = M_w \setminus \bigcup_{j=0}^{i} R[2j+1]|i=1,2,\ldots,\lceil k/2 \rceil,$$

where M_w and M_b are the sets of all mutually non-equivalent positions in which the corresponding color has the move. We denote by U' the set of positions that can be obtained from the positions in U by a backward step, i.e. those from which a move admissible in our game will arrive at a position in U. From any position in the set $R[2i+1]$ we may legally move to a position in $R[2i]$, and any position from which we may make such a move is won in no more than $2i+1$ moves. Therefore

$$R[2i+1] := R'[2i] \backslash \left(\bigcup_{j=0}^{i-1} R[2j+1] \right)$$

$$:= R'[2i] \cap M[2i-1].$$

In the same way we prove that the set $R[2i+2]$ consists of positions in the set $M[2i]$ from which all moves lead to positions of lower rank, at least one of them leading to a position of rank $2i+1$. But this means that it contains those and only those positions in $M[2i]$ to which no backward step can be made from positions in $M[2i+1]$:

$$R[2i+2] := M[2i] \backslash M'[2i+1];$$

moreover

$$M[i+1] := M[i-1] \backslash R[i+1] \ (i = 0, 1, \ldots).$$

We construct the sets $R[i]$ and $M[i]$ one after the other. Since for different i the sets $R[i]$ do not intersect, and the total number of positions is finite, the set $R[i]$ is empty for some i and $M[i] = M[i-2]$. Then no win for White is possible at positions in the sets $M[i]$ and $M[i-1]$.

The implementation of this algorithm for studying a game suffers from specific difficulties stemming from the fact that the sets $R[i]$ and $M[i]$ are not wholly stored in fast memory. The methods of coping with the corresponding problems are unrelated to the theme of this book.

The Construction of Models of a Game

Suppose given two game trees $\mathfrak{A}$ and $\mathfrak{A}'$ in which every move (A'_0, B') from the base position (A'_0) in $\mathfrak{A}'$ is mapped into some corresponding move (A_0, B) from the base position A_0 in $\mathfrak{A}$. Then we may choose a move from A_0 by finding a best move (A'_0, B') from the base position A'_0 in $\mathfrak{A}'$ and then choosing its image (A_0, B) in $\mathfrak{A}$. This process allows us to examine fewer positions if $\mathfrak{A}'$ is significantly smaller than $\mathfrak{A}$, and it is meaningful when we have some reason to expect that a best move in $\mathfrak{A}'$ maps into a good move in $\mathfrak{A}$.

Let us consider some techniques that, given a tree $\mathfrak{A}$, will construct a significantly smaller tree $\mathfrak{A}'$ which will to some degree satisfy our desire that its best moves map into good moves. Starting from the root A_0 of $\mathfrak{A}$, we add, one after another in $\mathfrak{A}'$, the moves belonging to $\mathfrak{A}$ from the positions already in $\mathfrak{A}'$, together with the positions to which they lead. At the newly added position the turn to move, and the score if the new position is terminal in $\mathfrak{A}$, are the same as they are in $\mathfrak{A}$. When no move $(A, B) \in \mathfrak{A}$ from a non-terminal position $A \in \mathfrak{A}'$ is contained in $\mathfrak{A}'$, we use some definite method to compute a score $\mathrm{sc}(A)$ that need not coincide with the score of this position in the original game.

Further, at some positions $A \in \mathfrak{A}'$ the side that has the move is allowed a choice: to make one of the moves $(A, B) \in \mathfrak{A}$ that are permissible in $\mathfrak{A}'$, or to accept a score (usually computed as for a non-terminal position in $\mathfrak{A}$) and take the position as terminal in $\mathfrak{A}'$. The possibility of scoring the position $A \in \mathfrak{A}$ is assumed to be due to a *blank move* $(A, A') \in \mathfrak{A}'$. Blank moves lead to terminal positions in $\mathfrak{A}'$ that are not found in $\mathfrak{A}$, and from a position in $\mathfrak{A}'$ only one blank move may be made. Since some move in $\mathfrak{A}$ must correspond to a move $(A_0, B) \in \mathfrak{A}'$, it is impossible to make a blank move from the base position $A_0 \in \mathfrak{A}'$.

The tree $\mathfrak{A}'$ that we construct in this way will be called a *model of the game tree* $\mathfrak{A}$. The construction process can be combined with a search. At each step in the search we may add the following modelling actions to those described in Chapter 1:

(1) deciding which moves $(A, B) \in \mathfrak{A}$ from the next position are admissible in $\mathfrak{A}'$:
(2) deciding whether a blank move (A, A') is admissible;
(3) computing the score at a position A that is non-terminal in $\mathfrak{A}$ and terminal in $\mathfrak{A}'$;
(4) computing the score at a position $A' \notin \mathfrak{A}$ arising from a blank move $(A, A') \in \mathfrak{A}$.

We shall ignore the problems of computing the scores, although we have said something about them in the first section of this chapter. We merely note that we take account of the features of the position $A \in \mathfrak{A}$ when we compute the score of a position A' arising from a blank move $(A, A') \in \mathfrak{A}$.

A model $\mathfrak{A}'$ of a game tree $\mathfrak{A}$ is said to be *equivalent* if it satisfies the following two conditions:

at every non-terminal position $A \in \mathfrak{A}'$ a non-blank best move (A, B) in $\mathfrak{A}'$ is also a best move in $\mathfrak{A}$;

when a blank move (A, A') in the game $\mathfrak{A}'$ is a best move, there exists a best move (A, B) in $\mathfrak{A}$ from the position A to a position $B \notin \mathfrak{A}'$.

For any game tree $\mathfrak{A}$ containing more than two positions we can construct a smaller equivalent model. Let $\mathfrak{A}'$ be a subtree of $\mathfrak{A}$ having its root at the base position A_0 and containing the first-rank position B such that (A_0, B) is a best move in $\mathfrak{A}$. It may be looked on as a game tree with scores at its terminal positions equal to their scores in $\mathfrak{A}$, and is equivalent to $\mathfrak{A}$. However, to make use of such a model, we need a method for determining the scores at its terminal positions.

In some cases we can find these scores without searching too many superfluous positions in the original tree $\mathfrak{A}$. In an endgame with King, Bishop, and Pawns against King and Bishop we may neglect variations in which the weaker side sacrifices a Bishop for a Pawn. In fact, there are no positions in an endgame with King and Bishop against a lone King in which either side can mate, and so all positions in such an endgame are drawn. Later we shall examine several somewhat more general methods for constructing equivalent models but, as we showed in Chapter 1, any such method must be founded on various properties of the rules of the game we are studying.

Game-playing algorithms often use models whose equivalence to the original game is either not proven or non-existent, i.e. they use heuristic models, whose plausibility is usually based on approximations. For some of these models we shall show later how we can compute the probabilities of correctly choosing a best move at positions in the original game $\mathfrak{A}$ or of computing their scores, whenever some definite information about the positions can be obtained without searching the corresponding subtree of the game $\mathfrak{A}$. Here we describe these models and the properties of the games that allow their use.

First we take up what we may call the *formal* methods of building models that require the least amount of analysis of the contemplated positions and moves. We may, for instance, simply set an *a priori* limit to the number of moves in $\mathfrak{A}$ that will be examined at the positions in the game $\mathfrak{A}'$. In the simplest case the limit depends only on the rank of the positions. We prescribe a function $\varphi(k)$ of the integer argument k, usually monotone decreasing and vanishing for some $k = n$. All moves (A, B_i) from White (Black) positions $A \in \mathfrak{A}'$ are arranged in order of decreasing (increasing) values of some function $f_{\text{ord}}(B_i)$ and we admit to the game $\mathfrak{A}'$ those and only those moves (A, B_i) for which $i \leq \varphi(k)$, where k is the rank of the position A.

We often use for the move-ordering function $f_{\text{ord}}(B)$ a function $f_{\text{sc}}(B)$ which determines the scores of the terminal positions in the game $\mathfrak{A}'$. In fact, however, these functions are subjected to variegated conditions. The function $f_{\text{sc}}(A)$ must yield values as close as possible to the true scores $\text{sc}(A)$ of the positions in the original game $\mathfrak{A}$. The values of $f_{\text{ord}}(A)$ should be such that the number of best moves (A, B_i) from positions $A \in \mathfrak{A}'$ has a high probability of being less than the prescribed limit. As we noted in Section 2, this implies that low numbers should be assigned not only to moves that lead to positions with extremal values of the evaluation function $f_{\text{sc}}(B_i)$ (maximal for White positions and minimal for Black) but also to moves leading to large expected values of the difference $|\text{sc}(B_i) - f_{\text{sc}}(B_i)|$.

Let $\mathfrak{A}''$ be a Shannon model of depth n of the same game $\mathfrak{A}$ with identical values of the evaluation function $f_{\text{sc}}(A)$ at its terminal positions. Then if we already have $\varphi(n) = 0$, the tree $\mathfrak{A}'$ constructed according to the above rules is a model of the game $\mathfrak{A}''$. If $\varphi(n-1) > 0$ and if at every non-terminal position $A \in \mathfrak{A}'$ at least one move (A, B) which is a best move in $\mathfrak{A}''$ is admissible in $\mathfrak{A}'$, then $\mathfrak{A}'$ is an equivalent model of $\mathfrak{A}''$ and the scores of all positions $A \in \mathfrak{A}'$ are the same in both games. When the best moves in $\mathfrak{A}''$ are not necessarily admissible in $\mathfrak{A}'$ we may ask about the probability of correctly estimating the score and choosing the best move by using the model $\mathfrak{A}'$.

We can estimate these probabilities for a uniform game $\mathfrak{A}_{m,n,s}$ of the type introduced above, having a tree depth n, alternate moves by White and Black, fanout m at positions A of rank $k < n$, s winning moves at positions where a win is possible, and scores $f_{sc}(A)$ having the values 0 or 1 at the terminal positions, all of which have rank n; all this provided we assume mutual independence among the probabilities that moves $(A, B) \in \mathfrak{A}_{m,n,s}$ are admissible in the model, and assume that for winning moves (A, B) these probabilities depend only on the rank k of the position A at which they can be made:

$$P\Big((A, B) \in \mathfrak{U}' | sc(A) = sc(B) = \begin{cases} 1, & \text{if } A \text{ is a White position} \\ 0, & \text{if } A \text{ is a Black position} \end{cases}\Big)$$

$$= \pi_k (0 \le k \le n)$$

Let p_k and q_k be the respective probabilities of obtaining a correct score for positions of rank k when we can or cannot make a winning move from them. The terminal positions will be correctly scored, i.e. $p_k = q_k = 1$. All moves from a losing position A lead to positions at which the opponent has a win. For the score at A to be correct, the scores of positions that can be reached from it by admissible moves in the game $\mathfrak{A}'$ must also be correct. Since the corresponding subtrees of $\mathfrak{A}'$ do not intersect, the probabilities of these events are independent, and we have the recursive relationships

$$q_k = p_{k+1}^{\phi(k)}, \qquad 0 \le k < n.$$

Let there be t winning and $\varphi(k) - t$ losing moves among the admissible moves from some winning position A. Then in order that the score at A be wrongly computed, the scores after winning moves must also be wrongly computed and after losing moves correctly computed. The probability of this event is equal to $(1 - q_{k+1})^t p_{k+1}^{\varphi(k)-t}$, where k is the rank of the position A. The probability that the contemplated case will occur is $C_s^t \pi_k^t (1 - \pi_k)^{s-t}$. Consequently the total probability of error $1 - p_k$ is equal to

$$\sum_{t=0}^{\min(s, \varphi(k)-s)} C_s^t \pi_k^t (1 - \pi_k)^{s-t} (1 - q_{k+1})^t p_{k+1}^{\varphi(k)-t}$$

and for $s \le \varphi(k) - s$

$$p_k = 1 - \big(\pi_k(1 - q_{k+1}) + (1 - \pi_k)p_{k+1}\big)^s p_{k+1}^{\varphi(k)-s}$$

$$(0 \le k < n).$$

Table 3. Probability that Scores Will Coincide When Some Moves are Not Considered

Notation

m—number of moves at non-terminal positions in the model,
s—number of winning moves at winning non-terminal positions,
n—depth of the uniform game,
π—probability that a winning move will have number $v \leq m$
P_n—probability of a win in the model game at a won position in the original game,
Q_n—probability of a win in the model game at a lost position in the original game.

	$s = 2,\quad m = 3,\quad \pi = 0.8$		$s = 2,\quad m = 5,\quad \pi = 0.8$		$s = 2,\quad m = 3,\quad \pi = 0.9$	
n	P_n	Q_n	P_n	Q_n	P_n	Q_n
0	1.0000	0.0000	1.0000	0.0000	1.0000	0.0000
1	0.9600	0.0000	0.9600	0.0000	0.9900	0.0000
2	0.9646	0.1153	0.9674	0.1846	0.9903	0.0297
3	0.9216	0.1025	0.8946	0.1528	0.9843	0.0288
4	0.9347	0.2173	0.9350	0.4270	0.9848	0.0463
5	0.8783	0.1835	0.7716	0.2852	0.9807	0.0450
6	0.9087	0.3224	0.9328	0.7265	0.9812	0.0569
7	0.8244	0.2497	0.5215	0.2937	0.9781	0.0554
8	0.8904	0.4397	0.9837	0.9614	0.9786	0.0642
9	0.7500	0.2942	0.1120	0.0790	0.9763	0.0627
10	0.8889	0.5781	0.99999	0.99998	0.9768	0.0694

	$s = 2,\quad m = 5,\quad \pi = 0.9$		$s = 2,\quad m = 5,\quad \pi = 0.95$		$s = 2,\quad m = 5,\quad \pi = 0.99$	
n	P_n	Q_n	P_n	Q_n	P_n	Q_n
0	1.0000	0.0000	1.0000	0.0000	1.0000	0.00000
1	0.9900	0.0000	0.9975	0.0000	0.9999	0.00000
2	0.9905	0.0490	0.9975	0.0124	0.9999	0.00050
3	0.9801	0.0466	0.9962	0.0123	0.9999	0.00050
4	0.9815	0.0957	0.9963	0.0187	0.9999	0.00055
5	0.9679	0.0889	0.9955	0.0185	0.9999	0.00055
6	0.9717	0.1506	0.9955	0.0224	0.9999	0.00056
7	0.9503	0.1339	0.9950	0.0222	0.9999	0.00056
8	0.9601	0.2249	0.9951	0.0247	0.9999	0.00056
9	0.9212	0.1841	0.9947	0.0245	0.9999	0.00056
10	0.9480	0.3366	0.9948	0.0261	0.9999	0.00056

Table 3 displays the values of p_k and $1 - q_k$ for various values of k, s, and φ when the values of the $\varphi(k)$ are identical for all non-terminal positions, i.e. $\varphi(0) = \varphi(1) = \ldots = \varphi(n-1) = \varphi$. Inspection of the table shows that except when π_k is very close to 1 the model should be used only for small depths n. If the game we are studying is nearly uniform, and the probabilities that moves will be admissible are nearly independent, it is inexpedient to increase the depth of the search on account of the uniform bound on the number of admissible moves. In fact, when the best move (A, B) at some position A is not admissible, it is highly probable that at nearby positions the best moves are also inadmissible. On this account the probability of erroneous scores only increases, and it is not surprising that programs with models of the type we have described play weakly.

The bounds $\varphi(A)$ on the number of admissible moves in the model $\mathfrak{A}'$ need not depend solely on the ranks of the corresponding positions A. However, up to now no such models have been studied theoretically nor have they been used in practice for game-playing programs. We shall look at a very simple case where:

(a) in all non-terminal positions A of the game $\mathfrak{A}$ White has the move and the moves from A are ordered, while the probabilities $P(l)$ that the move $(A, B_l) \in \mathfrak{A}$ with index l is a best move are identical and mutually independent;
(b) the positions of rank n, and only those, are terminal.

Thus, we are to find the critical path $(A_0, A_1, \ldots, A_n)$ leading from the base position A_0 to the terminal position A_n having the maximum score $sc(A_n)$.

Every position A_k of rank $k = 1, 2, \ldots, n$ can be defined by k coordinates $x_1, x_2, \ldots, x_k$, where x_i is the number of moves from A_i in the branch $(A_0, A_1, \ldots, A_k)$ leading to A_k from the base position A_0. Under our assumptions this branch has the probability $P(x_1) \times P(x_2) \times \cdots P(x_n)$ of being the beginning of the critical branch. Since the number of positions in the game tree $\mathfrak{A}'$ is bounded, we should first of all look at the terminal positions which have the greatest probability of being optimal. Accordingly, the game tree $\mathfrak{A}'$ must be constructed of branches leading to terminal positions having coordinates that satisfy the condition

$$P(x_1) \times P(x_2) \times \cdots \times P(x_n) \geq L,$$

where L is a suitably chosen constant.

We may often suppose that for the interesting values of x the function $P(x)$ is approximately linear, and that the linear function is nearly exponential, i.e.

$$P(x) \approx q - rx \approx qe^{-rx/q}.$$

Then the tree of the game $\mathfrak{A}'$ contains terminal positions with coordinates that satisfy the condition

$$P(x_1) \otimes P(x_2) \otimes \cdots \otimes P(x_n) \approx q^n e^{-\frac{r}{q}(x_1 + x_2 + \cdots + x_n)}$$

$$\leq L,$$

i.e. $x_1 + x_2 + \cdots + x_n \leq \ln \dfrac{L}{q^n} = C$

If we begin the numbering of the moves from non-terminal positions $A \in \mathfrak{A}'$ with 0, the condition for including the positions A_k of ranks $k = 1, 2, \ldots, n$ in $\mathfrak{A}'$ may be stated as the inequality

$$x_1 + x_2 + \cdots + x_k \leq C,$$

and we may specify the constant C immediately, without defining r, q, and L.

When we limit the number of admissible moves in a model of a two-person game $\mathfrak{A}$, the choice of the best move may be in error because the move may be inadmissible in the model $\mathfrak{A}'$, but also for other reasons. A bad move may be preferred to a best move because its refutation is inadmissible in $\mathfrak{A}'$. Since for a move (A, B_i) with a large value of i the chance of turning out to be bad is high, we should prolong the search for a refutation as far as possible unless it comes up among the earlier responses. In a model $\mathfrak{A}'$ of a game $\mathfrak{A}$ with alternating moves by White and Black, we might postulate, by analogy with the above argument, that the condition for including the positions A_k of ranks $k = 1, 2, \ldots, n$ is given by the inequality

$$\Phi(A_k) = x_k - \Phi(\Phi(A_{k-1})) \leq C,$$

where (A_{k-1}, A_k) is a move in the game $\mathfrak{A}$ leading to the position A_k and $\Phi(x)$ is a monotone increasing function of x. We have not, however, succeeded in formulating a probabilistic hypothesis that would support the use of this condition.

Let us now turn to the semantic method of constructing models, in which the admissibility of moves $(A, B) \in \mathfrak{A}$ is determined on the basis of a qualitative analysis of the moves or of the corresponding positions A and B. The notion of an *unstable* position makes sense in many games. In such a position there is a rather high probability that the value of the evaluation function $f(A)$ will differ significantly from the true score $sc(A)$. Moreover, either few moves $(A, B) \in \mathfrak{A}$ lead from it or the majority of them lead to bad positions. By analyzing some easily computed features of the position and of the moves leading to or from it that are admissible under the rules of the game $\mathfrak{A}$, we can determine whether or not the position is unstable.

For instance, in the game of noughts and crosses, those configurations consisting of m successive squares arranged in vertical, horizontal, or diagonal lines and occupied by one of the players with noughts or crosses, and having a continuation of $5\text{-}m$ contiguous free squares on one end or on both ends are called respectively half-open and open m-tuples. If the opponent has an open triplet all moves will lose except those that either immediately adjoin the triplet or make a threat to complete a quintuple of one's own. Thus positions in which the opponent has an open triplet are unstable.

In draughts—Russian, American, or international—positions in which a capture is possible are unstable. In fact, the material score in such positions may change in ways that cannot be predicted without analysis of the corresponding variations. (Moreover, in draughts captures are obligatory, so that the number of admissible moves in such positions is significantly less than in other positions.)

Chess positions arising after a capture may be considered to be unstable. As a rule, one may win back the value of a captured piece or even capture one of higher value. If one abstains from the capture, one cannot in general return to the former material balance at a later date. Thus the material score in such positions is unreliable and there are few non-losing moves. There is a substantial difference, however, between this example and the first two. In those, few of the moves from an unstable position are permitted by the rules of the game or fail to lead to obviously losing positions. In the chess example, however, it is merely highly probable that a 'quiet' move from an unstable position will be of poor quality.

Moves leading to an unstable position will be called *forcing* moves, and moves from an unstable position, except those leading to clearly or very probably lost positions, will be called *forced* moves. As is clear from the above examples, the features of forcing and forced moves are also easily computed (in game-playing algorithms for noughts and crosses, draughts, and chess we divide the unstable positions, forcing moves, and forced moves into different classes, but they all have easily computed features).

We can now define the concept of a *forced game*. Its positions are positions in the game $\mathfrak{A}$ and the admissible moves are forcing, forced, and blank moves. The admissibility of blank moves is related to the fact that in the games we are considering almost every position can be mapped into a position with the same configuration of the pieces, but with the move to be made by the side with the opposite color. If no such position exists (as in chess with a King in check), a blank move is inadmissible. Sometimes, however, a blank move is disallowed in other positions, as when a player under the threat of material loss will make an intermediate checking move (in models of chess a check is often regarded as a forcing move) against which his opponent defends himself; if a blank move were permitted, the program would not perceive a threat of material loss.

The notion of the forced game can be used to construct a model in the following way. As in Shannon's model the depth of search n is prescribed. In positions A_k of rank $k \leq n$ all moves in the game tree $\mathfrak{A}$ are admissible, but in contrast to Shannon's model, positions of rank n are taken to be non-terminal (unless they are terminal in $\mathfrak{A}$). At these and at all positions of higher rank, every move in the forced game is admissible. The ranks of positions in the model $\mathfrak{A}'$ may be either unbounded, or bounded by a search depth $n_1 \gg n$.

In many cases we can construct a model $\mathfrak{A}''$ equivalent to the model $\mathfrak{A}'$ described above but having fewer moves. To simplify the description of such a model we shall assume that: 1) in the original game $\mathfrak{A}$ White and Black move alternately; 2) the evaluation function $f(A)$ is the sum of material and positional components

$$f(A) = f_{\mathrm{m}}(A) + f_{\mathrm{p}}(A).$$

The material scoring function is linear:

$$f_{\mathrm{m}}(A) = \sum_{\mu=1}^{M} h_{\mu} P_{\mu}^{w}(A) - \sum_{\mu=1}^{M} h_{\mu} P_{\mu}^{b}(A),$$

where h_{μ} is the weight of the μ-th piece, M is the number of different pieces belonging to one of the sides, and $P_{\mu}^{w(b)}(A)$ is a predicate having the value 1 if the White (Black) piece is on the board and 0 if it is not. The values of the positional component $f_{\mathrm{p}}(A)$ of the evaluation function are non-negative and less than the minimum absolute difference $|f_{\mathrm{m}}(A') - f_{\mathrm{m}}(A'')|$ of the values of the material components of the scores.

Let A_k be the next position in the search tree being constructed for the game $\mathfrak{A}''$, with $k = n - 1$ and White (Black) to move. Let $\underline{\lim}$ ($\overline{\lim}$) as defined in Chapter 1 be the bounds for the scores

$$\underline{\lim} = \min \{ \mathrm{bd}(A_i) | A_i \in P_w \},$$

$$\overline{\lim} = \max \{ \mathrm{bd}(A_i) | A_i \in P_b \}$$

(we recall that P_w and P_b are the sets of positions in the branch $(A_0, A_1, \ldots, A_k)$ leading from the base position A_0 to A_k, the next position for White or Black, respectively; the bd A_i are the intermediate values of the parameters of the positions in this branch as calculated by the α, β-procedure instead of their scores). After the move $(A_k, A_{k+1}) \in \mathfrak{A}''$ there arises the position A_{k+1} with Black (White) to move and with the score

$$f(A_{k+1}) = f_{\mathrm{m}}(A_{k+1}) + f_{\mathrm{p}}(A_{k+1})$$
$$= f_{\mathrm{m}}(A_k) + (\pm) h + f_{\mathrm{p}}(A_{k+1}),$$

where h is the weight of the pieces captured in the move, or is 0 if (A_k, A_{k+1}) is a quiet move. The rank of A_{k+1} is not less than n, so that in general a blank move is admissible there.

The value of $\underline{\lim}$ ($\overline{\lim}$) is equal to the value of the evaluation function for some $A \in \mathfrak{A}$ and may be represented as a sum of material and positional components:

$$\underline{\lim} = \underline{\lim}_{\mathrm{m}} + \underline{\lim}_{\mathrm{p}},$$

$$\overline{\lim} = \overline{\lim}_{\mathrm{m}} + \overline{\lim}_{\mathrm{p}}.$$

If

$$f_{\mathrm{m}}(A_{k+1}) = f_{\mathrm{m}}(A_k) + h < \underline{\lim}_{\mathrm{m}} (f_{\mathrm{m}}(A_{k+1}))$$
$$= f_{\mathrm{m}}(A_k) - h > \overline{\lim}_{\mathrm{m}},$$

then

$$\underline{\lim} - f(A_{k+1}) = \underline{\lim}_m + \underline{\lim}_p - f_m(A_{k+1}) - f_p(A_{k+1})$$
$$\geq \underline{\lim}_m - f_m(A_{k+1}) - f_p(A_{k+1}) > 0,$$
$$f(A_{k+1}) - \overline{\lim} = f_m(A_{k+1}) + f_n(A_{k+1}) - \overline{\lim}_m - \overline{\lim}_p$$
$$\geq f_m(A_{k+1}) - \overline{\lim}_m - \overline{\lim}_p > 0,$$

since the absolute difference between the material scores of two positions is larger than the value $f_n(A)$ of the positional evaluation function for any position A.

After a blank move from the position A_{k+1} we reach a terminal position $\overline{A}_{k+1}$ in the game $\mathfrak{A}'$ described earlier, with the score

$$\mathrm{sc}_{\mathfrak{A}'}(\overline{A}_{k+1}) = f(A_{k+1}) < \underline{\lim} \ (\, > \overline{\lim})$$

Therefore, we also have $\mathrm{sc}_{\mathfrak{A}'}(\mathrm{A}_{k+1}) < \underline{\lim} \ (\, > \overline{\lim})$, i.e. after seeing the result of a blank move from A_{k+1} we step backward to the position A_k, and the value of $\mathrm{bd}(A_k)$ remains unchanged, as though the move (A_k, A_{k+1}) had not occurred. Thus we find that at positions with rank $k \geq n - 1$ and with material scores lying beyond the bounds $\underline{\lim}_m$ or $\overline{\lim}_m$, the only admissible moves are captures of pieces with weight $h > \underline{\lim}_m - f_m(A_k)(f_m(A_k) - \overline{\lim}_m)$ or moves in the game $\mathfrak{A}'$ after which a blank reply is prohibited. The method we have just described for constructing a model $\mathfrak{A}'$ will be called an absolute scheme.

Different branches of the model $\mathfrak{A}''$ may contain different numbers of quiet moves (A_k, A_{k+1}) that are inadmissible in the forced game. In fact, for $k < n$ some moves of this kind may be non-quiet. It has been suggested that in order to improve the quality of the move (A_0, B) selected at the base position A_0 it would be useful to equalize the numbers of quiet moves in the branches, and rules for constructing models called *quiet games* have been suggested for this purpose. There are two parameters—the search depth n and the minimum number d of quiet moves in a branch. In the positions A_k of rank k any move of the original game $\mathfrak{A}$ is admissible if $k < n$ or if the number of quiet moves (A_i, A_{i+1}) in the branch $(A_0, A_1, \ldots, A_k)$ is less than d. Otherwise, only forced moves are admissible.

A program using a quiet game is stronger than one using an absolute scheme with the same search depth n and the same scoring function $f(A)$. The time spent, however, is greater. We do not yet know how to use this extra time more effectively, nor even whether we can.

Similar considerations might explain the relative success of a program playing at the expert level on a rather slow machine ([36—38]). It allows quiet moves from a position A_k under the following condition: Let A_i be a position in the path $(A_0, A_1, \ldots, A_{k-1})$ leading from the base position A_0 to the immediate predecessor A_{k-1} of A_k, and let $\psi(A_i)$ be the number of moves from A_i that are admissible in the original game $\mathfrak{A}$. Then a quiet move from A_k is allowed if the $\psi(A_i)$ $(i = 0, 1, \ldots, k-1)$ satisfy the inequality

$$\psi(A_0) \cdot \psi(A_1) \cdots \psi(A_n) \leq M,$$

where M is a parameter determining the time taken by the search from the base position A_0. In unstable positions only captures are allowed (compulsory by the rules of draughts); the number of ways to capture is much smaller than the number of quiet moves allowed in other positions. Accordingly, if many unstable positions arise on the path its length is greater than that of a quiet path, in which such positions arise infrequently.

Semantic Models

The ultimate goal of a player is to reach a terminal position with a sufficiently favorable score. During the course of the game, however, he normally tries to attain intermediate goals—to reach positions that for one reason or other please him. In the semantic theory of games, for instance chess theory, there are many features of such positions and of the moves that lead to them, but these features lack precise definitions. In game programs some such features, which are given an exact definition (usually constricting their meaning), enter into the computation of the evaluation function $f(A)$. Thus, while studying a position in the course of the game, the program attempts to reach one or another intermediate goal.

In the search process, however, many moves are examined, which are meaningless with respect to reaching the goal. For this reason there is no time for a sufficiently deep search of the sensible variations. Moreover, different favorable features of a position do not always agree. As a result, the program may pursue a middle course in a game, making a sequence of discordant moves. This is especially true of positions in which neither player threatens to reach any intermediate goal whatsoever. Then paths that lead to positions with only slightly different values of the evaluation function compete for the choice of move by the program. Often the differences are due to random nuances. The program lacks a guiding thread —it has no plan and does not hinder its opponent's plans.

A deep search is infeasible unless we abandon a study of the results of various moves from positions of low rank, but prohibiting such moves on the basis of merely quantitative scores would not lead to good results. It has therefore been suggested that, in constructing a model of a given game, we should declare moves admissible or inadmissible with the help of a semantic analysis of the moves themselves or of the positions to which they lead. An example is afforded by the classification of moves in the forced game described in the preceding section.

From the technical point of view, there are two methods for carrying out such an analysis. The first consists in studying the position A from which a move is to be made, in order to generate only admissible moves and spend no time on the others. The second method consists in studying all moves leading from A (or at least all that are not obviously inadmissible) and then making a more or less laborious analysis of each to determine its admissibility. The first method offers a potential saving in time but selects moves on the basis of very simple formal features. To decide whether to use the second method one must estimate whether less time will be spent in analyzing the admissibility of moves than would be spent in searching the portion of the tree that is excluded by that analysis from the model being constructed.

Material gain is an important and easily examined intermediate goal in a game. In particular, forcing moves often represent the threat of winning material. Some non-forcing moves containing such a threat are easily defined; for instance, attacks on undefended major pieces (these are not included in the forced game because a formal definition of a sufficiently narrow class of forced replies is difficult). A wider class of moves posing the threat of material gain can be defined with the help of an auxiliary game model.

To study the move $(A, B) \in \mathfrak{A}$ we consider the models $\mathfrak{A}_{\text{force}}(B)$ and $\mathfrak{A}_{\text{force}}(\overline{B})$ with the initial configurations of the pieces on the board the same as that in the position B. At the non-terminal positions of the two models the moves of the forced game, and only those, are admissible. The move at the base position in the game $\mathfrak{A}_{\text{force}}(B)$ belongs to the side having it at B. Thus we may assume that this base position is simply the position B. The move at the base position $\overline{B}$ of the game $\mathfrak{A}_{\text{force}}(\overline{B})$ belongs to the side having it at A. (We assume that in the original game $\mathfrak{A}$ White and Black move alternately.)

If White (Black) has the move at A, the material component of the score of $\overline{B}$ in the game $\mathfrak{A}_{\text{force}}(\overline{B})$ is compared with the material component of the bound $\underline{\lim}_m$ ($\overline{\lim}_m$) determined when A is the next position in the search of the fundamental model of the game $\mathfrak{A}$. If in the first case this score is not less than $\underline{\lim}_m$, or in the second case not greater than $\overline{\lim}_m$, the move (A, B) is said to be non-losing in the forced variation. The material component of the score of $\overline{B}$ in the game $\mathfrak{A}_{\text{force}}(\overline{B})$ may be compared with the other bounds or with the value $f_m(A)$ of the material scoring function at A. (It is natural to choose the latter for positions of low rank, and the former for positions of high rank.) When this score attains the corresponding bounds or lies beyond them, (A, B) is called an *active* or a *second-level forcing* move.

This model, called an *active game* scheme, has been tested in practice. In positions A_k of rank $k < n - 1$ it admits all active moves that are non-losing in the forced variation. It also admits one so-called *best passive* move, which is non-losing in the forced variation but not active, and has the most favorable score (for its own side) among such moves in the game $\mathfrak{A}_{\text{force}}(\overline{B})$ (in a variant model, the game $\mathfrak{A}_{\text{force}}(B)$). If there are no active moves from

A that are non-losing in the forced variation, three best passive moves are admitted. The numbers 1 and 3 of best passive moves are program parameters. In positions A_k of rank $k \geq n$ only moves of the forced game are admitted, as in the absolute scheme.

Many decisions were taken for economy in the running time of the program. Thus the activity and safety tests of the moves (A_{n-1}, B) from positions A_{n-1} of rank $n-1$ took no less time than the search of the B-subtree consisting only of moves in the forced game, a search that might be obviated if the tests gave a negative answer (also, this search coincides with that of the tree $\mathfrak{A}_{force}(B)$ or differs from it only by somewhat wider initial bounds bd and Bd for the scores of the base position B). Therefore all moves in the original game $\mathfrak{A}$ were admissible at positions of rank $n-1$.

If the analysis of the trees in the games $\mathfrak{A}_{force}(B)$ or $\mathfrak{A}_{force}(\overline{B})$ shows that the move (A, B) lacks the necessary properties, i.e. loses in the forced game or is inactive, we may need to search the second of the two trees only to decide whether it relates to the number of best passive moves. But even this is not needed if we determine the best passive moves by comparing the scores at the base positions of whichever of the two trees was first selected for analysis. In one variant, the tree $\mathfrak{A}_{force}(B)$ is searched first and in the other $\mathfrak{A}_{force}(\overline{B})$, while the the non-selected tree is searched only if necessary. Both variants play almost identically.

In the forced game $\mathfrak{A}_{force}(\overline{A})$ the initial configuration of pieces on the board for the position $\overline{A}$ is the same as in the position A but with the move belonging to the opponent; if this game leads to a loss of material by the side having the move at A, it is highly likely that most of the quiet moves (A, B) in the game $\mathfrak{A}_{force}(B)$ will also lead to a loss of material. Clearly, such moves should first be tested for safety; then many of them will not need testing for activity. In the contrary case fewer such moves will lead to a loss (moves made from A under *zugzwang* are an exception, but this is not often found in the middle game). In this case moves should be tested first for activity and only the active and best passive moves should be tested for safety.

The rules for defining the admissibility of moves may be hybrids. For example, up to some rank n_1 all the moves in a game $\mathfrak{A}$ may be admissible, from ranks n_1 through n_2 all moves in the active game described above, and from n_2 onward only moves in the forced game. In the existing chess programs checking moves are counted as moves in the forced game only in positions up to a certain rank, or if the number of checks in the branch being examined is less than some given standard. In the program described above for the active game, all moves from the base position A_0 or from positions of rank 1 are admissible until a move is found that does not lose material when compared with the anticipated score established in an earlier search at lesser depth.

A program for the active game played rather weakly in the absolute scheme with the same depth of search n, but spent significantly less time (the variants of these programs used about the same amount of time, but were not compared). This program did not consider some moves that deserved attention, but no move included in the model tree was obviously senseless. However, variations identical from a chessplayer's viewpoint were repeatedly inspected at many positions in the search tree; their results would have been obvious to a player after they had been inspected once.

A typical example of superfluous work is the examination of so-called *pseudoactive* moves. Suppose that in the forced game $\mathfrak{A}_{force}(A)$ the opening side wins material, but does not win it in the active game following the forcing moves from A. Then many moves that have no effect on the variations in the forced game will seem to be active, but mistakenly so since after such a move (A, B) the game $\mathfrak{A}_{force}(\bar{B})$ turns out to be essentially the same as $\mathfrak{A}_{force}(A)$.

Active moves are instances of moves answering to a definite strategy, in this case the strategy of winning material. A definition of such moves is needed if we are to specify them. We might, as earlier, demand that an active move should not lose in the forced variation, but this demand seems too strong, in any case for positions of low rank. At the same time, it is scarcely worth while to call a move active if it puts a piece of its own in peril and simultaneously threatens to win material with the aid of the piece itself. Thus we must compare moves from different positions in the trees $\mathfrak{A}_{force}(\bar{B})$ and $\mathfrak{A}_{force}(B)$ among themselves.

To refine the notion of an active move we may use the fact that for many games we can define the concept of identical moves from different positions. After such a move, in all the positions where it is possible to make it, the placement of the pieces changes in the same way. An exact definition of sameness will be given in the next chapter. Here an intuitive notion will do. We may call a move (A, B) inactive if the opponent wins by the move (B, D) in the game $\mathfrak{A}_{force}(B)$ and the winning reply $(\bar{B}, C)$ is impossible in the game $\mathfrak{A}_{force}(\bar{B})$.

If a player in one of the variations has already won material it is useful for him to try defensive moves, after which his opponent cannot secure material equality. Defensive moves must be found also when the moves found earlier all lead to a loss of material. One test of the defensive qualities of a move (A, B) from the position A is the forced game $\mathfrak{A}_{force}(B)$ defined above. It may turn out, however, that a non-losing move in the forced variation will lead to a loss in the original game, which is richer in moves. A more precise definition of the conditions for a move to be defensive requires the use of concepts developed in the next chapter; nevertheless something can still be said here.

A threat of material loss is determined in some game tree, e.g. the B'-subtree $\mathfrak{A}'(B')$ of the contemplated model $\mathfrak{A}'$ of the original game $\mathfrak{A}$. Its base is at the position B', to which the preceding move (A, B') leads from the position A, the next position in the current state of the search. We ask what pieces were in fact lost in the game $\mathfrak{A}'(B')$. Sometimes it is easy to answer, as when the opponent has captured a piece standing at the instant of capture on the same square it occupied in the position A, while we could not reply by a capture of equal value. Then we could say that defensive moves would consist in retreating the piece to another square, protecting it by another piece, or blocking the path of the opponent.

The determination of intermediate goals other than the gain of material depends more substantially on the specifics of the game in question. Later we shall give an example of a certain class of model games in chess that offer a wider set of intermediate goals. These goals are the occupation of certain squares on the board by pieces in a given subset of one's own pieces, or attacks on these squares.To every intermediate goal there correspond strategies for reaching it, and to every strategy there correspond moves that answer to it. Any given move may answer to several strategies, and various strategies may be interconnected.

Suppose that to every chess piece there corresponds a board representing its potential—a 64-bit array representing the squares on the board. For those squares attacked by the piece and not occupied by pieces of its own color (these are the squares to which it can move), the corresponding bit is set to the value 1, and the remaining bits are set to 0. There are two boards per Pawn, one representing squares to which it can move, the other squares it can attack. Boards of this kind can specify the configurations of all pieces, of pieces of one color, pieces of a given kind, etc. The problems of setting up such boards and holding them in memory belong to programming methodology and are not of interest in this book.

To every elementary strategy there corresponds a square on the board which we call a null-rank square. The goal of a strategy is to to occupy the null-rank square with some (not arbitrary) piece of its own side or to attack it. Squares from which pieces can attack (for Pawns attack or move to) null-rank squares are called first-rank squares. Thus there will be first-rank squares for King, Queen, Rook, Bishop, Knight, and Black and White Pawns. Second-rank squares for a piece are those from which it can attack at least one first-rank square, not necessarily one of its own color, and so on through the higher ranks.

The general principle for defining moves responding to a given elementary strategy is this: the piece to be moved should go to a square of lower rank, or if the piece stands still the rank of the square on which it remains should decrease, else the rank of a square attacked by the opponent should increase. On the one hand, these requirements should be relaxed to permit a shift of the attack on a square of a given rank to another square of the same rank. On the other hand, they should be tightened to exclude commotions on the distant approaches to null-rank squares and to decrease the general number of strategic moves.

Let us look at a concrete example of a system for defining strategic moves. This system consists of a set of predicates $P(A, B)$ depending on the beginning and ending positions A and B of the move (A, B), or in some cases on the move itself. The move responds to the strategy if at least one of the predicates has the value 1 (is true). To widen the set of strategic moves we add to the original set one or more supplementary predicates of this kind. To narrow the set of strategic moves we can remove one of the predicates, or change it by imposing additional conditions on its truth value. In this way we can stipulate various means for representing the system in the course of its experimental trials.

First of all, moves of one's own pieces to chosen null-rank squares respond to elementary strategies. For instance, the strategy of the weak point requires the occupation of the point by minor pieces (Bishops or Knights); the strategy of the open file requires the incursion of major pieces (Rooks or Queens) into a square in the 7th or 8th rank of the file (with respect to one's own side); the strategy of the Pawn advance requires the movement of a Pawn to a null-rank square. Besides these, moves that result in new attacks on null-rank squares or on the opponent's pieces that are fighting on null-rank squares, or that result in the disappearance of the opponent's attack on null-rank squares or on one's own pieces standing on null-rank squares, are all responsive to an elementary strategy.

These features of strategic moves are easily defined with the aid of some redundant information about the positions, moves, and strategies. The positions are defined by two sets of boards: a) $\Phi_m[\mu]$ and $\Phi_e[\mu]$ for the positions of one's own and the opposing pieces; the bits are set to 1 for pieces of given types: Pawns for $\mu = 1$, Knights for $\mu = 2$, Bishops for $\mu = 3$, Rooks for $\mu = 4$, Queen (or Queens if there has been a promotion) for $\mu = 5$, and King for $\mu = 6$; and b) $M[i]$ ($i = 1, 2, \ldots, q_m$) and $E[i]$ ($i = 1, 2, \ldots, q_e$), where the bits are set to 1 for squares attacked by pieces of the corresponding type, one's own or the opponent's, and q_m and q_e are the respective numbers of these pieces. We shall examine these boards for the positions before and after a move. The first set will be marked with a prime, the second with a double prime.

A move will be described by 1) a six-bit array N in which a single bit is set to 1 in the place corresponding to the piece-type, 2) the board G' specifying the square from which the piece is moved, and 3) the board G'' specifying the destination square. Thus for a Pawn move $N = (000001)$, for a Knight move $N = (000010)$, etc. The move Bc1–g5, for instance, is described by $N = (000100)$, the board G' on which the square c1 is marked, and the board G'' on which the square g5 is marked. The strategy is described by 1) the board R_0 for the null-rank square, 2) the two boards $A_m[\mu]$, $A_e[\mu]$ specifying the squares from which, in the given position, the selected null-rank square is attackable by one's own or hostile pieces, respectively, (clearly $A_m[\mu] = A_e[\mu]$ for $\mu = 2, 3, 4, 5, 6$ but $A_m[1] \neq A_e[1]$), and by 3) the six-bit array S in which a single bit is set to 1 to specify the type of the piece trying to reach the target null-rank square.

We shall perform some set-theoretical operations on these boards: a bit in the board $P \cup Q$ is set to 1 if it is set to 1 in either P or Q or both; in

$P \cap Q$ if it is set to 1 in both P and Q; in $P \setminus Q$ if it is set to 1 in P but not in Q. The predicate $[P]$ is true if and only if P is not empty, i.e. at least one bit in it is set to 1; an empty board will be denoted by $\emptyset$. We use a similar notation for operations on sets with fixed numbers of bits and on their associated predicates.

With this notation, the predicate expressing occupancy of the null-rank square by the corresponding piece is

$$[R_0 \cap G'']\&[N \cap S],$$

The predicate expressing the appearance of new attacks by our pieces is

$$\left[\left(R_0 \cup \bigcup_{\mu=1}^{6} A_e''[\mu] \cap \Phi_e''[\mu]\right) \cap \bigcup_{i=1}^{q_m} (M''[i] \setminus M'[i])\right],$$

and that for the disappearance of former attacks by hostile pieces is

$$\left[\left(R_0 \cup \bigcup_{\mu=1}^{6} A_m''[\mu] \cap \Phi_m''[\mu]\right) \cap \bigcup_{i=1}^{q_e} (E'[i] \setminus E''[i])\right],$$

where $E''[i] = \emptyset$ if the opponent's i-th piece has been captured during the move in question.

Elementary strategies are compounded into non-elementary strategies by forming the union of their null-rank squares. First we define *active* strategies, aimed at reaching some intermediate target. For the strategy of White's center (we might also call it the development strategy) the null-rank squares are e4, e5, e6, d4, d5, and d6; for Black's development strategy they are e5,e4,e3, d5, d4, d3. For an attack on the opposing King the null-rank squares are the one on which he stands and the squares that he attacks. The strategy of the open file, mentioned above, is non-elementary, and so in essence are the strategies of Pawn advance, even though at any instant they have only one null-rank square. To every active strategy there corresponds a counter-strategy of the opponent, having the same null-rank squares.

To active strategies and counter-strategies there also correspond specific moves. For instance, the open-file strategy requires moves of major pieces into the file, and the preparation for these moves—new attacks by major pieces on the file, and the freeing up of squares for them (if such a square is occupied by another major piece, that piece must stay within the file). The strategy of the center requires castling and its preparation—clearing pieces from the line joining the King and the Rook, the flanking moves of the Bishop (Bc1—b2, Bf1—g2 for White; Bc8—b7, Bf8—g7 for Black) and the preparation for these, namely b2—b3 for a Bishop on c1, or g2—g3 for a Bishop on f1, and so on. There are also specific moves for the counterstrategies. The counter to an attack on the King may be an arbitrary move by the King; if the opponent has major pieces, the counterstrategy may construct escape hatches, as it may against the open file strategy. Specific moves are defined by using the same predicates as were used to define general moves. For example, the moves specified above for the open file strategy are defined by the predicates

$$([L \cap G''] \& \neg [L \cap G']) \& [N \cap S_l],$$

$$\bigvee_{i \text{ a major piece}} ([L \cap M''[i]] \& \neg [L \cap M'[i]]),$$

$$\left[L \cap G' \cap \bigcup_{i \text{ a major piece}} M_h'[i] \right] \& \neg [N \cap S_l],$$

$$[L \cap G''] \& \left[L \cap G' \bigcup_{i \text{ a major piece}} M_h[i] \right] \& [N \cap S_l],$$

where the index i ranges over the major pieces, L is the board for the file under consideration, $S_l = (011000)$ is a six-bit array in which the one-digits occupy the places of major pieces—Queen and Rook. The $M_h[i]$ are the boards for the horizontal potentials of the major pieces.

Strategic moves, as defined above, are basically those that lead to exchanges on squares of rank 0 or 1. When bringing a short-range piece (King, Knight) in from a distance, to a key square of a strategy, we use the natural metric of the board: a piece approaches if the sum of the absolute values of the horizontal and vertical differences between the coordinates of the square on which the piece stands and the null-rank square decreases. A Pawn approaches the scene of action in accordance with the strategy of Pawn advance. For transferring long-range pieces (Queen, Rook, Bishop) we use like Botwinnik the concept of the trajectory (cf. [7—9]). All these transfers, however, are considered as separate strategies connected in some way with the basic strategy that motivates them.

Most of the moves allowed by the rules of chess correspond to one or another strategy. Therefore, to shorten the search, not all strategies need be admitted. We need rules for prohibiting strategies and rules for admitting those previously prohibited. Such rules may be based on a semantic analysis of the position under study and on the amount of time the program has already spent on moves in the current game (under time pressure new active

strategies are wholly inadmissible). One of the aims in using a game model
with strategic moves is to prevent planless play, in which the choice of a
move depends on chance circumstances that influence the value of the
evaluation function in the terminal positions of the model.

One might hope that the following set of rules would meet the requirements: At the base position in the model, all strategies are admissible. After
a move A, B corresponding to one or more active strategies, all active
strategies to which it does not correspond are suppressed in the B-subtree,
but if the move corresponds to some admissible counter-strategy, no active
strategy is suppressed. In the C-subtree of the model, after the opponent's
reply B, C, counter-strategies are prohibited if they counter some active
strategy of the opponent to which his reply does not correspond. If however
the move B, C fits one of the opponent's admissible counter-strategies, our
own counter-strategies are not suppressed.

It turns out, however, that almost all moves fit some or other counter-
strategy. In fact, most of one's own side of the board consists of low-rank
squares for counter-strategies, and new attacks on one's own pieces often
result from one's own moves. Therefore the above mechanism for excluding
moves is not strong enough. One might try to limit the number of
counter-strategic moves in a branch, separately for White and Black, or to
admit such moves only as replies to moves by the opponent that correspond
to one of his active strategies. To be sure, what to do about counter-strategic moves at the base position is not obvious, but one might for example
allow all of them. There is a more complex, 'semantic', solution. When the
search reaches a position A for the first time, all moves corresponding to
active admissible strategies are allowed, plus perhaps a few moves that are
counter-strategic with respect to an admissible active strategy of the opponent, provided these are highly enough regarded by the best move routine.
While searching the B-tree after a move (A, B) corresponding to various
active strategies, one suppresses active strategies to which that move does
not correspond. If, however, one is studying a counter-strategic move, no
new prohibitions of active strategies occur in the search of the corresponding subtree.

Now suppose that we have returned to A after investigating the move
(A, B). If the move turned out to be a refutation, nothing remains to be
done, since we immediately step backward. If it is a bad move, we
investigate the next of the moves permitted earlier. If, however, it is an
improving move, that is, if it yields a partial score for A lying between the
old bounds $\underline{\lim}$ and $\overline{\lim}$, we change the set of admissible counter-strategies.
Initially the set is empty; the admissibility of a counter-strategic move in
the array of best moves does not mean that other moves corresponding to
the same counter-strategies are admissible. In the case we are now considering there is a critical branch from the position A beginning with the move
(A, B) just investigated. From there onward, counter-strategies to active

strategies of the opponent are admissible if they remain so at the end of the critical branch; if others were admissible earlier, they are now suppressed. However, if all the opponent's moves on the critical branch are counter-strategic, all our counter-strategies are suppressed.

If all the available moves have been investigated and have turned out to be bad, there exists a left W- or B-tree with base position at A, depending on whether White or Black has the move there. Then we also admit counter-strategies to any active strategies of the opponent that are admissible at the termini of the tree. However, if there are enough counter-strategic moves in the branch leading from the base position A_0 to the position A, and if we have found a satisfactory move at one of the lower rank positions in the branch, at which we have the move, no counter-strategies are admitted and we step backward from A.

The difficulty in carrying out such a plan is that at the precise moment when we admit some counter-strategy many moves (A, B) from A may have been studied and even rejected because they did not correspond to the strategies then admissible, or these moves may have been admitted but with a restriction on the set of admissible strategies. Therefore, after the set of admissible counter-strategies is changed, we must inspect these moves again. Those that correspond to newly admitted counter-strategies must be let into the search; some may have been judged admissible earlier, and if they were then held to be counter-strategic there is no need to inspect them again. However, they may have corresponded to some active strategies, and in this case such strategies must be prohibited and only those active strategies at A to which the moves in question do not correspond may be admitted. In any case, we cannot avoid repeated inspection of some variations in this model.

On the other hand, some of the strategies link together. Advancing a Pawn is useful for supporting neighboring Pawns (but not to outrun them); an attack on the King is supported by a corresponding Pawn advance; the strategy of the open file by the advance of the bordering Pawns, often necessary when the opponent has a Pawn on the file in question. Such a linkage can be realized by means of the graph Ind of induced strategies. If the arc $(S_1, S_2) \in$ Ind, the admissibility of the strategy S_1 implies the admissibility of S_2. If S_2 is induced, the moves that answer to it must be counted as also answering to the inducing strategy S_1.

Transformations of strategies are also useful. For example, if the strategy of the open file leads to an invasion by major pieces, we should adopt a supplementary strategy of an attack on the King; or, after stationing our pieces on a null-rank square that is the next point on a trajectory, we should consolidate our elementary strategy and go over to the strategy of reaching the next point on the trajectory. The conditions for admissibility of a new strategy may also depend on changes in the position that are not directly connected with the aims of our strategies. The open file strategy may be admitted if a file opens up; a Pawn advance strategy when a passed Pawn eventuates, and so on. A change in the material balance may induce a review of formerly active or protective strategies, and the review may generate new admissions and suppressions in the subtrees rooted at the positions where the material balance changes.

Some of the above proposals have been implemented. A small practical experiment has shown that there are prospects for semantic models of the kind we have described, but fundamental research is still in the future.

CHAPTER 3

The Method of Analogy

Identical Moves in Different Positions

We have said that a game-playing algorithm often inspects the same thing many times over. A human, having studied a situation once, will in the future draw conclusions by the use of analogy. But, it often happens that seemingly insignificant changes in the position alter the course of the game and lead to substantially different outcomes. Such changes are said to be essential with respect to the contemplated variations. A human decides, well or poorly, whether a position that has been studied differs essentially from one that has not, and accordingly does or does not investigate variations starting from the latter. If we are to devise algorithms that use this method, we must analyze a) the notion of analogous moves (later we shall often use the term 'the same' rather than 'analogous') and b) the notion of the difference between positions essential for given variations.

We begin with the notion of similar moves. In the games that we have used and shall use as examples, pieces are moved about on a board and the moves are elementary displacements determined by a finite set of rules which is small compared to the number of positions, or even compared to the number of classes of mutually similar moves. To make the definition of a move more universal, we may consider an auxiliary square in addition to the customary squares of the board; in chess, for example, or draughts, captured pieces will move to the auxiliary square. The examples of moves that we have just given are taken from chess or draughts, but we may also imagine noughts-and-crosses as played on a board divided into squares; the pieces do not move on this board, but move to squares on it from an auxiliary square. Card games may be described in a similar way, e.g. bridge is played on six squares—N, S, E, W, the 'table', and an auxiliary square to which pieces move on quitting further play.

We may use a standard form for prescribing the rules of a game, by defining the admissible moves. Positions in the game are described by a set of incidence relations that hold among the elements of finite sets: T—the board (its elements are called squares); F—the set of pieces; and C—the set of players of the game. The relationship f/t is expressed in words as 'the piece f occupies the square t'; f/c as 'the piece f belongs to player c' (in many cases the relationships f/c do not change, i.e. each piece has its own 'color'). Each piece $f \in F$ in a given position stands on one square of the board and belongs to one of the players in the game. An arbitrary admissible move is the realization of a *virtual move*—a change in one of the incidence relationships. The lowest level in the hierarchy of rules describes the set of virtual moves.

A virtual move may be prescribed by specifying the following sets of incidence relationships:

(1) The set $\{f_i/t_i', f_i/c_i'\}$ which are satisfied before the move and not satisfied after it;
(2) the set $\{f_i/t''_i, f_i/c''_i\}$, which are satisfied after the move but not before it;
(3) the set $\{f_i/t'''_i, f_i/c'''_i\}$, which must be satisfied before the move (else the contemplated virtual move from the given position is inadmissible) and continue to be satisfied after the move.

Often the third set of relationships expresses the condition that some set of squares on the board must be empty if the virtual move is to be admissible. The squares c_i that satisfy the incidence relationships in the third set form a subset $L \subset T$ which we call the 'line of the move' or the 'trajectory'; (the line contains no squares when the third set is empty). In the first approximation virtual moves are admissible from positions at which the three given sets of incidence relationships are satisfied, the remaining incidence relationships being arbitrary. A real move is an admissible virtual move. One of the admissibility conditions relates to the turn to move. others are specified by using the notion of virtual move and some other notions found at higher levels in the hierarchy of rules.

Intuitively we would say that real moves are similar if they are realizations of the same virtual move. In specific games, however, we must take account of various circumstances belonging to higher levels, e.g. whether a check to a King arises or vanishes in consequence of a virtual move. Also, we may find it desirable to regard certain real moves as similar when the virtual moves they correspond to are different but have lines that are in some sense near together. In general, we shall hold to the intuitive definition.

Before we study the notion of essential differences between positions we pause to ask what kinds of situations make it desirable to reason by analogy in reaching a decision. Let B be the current position in a search. If (B, B') is a best move, we want to ascertain its consequences as precisely as we can. If it is a refutation, we must establish the fact that it is, in order to step backward from B without investigating other moves (B, C) (if we know that we may step back from B without studying any of the moves from it, then very probably we may omit a forward step to it). Moreover, since best and refutation moves can be relatively well guessed, the majority of the moves found in the search tree are bad. Therefore, to shorten the search, methods for mass pruning of bad moves are extremely valuable, and among these is the use of analogy.

To this end we consider the following scheme for reasoning by analogy. Suppose that at two positions B and C in which the move belongs to the same color, say White, we may make the similar moves (B, B_1) and (C, C_1). A search of the B_1-subtree of the game $\mathfrak{A}$ has shown that $\mathrm{sc}(B) \leq M$, and the corresponding search subtree $\tilde{\mathfrak{A}}(B)$ includes the move (B, B_1). The positions B, C and the tree $\tilde{\mathfrak{A}}(B)$ determine the value of the *influence* predicate $\mathrm{Inf}(B, C, \tilde{\mathfrak{A}}(B))$. The value 1 means that the difference between B and C influences the search tree $\tilde{\mathfrak{A}}(B)$. Then we must investigate the consequences of the move $C, C_1)$. The value 0 means that there is no influence, and the argument by analogy correctly concludes that $\mathrm{sc}(C_1) \leq m$.

To define the influence predicate so that its value can be determined without a search of the C_1-subtree, we examine the case $\mathrm{sc}(C_1) > m$. We assume for simplicity that White and Black move alternately. We also call positions with scores greater than m won positions, those with scores not greater than m lost, and regard similar moves as identical. Since a search

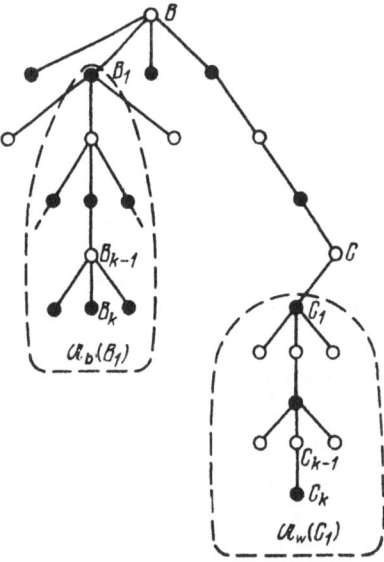

Figure 18

has shown that B_1 is lost, there exists a minimally inclusive B-pruned tree $\mathfrak{A}_b(B_1)$ in which all terminal positions are lost. So also there exists a W-pruned tree $\mathfrak{A}_w(C_1)$ with base C_1, in which the terminal positions are won, and in which there is at least one move from every non-terminal White position. To be sure, in contrast to the tree $\mathfrak{A}_b(B_1)$, this one is unknown, but for the present we do not need to know it.

The tree $\mathfrak{A}_b(B_1)$ contains a move $\Psi_1 = (B_1, B_2)$ which wins for Black and leads to a White position B_2. Suppose this same move is admissible at C_1 under the rules of the game $\mathfrak{A}$. It leads to some C_2, also a White position. Since $\mathfrak{A}_w(C_1)$ is a W-pruned tree and C_1 is a Black position this tree contains all moves in the game $\mathfrak{A}$ from C_1, including the move $\Psi = (C_1, C_2)$, and so contains the position C_2. If this is non-terminal the tree $\mathfrak{A}_w(C_1)$ contains a move $\Psi_2 = (C_2, C_3)$ leading from it to a Black position just as the move (B_2, B_3) does from the position $B_2 \in \mathfrak{A}_b(B_1)$ provided the latter move exists. As an arbitrary move from a White position in the tree $\mathfrak{A}_b(B_1)$, $(B_2, B_3) \in \mathfrak{A}_b(B_1)$ (see Figure 18).

We may repeat this argument to show that there exist branches $(B, B_1, \ldots, B_k)$ and $(C, C_1, \ldots, C_k)$ beginning at the positions B and C, respectively, and containing the same moves $\Psi_0, \Psi_1, \Psi_{k-1}$ in the same order (Ψ_0 is the move (B, B_1) which by hypothesis is identical to the move (C, C_1)), and the branch $(B, B_1, \ldots, B_k)$ belongs to the tree $\tilde{\mathfrak{A}}(B)$ and to its subtree $\mathfrak{A}_{\Psi_0}(B) = (B, B_1) \cup \mathfrak{A}_b(B_1)$. Since the trees $\mathfrak{A}_b(B_1)$ and $\mathfrak{A}_w(C_1)$ are finite, the process of constructing the sequence of moves $\Psi_0, \Psi_1, \ldots, \Psi_{k-1}$ must come to an end. The impossibility of continuing can arise only from one of the following causes:

(1) a Black move $J_k = (B_k, B_{k+1})$ which wins for Black at the position $B_k \in \mathfrak{A}_b(B_1)$, is impossible at the corresponding position;
(2) A White winning move $\Psi_k = (C_{k-1}, C_k) \in \mathfrak{A}_w(C_1)$ is impermissible at the position B_k;
(3) the position B_k is terminal and C_k is not;
(4) the position C_k is terminal and B_k is not;
(5) both B_k and C_k are terminal.

Our main interest lies in the first two cases, depicted in Figure 19. The last three cases may be excluded by a formal method. For games with two outcomes (which we are now studying since we regard all positions with scores not exceeding m as lost and positions with scores exceeding m as won) we can construct an equivalent model in which terminal positions are always lost for the side that has the move there. To do this we need only add to the game a fictitious move $(F, \bar{F})$ at all terminal positions F that are won for our own side; the new position $\bar{F}$ has the same score as F, i.e. is lost for the side having the move.

If the game $\mathfrak{A}$ satisfies our condition, C_k is a White position in Case 3 and there is a winning move $\Psi_{k+1} = (C_k, C_{k+1})$ from it. But B_k is terminal, and all moves, Ψ_{k+1} included, are prohibited from it. Thus in Case 3, Case

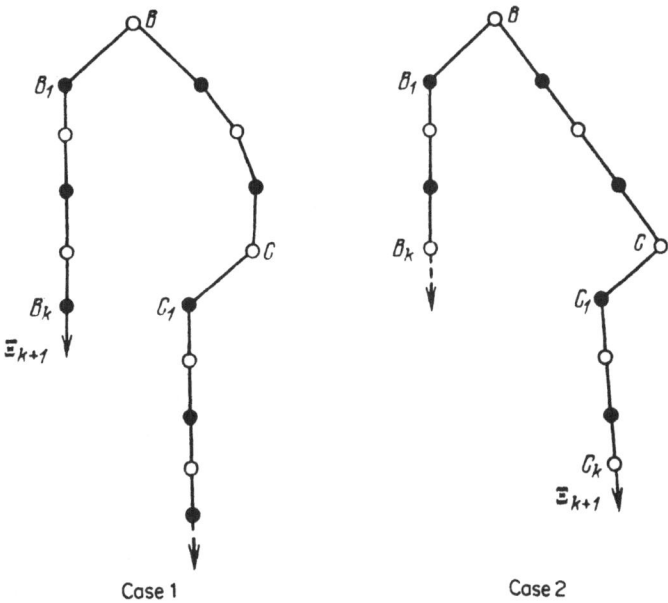

Case 1   Case 2

Figure 19

2 also holds; similarly, Case 1 holds when Case 4 does. Case 5 is impossible, since it implies different scores at the terminal positions B_k and C_k, where the same side has the move. This formal exclusion of Case 5, however, is useless for our investigation, since the fictitious move is defined quite differently from the genuine moves, and does not lend itself to the use of our notion of virtual move and our method of defining it.

But, it is especially important to exclude Case 5. Cases 3 and 4 may be studied by methods like those we shall use later for Cases 1 and 2. We shall admit some virtual moves from the positions B_k and C_k but not from other positions. In Case 5 we would have to use rules for determining the score at terminal positions; these would be complex, involving both real and virtual moves. Therefore we shall for the moment consider some more or less general examples of rules that result in the exclusion of Case 5 (and partly of Cases 3 and 4).

Suppose that a win or loss is determined by the configuration of only some of the pieces, and the pieces playing a role in the final configurations that determine a loss for White at terminal positions of $\mathfrak{A}_b(B_1)$ are identically placed in the positions B and C. Then the sequence of moves that leads from B_1 to the terminal position also leads from C_1 to a terminal position with the same score. For instance, in the game of 'Wolf and Sheep' the Wolf wins if he reaches the opponent's shelter in the first rank, just as in Russian draughts. Thus for this game both Cases 3 and 5 are impossible.

Furthermore, winning terminal configurations generally arise only from a move made by the winning side. In the game of noughts-and-crosses, for instance, such a configuration exists after the winner's move—five of his pieces placed sequentially in a vertical, horizontal, or diagonal line. We may modify the rules of this game to some extent, so that the move number becomes uniquely related to the score of the corresponding terminal positions, and Cases 3, 4, and 5 become impossible. It suffices that after the move number $L+1$ an arbitrary position is drawn if neither side has won earlier. In games such as draughts and chess a single-valued relationship between the turn to move and the score of the corresponding terminal position is prevented by the possibility of a draw based on a negative definition (neither side can force a win).

Another cause for the impossibility of Case 5 in many games, including of course model games, is that the score of their terminal positions is equal to the sum of the material gains achieved at each move in the sequence $\Psi_1, \Psi_2, \ldots, \Psi_k$ leading from the base position A_0 to the terminal position F:

$$\mathrm{sc}(F) = \sum_{i=1}^{k} h(\Psi_i).$$

For example, in many card games (and in *odnomastka*, a similar model game with complete information,) the scores are equal to the weighted sum of the tricks—sets of pieces simultaneously laid on the table. The weights of such tricks have opposite values for White and Black, and in some games depend on the pieces they contain.

In games such as draughts and chess the scores at terminal positions are immediately determined, in another way. Nevertheless, to achieve a winning position in these games it is often necessary to gain a preliminary material preponderance. Therefore, in computing the scores at terminal positions in models of these games, an important component is the material score discussed in the preceding chapter:

$$f_{\mathrm{m}}(C) := \sum_{\mu=1}^{M} h_\mu(C) P_\mu^w(C) - \sum_{\mu=1}^{M} h_\mu(C) P_\mu^b(C).$$

If the sequence of moves $\Psi_1, \Psi_2, \ldots, \Psi_k$ leads from the posiition B to the terminal position F, the material score of the latter is given by

$$f_M(F) = f_M(B) + (f_M(F) - f_M(B))$$

$$= f_M(B) + \sum_{\mu=1}^{M} \left(P_\mu^b(F) - P_\mu^b(B) \right) h_\mu - \sum_{\mu=1}^{M} \left(P_\mu^w(F) - P_\mu^w(B) \right) h_\mu$$

$$= f_M(B) + \sum_{i=0}^{\lceil k/2 \rceil} h(\Psi_{2^i+1}) - \sum_{i=1}^{\lceil k/2 \rceil} h(\Psi_{2^i}),$$

where $h(\Psi)$ is the weight of the move Ψ. This is equal to the sum of the weights of the opponent's pieces that have been captured and the difference of the new and former weights of our own pieces that have changed position. This latter quantity differs from 0 for moves with promotion to King (draughts), Queen (chess), etc. For moves Ψ without capture or promotion, $h(\Psi) = 0$.

Thus if the variation $(\Psi_0, \Psi_1, \ldots, \Psi_{k-1})$ leads from the position B to a material loss $(f_m(B_k) < f_m(B))$ it leads to the same loss from the position C. If the change in the positional score cannot compensate for White's material loss after the move $\Psi_0 = (B, B_1)$ from B, and $f_m(B) = f_m(C)$, then C_k, which corresponds to the terminal position $B_k \in \mathfrak{A}_{\Psi_0}(B)$, is necessarily lost for White. This means that Case 5 is impossible in our models. Cases 3 and 4 are excluded for these models because all terminal positions are lost for their own side, except those arising after the so-called *blank* moves. We shall discuss this point in more detail in the next section.

Thus, for a rather broad class of games the exclusion of Cases 1 and 2 at pairs of positions $B_k \in \mathfrak{A}_{\Psi_0}(B)$ and C_k belonging to the 'parallel' tree $\mathfrak{A}_{\Psi_0}(C)$ implies that $\mathrm{Inf}(B, C, \mathfrak{A}_{\Psi_0}(B)) = 0$. Therefore we may construct the influence predicate by studying the conditions that allow the difference in the simultaneous admissibility of the same move in parallel positions. Let $\mathrm{Inf}(B, C, D)$ be a predicate having the value 1 if a) Black has the move at $D \in \mathfrak{A}_{\Psi_0}(B)$ and the move $\Psi = (D, G) \in \mathfrak{A}_{\Psi_0}(B)\}$ *is inadmissible at the parallel position* $E \in \mathfrak{A}_{\Psi_0}(C)$, or b) White has the move at D and there exists a virtual move $\Psi(E, H)$ which is admissible at the parallel position $E \in \mathfrak{A}_{\Psi_0}(C)$ and inadmissible at D. The influence predicate $\mathrm{Inf}(B, C, \mathfrak{A}_{\Psi_0}(B))$ is easily expressed by means of such a 'local' predicate:

$$\mathrm{Inf}\left(B, C, \mathfrak{A}_{\Psi_0}(B)\right) = \bigvee\left\{\mathrm{Inf}(B, C, D) \mid D \in \mathfrak{A}_{\Psi_0}(B)\right\}.$$

Since an arbitrary position is fully defined by a finite collection of features, i.e. by the values of elementary predicates belonging to a finite set $\{\pi_\alpha\}$, the predicate $\mathrm{Inf}(B, C, D)$ can be represented in disjunctive normal form with components $\pi_\alpha(B)$, $\pi_\alpha(C)$, and $\pi_\alpha(D)$. (We may assume that the set $\{\pi_\alpha\}$ contains the negation of each of its predicates.) This form can be written in terms of the non-elementary predicates of the positions B, C, and D:

$$\mathrm{Inf}(B, C, D) = \bigvee_{i=1}^{s} \left(\rho_i(B) \& \sigma_i(C) \& \tau_i(D)\right).$$

Consequently

$$\mathrm{Inf}\left(B, C, \mathfrak{A}_{\Psi_0}(B)\right) = \bigvee_{D \in \mathfrak{A}_{\Psi_0}(B)} \left(\bigvee_{i=1}^{s} \left(\rho_i(B) \& \sigma_i(C) \& \tau_i(D)\right)\right)$$

$$= \bigvee_{i=1}^{s} \left(\rho_i(B) \& \sigma_i(C)\right) \& \left(\bigvee_{D \in \mathfrak{A}_{\Psi_0}(B)} \tau(D)\right)$$

$$= \bigvee_{i=1}^{s} \left(\rho_i(B) \& \sigma_i(C) \& \tau\left(\mathfrak{A}_{\Psi_0}(B)\right)\right).$$

for an arbitrary set of positions $\mathfrak{W}$

$$\tau_i(\mathfrak{W}) = \vee\{\tan_i(D) \mid D \in \mathfrak{W}\}, \qquad i = 1, 2, \ldots, s.$$

The predicates $\tau_i(\mathfrak{W})$ have an important property:

$$\tau_i(\mathfrak{W}_1 \cup \mathfrak{W}2) = \tau_i(\mathfrak{W}1) \vee \tau_i(\mathfrak{W}_2).$$

Thus the predicates $\rho_i(B)$ and $\sigma_i(C)$ may be determined for the positions B and C, while the values of the predicates $\tau_i(\mathfrak{A}_{\Psi_0}(B))$ can be calculated during the search of the tree $\mathfrak{A}_{\Psi_0}(B)$ and the amount of information about them does not depend on the number of positions in the tree. If the predicate $\text{Inf}(B, C, D)$ is *exact*, i.e. if it is a necessary and sufficient condition for the occurrence of Cases 1 and 2, the number s of predicates may be very large, so that the computation of the predicates $\tau_i(\mathfrak{A}_{\Psi_0}(B))$ consumes large amounts of time and memory. However, as we shall see, we can develop predicates for specific games that can be computed quickly enough, if we replace the necessary and sufficient conditions for the existence of Cases 1 and 2 by necessary conditions only.

Suppose that a search of the tree $\mathfrak{A}_{\Psi_0}(B)$ has shown that the move $\Psi_0 = (B, B_1)$ leads to a lost position, and after the moves $\Theta_1, \Theta_2, \ldots, \Theta_{2l}$ we arrive from B at a position S which is also White. Fairly often some new moves $\Omega_1, \Omega_2, \ldots, \Omega_r$ can be made from C, although they were inadmissible at B. Then the influence predicate $\text{Inf}(B, C, \mathfrak{A}_{\Psi_0}(B))$ is equal to 1, since the position C in the tree $\mathfrak{A}_{\Psi_0}(C)$ corresponds to the position B in the tree $\mathfrak{A}_{\Psi_0}(B)$ and we are in the presence of Case 2. Replacing the trees $\mathfrak{A}_{\Psi_0}(B)$ and $\mathfrak{A}_{\Psi_0}(C)$ by $\mathfrak{A}_{\Psi_0}(B_1)$ and its parallel tree $\mathfrak{A}_{b_{\text{II}}}(C_1)$ normally does not help us, since some at least of the new moves Ω_λ remain admissible at White positions in the tree $\mathfrak{A}_{b_{\text{II}}}(C_1)$. Thus if we use the notion of influence as defined above we cannot shorten the search of the C-subtree of $\mathfrak{A}$.

Also, the new moves Ω_λ $(1 \leq \lambda \leq r)$ may be irrelevant to the play and will therefore lose quickly. Common sense tells us that in this case our earlier conclusion about the move Θ_0, which is admissible at C also, will probably still be valid. We need to reconsider it only if the sequence of moves $\Lambda(\Theta_1, \Theta_2, \ldots, \Theta_{2l})$—or variants of it—which demonstrate that White loses after the moves $\Omega_\lambda (1 \leq \lambda \leq r)$ turn out to influence variants of the tree $\mathfrak{A}_{\Psi_0}(B)$. But this is a different concept of influence. It is narrower, since now it prohibits not all instances of Case 2, but only those in which White moves that have not yet been inspected are admissible . The new influence predicate $\text{Inf}'(B, C, \mathfrak{A}_b(B))$ is determined by the conditions that permit new moves, not in Ω_λ, to arise at positions in the tree $\mathfrak{A}_{\Psi_0}(C)$.

A scheme for shortening the search with this notion of influence is shown in Figure 20. First we investigate the new moves $\Omega_1, \Omega_2, \ldots, \Omega_r$. Let $\mathfrak{A}_{\{\Omega_\lambda\}}(C)$ be the corresponding search tree, and let $\mathfrak{A}_{\{\Omega_\lambda\}}(B) = \Lambda \cup \mathfrak{A}_{\{\Omega_\lambda\}}(C)$. If the tree $\mathfrak{A}_{\{\Omega_\lambda\}}(C)$ has no influence on the tree $\mathfrak{A}_{\Psi_0}(B)$, the move $\Psi_0 = (B, B_1)$ $= (C, C_1)$ from the position C loses just as it does from the position B. (A more precise description of this method for shortening the search will be

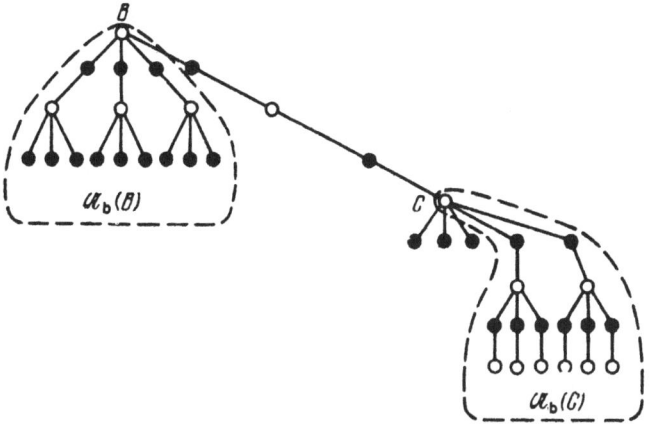

Figure 20

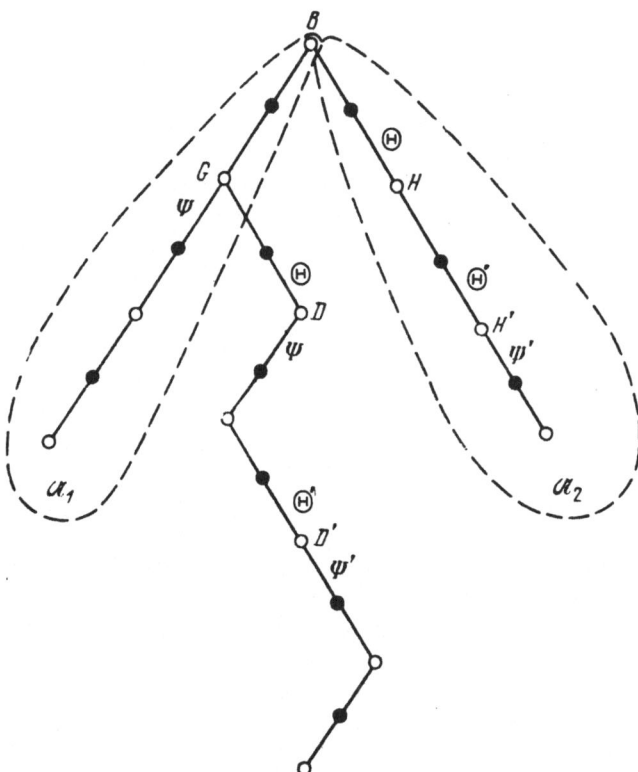

Figure 21

given in Section 4 of this chapter.) Thus we need to determine the influence of a subtree $\mathfrak{A}_1$ with base B on another subtree $\mathfrak{A}_2$ with the same base, rather than the influence of two different positions B and C on the search tree $\mathfrak{A}_{\Psi_0}(B)$. An influence predicate $\mathrm{Inf}(B, \mathfrak{A}_1, \mathfrak{A}_2)$ of this kind is the disjunction of the influence predicates of the pairs of branches Λ_1 and Λ_2 in these subtrees:

$$\mathrm{Inf}(B, \mathfrak{A}_1, \mathfrak{A}_2) = \vee \left\{ \mathrm{Inf}(B, \Lambda_1, \Lambda_2) \mid gL_1 \in \mathfrak{A}_1, \Lambda_2 \in \mathfrak{A}_2 \right\}.$$

The B-subtree of the game $\mathfrak{A}$ contains compound branches, in which the moves of some branch $\Lambda_1 \subset \mathfrak{A}_1$ alternate with the moves of some other branch $\Lambda_2 \subset \mathfrak{A}_2$, while the relative order of the moves in the two branches is preserved (cf. Figure 21). Only one of the branches Λ_1, Λ_2 may have an odd length, else the numbers of White and Black moves in the compound branch Λ would differ by 2, which is impossible in a game where White and Black move alternately. The positions D in Λ that are incident on the moves $\Psi \in \Lambda_1$ and $\Theta \in \Lambda_2$ correspond to the positions $G \in \mathfrak{A}_1$ and $H \in \mathfrak{A}_2$ that are incident on the moves Ψ and Θ in the respective branches Λ_1 and Λ_2. When both the incident positions D' of a move belong to the same branch, Λ_1 or Λ_2, the position H' corresponding to D' lies in only one of the trees $\mathfrak{A}_1$, $\mathfrak{A}_2$. The final position in the branch Λ corresponds to the final position of the same color in either Λ_1 or Λ_2 when the lengths of these branches have unequal parities, or to the final positions in both branches when their lengths have even parity.

In this more general situation, Case 1 indicates that at some position D lying in the branch Λ, with color different from that of B, some move Ψ is inadmissible under the rules of the game $\mathfrak{A}$ but is admissible at the corresponding position G in one or other of the branches Λ_1 or Λ_2 and belonging to that branch; Case 2 indicates that some move Ω is admissible at the position D (of the same color as B) and inadmissible at both of the corresponding positions.

Axioms of Influence and Possibilities for Shortening the Search

We suppose now that the influence predicate $\mathrm{Inf}(B, \Lambda_1, \Lambda_2)$ has been defined as a function of the position B and two issuing branches Λ_1 and Λ_2. Its properties, as used to prove the theorem on the possibility of shortening the search, may be stated axiomatically, but before we state them we need some preliminary definitions.

The branch $\Lambda(\Psi_1, \Psi_2, \ldots, _\Lambda)$ is said to be *composed of the two branches* $\Lambda_1(\Theta_1, \Theta_2, \ldots, \Theta_k)$ and $\lambda_2(\Omega_1, \Omega_2, \ldots, \Omega_{1-k})$:

$$\Lambda = \Lambda_1 + \Lambda_2,$$

if every move $\Psi_h \in \Lambda$ coincides with one of the moves $\Theta \in \Lambda_1$ or $\Omega \in \Lambda_2$, while the correspondence between the moves in the branch Λ and the moves in Λ_1 and Λ_2 is one-to-one, and the ordering of the moves in these branches is compatible, i.e.

$$\Theta_i = \Psi_h \& \Theta_{i+1} = \Psi_g \Rightarrow h < g, \qquad i = 1, 2, \ldots, k-1,$$

$$\Omega_j = \Psi_m \& \Omega_{j+1} = \Psi_n \Rightarrow m < n, \qquad j = 1, 2, \ldots, l-k-1.$$

The branch $\Lambda(\Psi_1, \Psi_2, \ldots, \Psi_l)$ is said to be *strictly composed of the two branches* $\Lambda_1(\Theta_1, \Theta_2, \ldots, \Theta_k)$ and $\Lambda_2(\Omega_1, \Omega_2, \ldots, \Omega_{l-k})$:

$$\Lambda = \Lambda_1 * \Lambda_2,$$

if

$$\Psi = \begin{cases} \Theta_i, & i = 1, 2, \ldots, k, \\ \Omega_{i-k}, & i = k+1, k+2, \ldots, 1. \end{cases}$$

The non-commutative operation of strict composition, as so defined, may at times fail to define a branch in $\mathfrak{A}$ with its origin at B, since some move Ψ_i may be inadmissible under the rules of $\mathfrak{A}$ at the position arising after the sequence of moves $\Psi_1, \Psi_2, \ldots, \Psi_{i-1}$ (in this case clearly $i > k$). Nevertheless, the branch $\Lambda = \Lambda_1 * \Lambda_2$ may be admissible even when the branch Λ_2 issuing from B consists of moves not necessarily admissible at the corresponding positions. The non-strict composition $\Lambda = \Lambda_1 + \Lambda_2$ is not uniquely defined (the moves in the branches may alternate in various ways) and may also define an inadmissible branch. It may be regarded as commutative, in the sense that the relationship $\Lambda = \Lambda_1 + \Lambda_2$ among the branches is commutative in the second and third arguments. The strict composition of the branch Λ and a move Θ is defined as for branches:

$$\Lambda(\Psi_1, \Psi_2, \ldots, \Psi_l) * \Theta := \Lambda'(\Psi_1, J_2, \ldots, \Psi_l, \Theta).$$

Let B be a fixed position which will serve as the origin of all branches to be considered. If $\mathrm{Inf}(B, \Lambda_1, \Lambda_2) = 1$, we shall say that the branch Λ_1 influences the branch Λ_2 and we write $\Lambda_1 \sim \Lambda_2$. We shall also consider the influence relation $\Lambda \sim \Psi$ of the branch Λ on the move Ψ; this is not the same as the influence $\Lambda \sim L(\Psi)$ of Λ on the branch $L(\Psi)$ which consists of the single move Ψ. If the move Ψ is admissible at B, the relation $\Lambda \sim \Psi$ implies that $\Lambda \sim L(\Psi)$, but the converse is not true.

The position D to which the branch $\Lambda(\Psi_1, \Psi_2, \ldots, \Psi_l)$ leads from B will be denoted by $\mathrm{fin}(B, \Lambda)$. The color of the move Ψ will be denoted by $\mathrm{col}\,\Psi$; the color of the side having the move at A will be denoted by $\mathrm{col}\,A$, and the color of the side having the move at the position $\mathrm{fin}(B, \Lambda)$ by $\mathrm{col}\,\Lambda$. (Since we are currently considering games $\mathfrak{A}$ in which White and Black move alternately, $\mathrm{col}\,\Lambda$ is uniquely defined by $\mathrm{col}\,B$ and the parity of the length of the branch Λ.) When B is a White position we write $\mu \underset{\mathrm{col}\,B}{\preccurlyeq} \nu$ to mean that $\mu \leq \nu$, and $\mu \underset{\mathrm{col}\,B}{\prec} \nu$ to mean that $\mu + \delta \leq \nu$, where δ is some preselected quantity; when $\mathrm{col}\,B$ is Black, we use the same notation to mean that $\mu \geq \nu$ and $\mu - \delta \geq \nu$, respectively. If $\mathrm{sc}(D) \underset{\mathrm{col}\,B}{\prec} \mathrm{sc}(E)$ we shall say that the score at the position D is significantly worse for $\mathrm{col}(B)$ than the score at E.

The set of virtual moves admitted by the rules of $\mathfrak{A}$ at some position A will be denoted by $M(A)$.

The Influence Axioms. In the following formulations we shall suppose that in the game $\mathfrak{A}$ White and Black move alternately, and that all branches issue from a fixed (arbitrary) position B and consist of moves admissible in $\mathfrak{A}$ at the corresponding positions.

Axiom 1. Axiom on the relationship between influence on a move and influence on a branch: If the branch Λ_1 influences the move Ψ with the color $\mathrm{col}\,B$ and Ψ belongs to the branch Λ_2, then Λ_1 influences Λ_2:

$$\Lambda_1 \sim \Psi \& \mathrm{col}\,\Psi = \mathrm{col}\,N \& \Psi \in \Lambda_2 \Rightarrow \Lambda_1 \sim \Lambda_2.$$

Axiom 2. Axiom of symmetry: If the branch Λ_1 influences the branch Λ_2, then Λ_2 influences Λ_1:

$$\Lambda_1 \sim \Lambda_2 \Rightarrow \Lambda_2 \sim \Lambda_1.$$

Axiom 3. First axiom on the composition of branches: If the branch Λ is composed of the branches Λ_1, Λ_2 and influences the move Ψ, then either Λ_1 influences Ψ or Λ_2 influences Ψ or Λ_1 influences Λ_2.

$$\Lambda = \Lambda_1 + \Lambda_2 \& \Lambda \sim \Psi \Rightarrow \Lambda_1 \sim \Psi \vee \Lambda_2 \sim \Psi \vee \Lambda_1 \sim \Lambda_2.$$

Axiom 4. The second axiom on the composition of branches: If the branch Λ is composed of the branches Λ_1 and Λ_2 and influences the branch Λ_3, then Λ_1 influences Λ_3 or Λ_2 influences Λ_3 or Λ_1 influences Λ_2:

$$\Lambda = \Lambda_1 + \Lambda_2 \& \Lambda \sim \Lambda_3 \Rightarrow \Lambda_1 \sim \Lambda_3 \vee \Lambda_2 \sim \Lambda_3 \vee \Lambda_1 \sim \Lambda_2.$$

Axiom 5. Axiom on Case 1: Let Λ_1, Λ_2 be branches with even and odd lengths, respectively. Let $\Lambda = \Lambda_1 + \Lambda_2$, and let the move Ψ be admissible under the rules of the game $\mathfrak{A}$ at the position $\mathrm{fin}(B, \Lambda_2)$ and inadmissible at $\mathrm{fin}(B, \Lambda)$ (see Figure 22). Then the branch Λ_1 influences $\Lambda_2 * \Psi$:

$$\mathrm{col}\,\Lambda_1 = \mathrm{col}\,B \& \mathrm{col}\,\Lambda_2 \neq \mathrm{col}\,B \& \Lambda$$
$$= \Lambda_1 + \Lambda_2 \& \Psi \in M(\mathrm{fin}(B, \Lambda_2)) \&$$
$$\Psi \notin M(\mathrm{fin}(B, \Lambda)) \Rightarrow \Lambda_1 \sim \Lambda_2 * \Psi.$$

Axiom 6. First axiom on Case 2. Let Λ_1 and Λ_2 be two branches of even length, let $\Lambda = \Lambda_1 + \Lambda_2$, and let the move Ψ be inadmissible under the rules of the game $\mathfrak{A}$ at the positions $\mathrm{fin}(B, \Lambda_1)$ and $\mathrm{fin}(B, \Lambda_2)$ but admissible at $\mathrm{fin}(B, \Lambda)$ (see Figure 23). Then Λ_1 influences Λ_2:

$$\mathrm{col}\,\Lambda_1 = \mathrm{col}\,\Lambda_2 = \mathrm{col}\,B \& \Lambda = \Lambda_1 + \Lambda_2$$
$$\& \Psi \notin (M(\mathrm{fin}(B, \Lambda_1)) \cup M(\mathrm{fin}(B, \Lambda_2)))$$
$$\& \Psi \in M(\mathrm{fin}(B, \Lambda)) \Rightarrow \Lambda_1 \sim \Lambda_2.$$

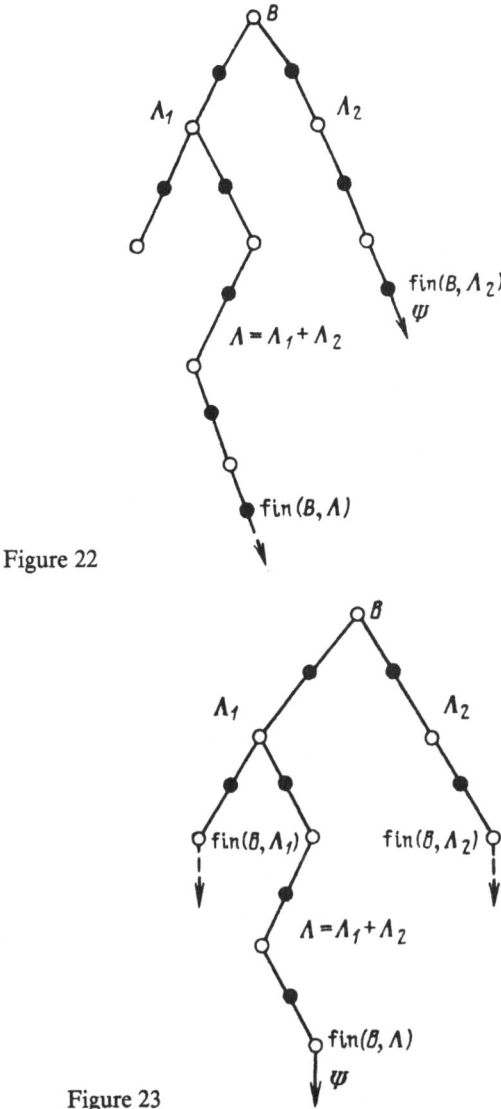

Figure 22

Figure 23

Axiom 7. Second axiom on Case 2. Let Λ_1 and Λ_2 be branches of even length, let $\Lambda = \Lambda_1 + \Lambda_2$, and let the move Ψ be admissible under the rules of the game A at the positions B, $\mathrm{fin}(B, \Lambda_1)$, and $\mathrm{fin}(B, \Lambda)$, but not admissible at the position $\mathrm{fin}(B, \Lambda_2)$ (see Figure 24). Then Λ_1 influences Λ_2 or Ψ:

$\mathrm{col}\,\Lambda_1 = \mathrm{col}\,\Lambda_2$

$\quad = \mathrm{col}\,B \& \Lambda = \Lambda_1 + \Lambda_2 \& \Psi \in (M(B) \cap M(\mathrm{fin}(B, \Lambda_1)) \cap M(\mathrm{fin}(B, \Lambda)))$

$\quad \& \Psi \notin M(\mathrm{fin}(B, \Lambda_2)) \Rightarrow \Lambda_1 \sim \Lambda_2 \vee \Lambda_1 \sim \Psi.$

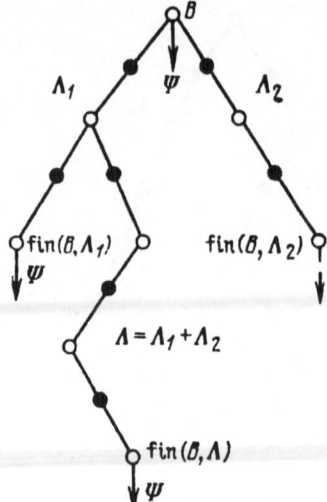

Figure 24

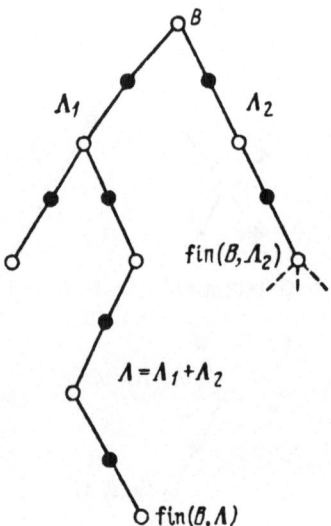

Figure 25

Axiom 8. Axiom on Cases 3 and 4. Let Λ_1 be a branch of even length, let $\Lambda = \Lambda_1 + \Lambda_2$, and let one and only one of the positions $\text{fin}(B, \Lambda)$ and $\text{fin}(B, \Lambda_2)$ be terminal (see Figure 25). Then Λ_1 influences Λ_2:

$$\text{col } \Lambda_1 = \text{col } B \& \Lambda = \Lambda_1 + \Lambda_2$$
$$\& M(\text{fin}(B, \Lambda)) = \varnothing \& M(\text{fin}(B, \Lambda_2)) \neq \varnothing$$
$$\vee M(\text{fin}(B, \Lambda)) \neq \varnothing \& M(\text{fin}(B, \Lambda_2)) = \varnothing \Rightarrow \Lambda_1 \sim \Lambda_2.$$

Axiom 9. Axiom on Case 5. Let Λ_1 be a branch of even length, let $\Lambda = \Lambda_1 + \Lambda_2$, and let both positions $\text{fin}(B, \Lambda)$ and $\text{fin}(B, \Lambda_1)$ be terminal

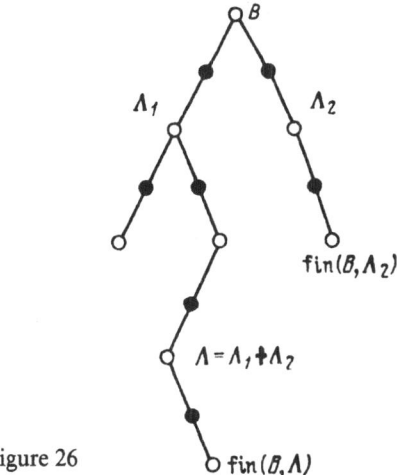

Figure 26

(see Figure 26). Then either Λ_1 influences Λ_2 or the score at the position $\text{fin}(B, \Lambda)$ is not better for col B than the score at $\text{fin}(B, \Lambda_2)$:

$$\text{col } \Lambda_1 = \text{col } B \& \Lambda = \Lambda_1 + \Lambda_2 \& M(\text{fin}(B, \Lambda))$$
$$= M(\text{fin}(B, \Lambda_2)) = \varnothing$$
$$\Rightarrow \Lambda_1 \sim \Lambda_2 \lor \text{sc}(\text{fin}(B, \Lambda)) \underset{\text{col } B}{\preccurlyeq} \text{sc}(\text{fin}(B, \Lambda_2)).$$

Constructively defined influence relations for specific games satisfy these axioms so to speak 'essentially', that is, when their assertions are invalid then the contemplated branches, their ends, or the moves will satisfy supplementary conditions that we can specify or exploit. Thus the formulations of the theorems we shall be proving must be sharpened, and their proofs must be supplemented by investigations of special cases. Nevertheless the leading ideas of our proofs will remain the same as those in this chapter. Of course, this system of concepts and axioms connected with the influence relation is not in final form.

In particular, for draughts, chess, similar games, and many models of similar games, we may replace the axioms for Cases 3, 4, and 5 by conditions that hold when these cases arise. The conditions are a consequence of certain properties of such games that we shall now discuss.

(1) For every position A in the contemplated games there is a function yielding the material balance $f_m(A)$.
(2) For any position $B \in \mathfrak{A}$ and for any branch Λ issuing from it that consists of moves admissible under the rules of $\mathfrak{A}$ at the corresponding positions, the condition

$$f_m(\text{fin}(B, \Lambda)) = f_m(B) + \sum_{\Psi \in \Lambda} h(\Psi)$$

holds, where the weights $h(\Psi)$ are defined for all elements Ψ of the set V of virtual moves and do not depend on the positions at which these moves are made.

(3) The set V may contain blank moves of both White and Black, which we shall denote indifferently by the symbol $m_\varnothing$, since this leads to no confusion. The weight $h(m_\varnothing)$ of a blank move is equal to 0, and a blank move always leads to a terminal position.

(4) If a terminal position $F \in \mathfrak{A}$ is the end of a non-blank move $(D, F) \neq m_\varnothing$, it either has a score which we may take as losing for col F (we recall that White and Black move alternately in $\mathfrak{A}$ and therefore the concept of the color of a terminal position makes sense), or it is a position of a special type that may in some approximations be ignored. For instance, White or Black wins at draughts when the opponent is either bereft of pieces or has all his pieces 'locked', unable to make a legal move. These cases arise when the opponent has the move. In exactly the same way White or Black is mated in chess, i.e. loses, in positions where the loser has the move.

The existence of a third outcome of the game—namely a draw—prevents a complete satisfaction of Condition 4; however, we may take as terminal positions with score 1/2 only those in which the side that was seeking a win has the move. We may always do this in draughts, where a draw occurs when it can be proved that neither side can win. The same is true of chess, with a reservation excluding stalemated positions, either ignoring them completely or subjecting them to specific investigation (such investigations will enable us to prove even for chess the theorems we shall be stating later).

If there are no blank virtual moves in the game $\mathfrak{A}$, Property 4 suffices to exclude Cases 3, 4, and 5.

(5) If the blank move $(D, F) = m_\varnothing$ leads to the terminal position F, the score there is equal to the material balance:

$$(D, F) = m_\varnothing \Rightarrow \mathrm{sc}(F) = f_m(F).$$

Let us see how this property is used in the investigation of Cases 3, 4, and 5. (To prove the theorem on the carry-over of the results of an estimate of a group of moves we need more complex arguments, but these are essentially close to those we shall adduce below.)

Case 3 means that $\Lambda = \Lambda_1 + \Lambda_2$, the position $\mathrm{fin}(B, \Lambda_1)$ is terminal and $\mathrm{fin}(B, \Lambda)$ is not. Accordingly neither of the branches Λ and Λ_1 contains a blank move, since by Property 3 a blank move leads to a terminal position. In such a situation, certain conditions on Case 2 are satisfied if Λ_1 has even length, and Case 2 may be investigated with the aid of these axioms. If Λ_1 has odd length Property 4 implies that the terminal position $\mathrm{fin}(B, \Lambda_1)$ is lost for col $\Lambda_1 \neq$ col B. In the situations we shall be studying this turns out to be impossible.

In Case 4 $\Lambda = \Lambda_1 + \Lambda_2$, the next move Θ is the same in both branches Λ and Λ_1, and the position $\text{fin}(B, \Lambda)$ is terminal, while $\text{fin}(B, \Lambda_1)$ is not. Then the move Θ, which leads to $\text{fin}(B, \Lambda_1)$ is non-blank. When the branch Λ_1 has an odd length the conditions of the axiom on Case 1 are satisfied. But if this length is odd, Property 4 implies that the position $\text{fin}(B, \Lambda)$ is lost for col $\Lambda = $ col B, which is impossible in the cases that we considered in proving the theorem on the transfer of scores.

Finally, suppose that $\Lambda = \Lambda_1 + \Lambda_2$, that the next moves in the branches Λ and Λ_1 coincide, and that the positions $\text{fin}(B, \Lambda)$ and $\text{fin}(B, \Lambda_1)$ are both terminal; i.e. we have Case 5. If the next move Θ in the branches Λ and Λ_1 is non-blank, Property 4 implies that the positions $\text{fin}(B, \Lambda)$ and $\text{fin}(B, \ldots, \Lambda_1)$ are lost for the same side. If, however, $\Theta = m_\varnothing$, Property 5 implies that

$$
\begin{aligned}
\text{sc}\left(\text{fin}(B, \Lambda)\right) &= f_{\text{m}}\left(\text{fin}(B, \Lambda)\right) \\
&= f_{\text{m}}(B) + \sum_{\Psi \in \Lambda} h(\Psi) \\
&= f_{\text{m}}(B) + \sum_{\Psi \in \Lambda_1} h(\Psi) + \sum_{\Psi \in \Lambda_2} h(\Psi) \\
&= f_{\text{m}}\left(\text{fin}(B, \Lambda_1)\right) + \sum_{\Psi \in \Lambda_2} h(\Psi) \\
&= \text{sc}\left(\text{fin}(B, \Lambda_1)\right) + \sum_{\Psi \in \Lambda_2} h(\Psi).
\end{aligned}
$$

So, if

$$
\sum_{\Psi \in \Lambda_2} h(\Psi) \underset{\text{col } B}{\preccurlyeq} 0 \quad \text{and} \quad \text{sc}\left(\text{fin}(B, \Lambda_1)\right) \underset{\text{col } B}{\prec} m,
$$

we have

$$
\text{sc}\left(\text{fin}(B, \Lambda)\right) \underset{\text{col } B}{\prec} m.
$$

Thus we cannot have the position $\text{fin}(B, \Lambda_1)$ lost for the side $\text{col}(B)$ and $\text{fin}(B, \Lambda)$ not lost for that side; however we must impose the supplementary constraint

$$
\sum_{\Psi \in \Lambda_2} h(\Psi) \preccurlyeq 0.
$$

We now introduce yet more notation. Let $\mathfrak{B}$ be a subtree of the game $\mathfrak{A}$, B its base, and $L(\Psi_1, \Psi_2, \ldots, \Psi_l)$ a sequence of virtual moves. We define the relation $L \lozenge B$; the sequence L defines a branch with base B consisting of moves admissible under the rules of $\mathfrak{A}$ at the corresponding positions and belonging to the subtree B, i.e.

$$
\begin{aligned}
L \lozenge B := {}& \Psi_1 = (B, B_1) \in \mathfrak{B} \,\& \\
& \Psi_2 = (B_1, B_2) \in \mathfrak{B} \,\& \ldots \& \\
& \Psi_l = (B_{l-1}, B_l) \in \mathfrak{B}.
\end{aligned}
$$

Then we say that the branch L is incident on the subtree $\mathfrak{B}$. Using the notion of incidence we can define the influence of a branch L on a subtree $\mathfrak{B}$ of the game tree $\mathfrak{A}$, and the influence of a subtree $\mathfrak{B}_1$ on a subtree $\mathfrak{B}_2$ (these subtrees must have a common root):

$$L \sim \mathfrak{B} := \exists \Lambda (\Lambda \diamondsuit \mathfrak{B} \& L \sim \Lambda);$$

$$\mathfrak{B}_1 \sim \mathfrak{B}_2 := \exists \Lambda_1 \exists \Lambda_2 (\Lambda_1 \diamondsuit \mathfrak{B}_1 \& \Lambda_2 \diamondsuit \mathfrak{B}_2 \& \Lambda_1 \sim \Lambda_2).$$

We shall consider a certain subtree $\mathfrak{S}(B)$ of the tree for our game $\mathfrak{A}$, and the corresponding model games. The position B is the root of the subtree and is the base position of the model game. For every position $D \in \mathfrak{S}(B)$ we denote by $S(D)$ the set of virtual moves to which the arcs $(D, E) \in \mathfrak{S}(B)$ correspond. Clearly $S(D) \subset M(D)$. The side $\mathrm{col}_{\mathfrak{S}(B)}(D)$ having the move at the position D in the game $\mathfrak{S}(B)$ and the scores at the positions $F \in \mathfrak{S}(B)$ that are terminal in the original game and therefore terminal in $\mathfrak{S}(B)$ are defined in the natural way:

$$\mathrm{col}_{\mathfrak{S}(B)}(D) := \mathrm{col}(D);$$

$$\mathrm{sc}_{\mathfrak{S}(B)}(F) := \mathrm{sc}(F).$$

If, however, F is terminal in $\mathfrak{S}(B)$ only, and not in $\mathfrak{A}$, its score in the model game is the least of those possible in $\mathfrak{A}$ when F is a White position and the greatest when F is a Black position.

$$F \in \mathfrak{S}(B) \& M(F) \neq \varnothing \& S(F) = \varnothing$$
$$\Rightarrow \forall D \subset \mathfrak{A} \left\{ \mathrm{sc}_{\mathfrak{S}(B)}(F) \underset{\mathrm{col}\,B}{\preccurlyeq} \mathrm{sc}(D) \right\}.$$

Let $\mathfrak{S}(C)$ be such a model game. In particular, we shall allow all its sets $S(D)$ to coincide with the set $M(D)$ of virtual moves admissible in $\mathfrak{A}$ at the corresponding positions. Then $\mathfrak{S}(C)$ is defined by a C-subtree of $\mathfrak{A}$, which we shall denote by $\mathfrak{A}(C)$. The model game $\mathfrak{S}_T(B)$ is called a *test model* of the game $\mathfrak{S}(C)$ if $\mathrm{col}\,B = \mathrm{col}\,C$ and the sets of allowed moves at corresponding positions in the two games satisfy the following conditions:

$$S(C) \subset S(B),$$

$$\left.
\begin{aligned}
& S(\mathrm{fin}(C, \Lambda)) \cap M(\mathrm{fin}(B, \Lambda)) \subset \\
& \quad \subset S(\mathrm{fin}(B, \Lambda)) | \mathrm{col}\,\Lambda = \mathrm{col}\,C; \\
& S(\mathrm{fin}(B, \Lambda)) \cap M(\mathrm{fin}(C, \Lambda)) \subset \\
& \quad \subset S(\mathrm{fin}(C, \Lambda)) | \mathrm{col}\,\Lambda \neq \mathrm{col}\,C
\end{aligned}
\right\} \Lambda \diamondsuit \mathfrak{S}(C), \Lambda \diamondsuit \mathfrak{S}_T(B)$$

If the branch Λ is incident on both subtrees $\mathfrak{S}_T(B)$ and $\mathfrak{S}(C)$, we shall say that the position $\mathrm{fin}(B, \Lambda)$ $(\mathrm{fin}(C, \Lambda))$ is *obtained from* $\mathrm{fin}(C, \Lambda)$ $(\mathrm{fin}(B, \Lambda))$ *by parallel transfer*.

If $\mathrm{col}\,B = \mathrm{col}\,C$ and $S(C) \subset M(B)$ we may construct the test model $\mathfrak{S}_T(B)$ for a given game $\mathfrak{S}(C)$ recursively, defining the moves allowed at the test positions as follows:

$$S(B) := S(C);$$

$$S(\operatorname{fin}(B, \Lambda)) := M(\operatorname{fin}(B, \Lambda)) \cap$$

$$\cap S(\operatorname{fin}(C, \Lambda)) | \Lambda \diamondsuit \mathfrak{S}(C), \Lambda \diamondsuit \mathfrak{S}_T(B).$$

Normally, however, we do not know the game $\mathfrak{S}(C)$ nor even the position C itself when we specify the base B of the test model $\mathfrak{S}_T(B)$. If, however, we do know that the sets of allowed moves at the positions of $\mathfrak{S}(C)$ must belong to the subsets of virtual moves $R(\Lambda) \subset V$ defined by the branches Λ leading to those positions, we may attempt to construct the test model $\mathfrak{S}_T(B)$.

The test game is defined recursively by its sets of allowable moves:

$$S(B) := M(B) \cap R(\varnothing);$$

$$S(\operatorname{fin}(B, \Lambda)) := M(\operatorname{fin}(B, \Lambda)) \cap R(\Lambda) | \Lambda \diamondsuit \mathfrak{S}_T(B).$$

Here $R(\varnothing)$ is the set of virtual moves to which the set $S(C)$ of moves allowed at the base C of $\mathfrak{S}(C)$ must belong. In some model games used in algorithms for game programming the subsets $R(\Lambda)$ of the set V of all virtual moves coincide for all the subtrees $\mathfrak{A}(C)$ for positions C of sufficiently high rank. In the extreme case, $R(\Lambda) \equiv V$. If B is of lower rank, the model $\mathfrak{S}_T(B)$ with base B is a test model for those games $\mathfrak{A}(C)$ in which virtual moves that do not exist at B are disallowed at the base C.

When $\mathfrak{S}_T(B)$ is a test model for a number of the games $\mathfrak{A}(C_i)$ ($i = 1, 2, \ldots, k$) there is a potential for shortening the search of the original game $\mathfrak{A}$, as a consequence of a theorem on the transfer of scores of moves which we now state and prove.

Theorem on the Transfer of Scores of Moves. *Let the following conditions be satisfied:*

(1) *In the game $\mathfrak{A}$: White and Black move alternately; a set V of virtual moves is defined, as is an influence relation of branches on moves, branches, and trees that satisfies the axioms and definitions given in this section;*

(2) *$\mathfrak{S}(C) \subset \mathfrak{A}$ is a model game with base C in which an arbitrary move Θ that is admissible under the rules of $\mathfrak{A}$ at C but not admissible there under the rules of $\mathfrak{S}(C)$ is inadmissible at the other positions of $\operatorname{col} C$ in the model game;*

(3) *L is a branch of even length, beginning at B and ending at C, and $\mathfrak{S}_T(B)$ is a test model game for $\mathfrak{S}(C)$ with base B;*

(4) *$\Psi_1 \in S(C) \subset S(B)$ is a virtual move, $\Psi_1 \in (B, B_1) = (C, C_1)$. $\mathfrak{S}_T(B, \Psi_1)$ and $\mathfrak{S}(C, \Psi_1)$ are subtrees of $\mathfrak{S}_T(B)$ and $\mathfrak{S}(C)$ respectively, consisting of positions and arcs in the branches $K(\Psi_1, \ldots)$ having Ψ_1 as first move (see Figure 27).*

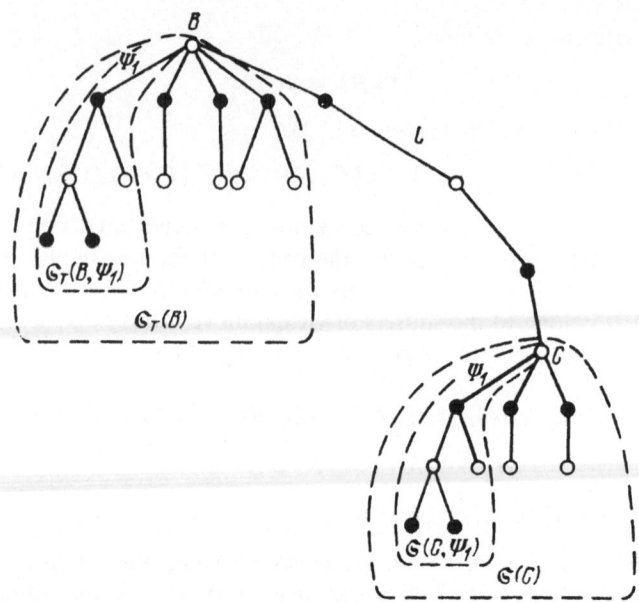

Figure 27

(5) *The position B_1 has the score* $\mathrm{sc}_{\mathfrak{S}_T(B)}(B_1) \underset{\mathrm{col}\, C}{\prec} m$ *in the game* $\mathfrak{S}_T(B)$.
$\mathfrak{S}'_T(B)$ is a minimal col B_1-*pruned tree of the game* $\mathfrak{S}_T(B, \Psi_1)$ *for whose every terminal position F we have* $\mathrm{sc}_{\mathfrak{S}_T(F)}(B_1) \underset{\mathrm{col}\, C}{\prec}$.
   Then either the position C has a score $\mathrm{sc}_{\mathfrak{S}(C)}(C) \underset{\mathrm{col}\, C}{\prec} m$ *in the game* $\mathfrak{S}(C)$,
or the branch L influences the tree $\mathfrak{S}'_T(B, \Psi_1)$, *or it influences some move* Θ
admissible under the rules of the game $\mathfrak{A}$ *and allowed at the positions B and C
but either inadmissible or disallowed in the game* $\mathfrak{S}_T(B)$ *after the move* Ψ_1
and some reply to it Ψ_2.

We may exclude the axioms on Cases 3, 4, and 5 from our axiom system.
Then we must add the following item to the hypothesis of our theorem:

(6) *If K is a branch incident on the trees* $\mathfrak{S}_T(B)$ *and* $\mathfrak{S}(C)$, *and the
positions* $\mathrm{fin}(B, K)$ *and* $\mathrm{fin}(C, K)$ *are terminal, then*

$$\mathrm{sc}_{\mathfrak{S}(C)}(\mathrm{fin}(C, \Lambda)) \underset{\mathrm{col}\, \Lambda}{\preceq} \mathrm{sc}_{\mathfrak{S}_T(B)}(\mathrm{fin}(B, \Lambda));$$

if, however, only one of these positions is terminal, then

$$\mathrm{sc}_{\mathfrak{S}_T(B)}(\mathrm{fin}(B, \Lambda)) \underset{\mathrm{col}\, \Lambda}{\prec} m \,|\, S(\mathrm{fin}(B, \Lambda)) = \varnothing,$$

$$\mathrm{sc}_{\mathfrak{S}(C)}(\mathrm{fin}(C, \Lambda)) \underset{\mathrm{col}\, \Lambda}{\prec} m \,|\, S(\mathrm{fin}(C, \Lambda)) = \varnothing,$$

From now on we shall say that a position D in the trees $\mathfrak{S}_T(B)$ and
$\mathfrak{S}(C)$ is a *losing*, or *lost*, *position* if it has scores in the corresponding

games that are worse than m for the color $col(C)$, and moves that lead to a losing position will be called *losing moves*.

The basic ideas of the proof were set out in the preceding section. We repeat them here and mark the fundamental steps in the proof, which in an altered form appears in the proofs of other theorems on the possibility of shortening the search in the game tree by using the influence relation.

The first step is the construction of the branch Λ, which is similar to the critical branch described in Chapter 1. The latter is the intersection of minimal W-pruned and B-pruned trees coordinated with respect to scores.

We take the intersection of the trees $\mathfrak{S}'_T(B, \Psi_1)$ and $\mathfrak{S}''(C, \Psi_1)$. We recall that $\mathfrak{S}'_T(B, \Psi_1)$ is a minimal pruned subtree of the game tree $\mathfrak{S}_T(B, \Psi_1)$ of the color opposite to $col\,B = col\,C$; the terminal positions of the intersecting trees have scores $\underset{col\,C}{\prec} m$, by the hypothesis of the theorem; $\mathfrak{S}''(C, \Psi_1)$ is a minimal $col\,C$-pruned subtree of the game $\mathfrak{S}(C, \Psi_1)$, in which it defines the score of the position C. Although these trees have different roots, their intersection can be defined by parallel transfer of branches.

If the position C_1 is lost in the game $\mathfrak{S}(C)$, one of the alternatives in our theorem is satisfied. We therefore assume that it is not lost. Then no move $(D, E) \in \mathfrak{S}''(C, \Psi_1)$ loses, and no terminal position of $\mathfrak{S}''(C, \Psi_1)$ is lost.

The first move in the branch Λ is $\Psi_1 = (B, B_1) = (C, C_1)$. If B_1 is non-terminal in the game $\mathfrak{S}_T(B)$, the tree $\mathfrak{S}'_T(B, \Psi_1)$ contains a move Ψ_2 leading from B_1 to a lost position $B_2 \in \mathfrak{S}'_T(B, \Psi_1)$. Such a move is unique since $\mathfrak{S}'_T(B, \Psi_1)$ is minimal. Suppose that it is admissible under the rules of the game $\mathfrak{A}$ and allowed under the rules of the model game $\mathfrak{S}(C)$ at the position $C_1 \in \mathfrak{S}''(C, \Psi_1)$ obtained from B_1 by parallel transfer. Since the color of the move $\Psi_2 = (C_1, C_2)$ is opposite to $col\,C$, both the move and the position C_2 to which it leads belong to the $col\,C$-pruned tree $\mathfrak{S}''(C, \Psi_1)$.

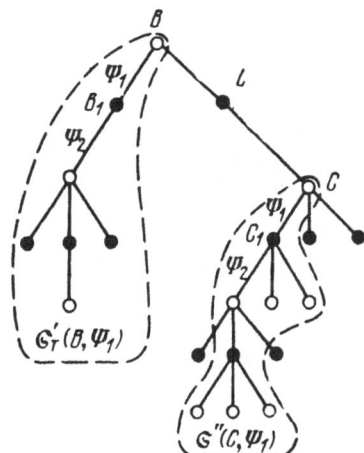

Figure 28

In the same way we transfer to the tree $\mathfrak{S}''(B, \Psi_1)$ the unique non-losing move $\Psi_3 = (C_2, C_3) \in \mathfrak{S}''(C, \Psi_1)$ if it exists, and so on. (See Figure 28.)

Thus we have uniquely defined the branch Λ incident on the trees $\mathfrak{S}'_T(B, \Psi_1)$ and $\mathfrak{S}''(C, \Psi_1)$ and not further extendable. It is the intersection of $\mathfrak{S}'_T(B, \Psi_1)$ and the tree $\mathfrak{S}''(C, \Psi_1)$ obtained by parallel transfer of the base B of the former. We call the branch Λ *pseudocritical*.

In the second step of the proof we see why the branch Λ cannot be extended. There may be two reasons:

(1) the branch Λ as constructed from B or C may have reached a terminal position in the corresponding model game;
(2) the next move Ψ_k chosen at one of the branch ends $\mathrm{fin}(B, \Lambda)$ or $\mathrm{fin}(C, \Lambda)$ is not allowed under the rules of the game $\mathfrak{A}$ or is inadmissible in the other branch.

As we showed in the preceding section, five different cases may arise at the ends $\mathrm{fin}(B, \Lambda)$ and $\mathrm{fin}(C, \Lambda)$.

We now go to the third step in the proof—the investigation of each of these cases and the proof that the branch Λ influences some branch of the tree $\mathfrak{S}'_T(B, \Psi_1)$ or influences a move satisfying the conditions of the theorem.

(1) the branch Λ has odd length $2j-1 (j \geq 1)$ and the move Ψ_{2j} at its end (the position $B_{2j-1} = \mathrm{fin}(B, \Lambda) \in \mathfrak{S}'_T(B, \Psi_1)$) that would extend it in the tree $\mathfrak{S}'_T(B, \Psi_1)$ is inadmissible under the rules of the game $\mathfrak{A}$ at the corresponding position $C_{2j-1} = \mathrm{fin}(C, \Lambda) \in \mathfrak{S}''(C, \Psi_1)$ or is not allowed in the model game $\mathfrak{S}(C)$. In particular, this case arises when the position C_{2j-1} is terminal and B_{2j-1} is not. Since $\mathrm{col}\, C_{2j-1} \neq \mathrm{col}\, C$, moves that are allowed at B_{2j-1} and admissible at C_{2j-1} are allowed there. Of course, the move Ψ_{2j} would be allowed at C_{2j-1} if it were admissible there, i.e. it is inadmissible under the rules of the game $\mathfrak{A}$. By the axiom on Case 1, $L \sim \Lambda * \Psi_{2j}$. But the branch $\Lambda * \Psi_{2j}$ is incident on the tree $\mathfrak{S}'_T(B, \Psi_1)$ and therefore $L \sim \mathfrak{S}'_T(B, \Psi_1)$.
(2) The branch Λ has even length $2j (j \geq 1)$ and the move Ψ_{2j+1} at its end

$$C_{2j} = \mathrm{fin}(C_1, \Lambda) \in \mathfrak{S}''(C, \Psi_1),$$

which would extend it in the tree $\mathfrak{S}''(C, \Psi_1)$, is either inadmissible under the rules of the game $\mathfrak{A}$ or is not allowed in the game $\mathfrak{S}(C)$ at the position

$$B_{2j} = \mathrm{fin}(B, \Lambda) \in \mathfrak{S}'_T(B, \Psi_1).$$

Then B_{2j} also may be a terminal position. Since $\mathrm{col}\, C_{2j} = \mathrm{col}\, C$, moves allowed at C_{2j} and admissible at B_{2j} are allowed there. Accordingly the move Ψ_{2j+1} is inadmissible under the rules of the game $\mathfrak{A}$. If it is not admissible at $C = \mathrm{fin}(B, L)$, the first axiom on Case 2 implies that $L \sim \Lambda$ and so $L \sim \mathfrak{S}'_T(B, \Psi_1)$. If, however, it is admissible at C it is allowed there, by the conditions of the theorem (else it would not be allowed at $C_{2j} = \text{fin}(C, \Lambda) \in \mathfrak{S}(C)$). It is also allowed at B, since $S(C) \subset S(B)$. By the second axiom on Case 2, $L \sim \Lambda$ or $L \sim \Psi_{2j+1}$. In the first alternative, as before, $L \sim \mathfrak{S}'_T(B, \Psi_1)$.

In the second alternative, if the move Ψ_{2j+1} is admissible and allowed at the position $B_2 \in \mathfrak{S}'_T(B, \Psi_1)$ after the moves $\Psi_1 = (B, B_1)$ from B and $\Psi_2 = (B_1, B_2)$ from B_1, the move $\Psi_{2j+1} = (B_{2j}, D)$ and the position D belong to the col B_1-pruned tree $\mathfrak{S}'_T(B, \Psi_1)$ since

$$\text{col } \Psi_{2j+1} = \text{col } C_{2j} = \text{col } C = \text{col } B \neq \text{col } B_1.$$

Accordingly, the branch $\Lambda'(\Psi_1, \Psi_2, \ldots, \Psi_{2j+1})$ is incident on the tree $\mathfrak{S}'_T(B, \Psi_1)$. By the axiom on influences on moves and branches, $L \sim \Lambda'$ and so $L \sim \mathfrak{S}'_T(B, \Psi_1)$. Now there remains only the case in which the branch L influences the move Ψ_{2j+1}, which is admissible and allowed at the beginning B and end C of the branch, but not admissible or not allowed after a move from B and the reply Ψ_2. In concrete games we may formulate supplementary conditions to be satisfied by the tree $\mathfrak{S}'_T(B, \Psi_1)$ in order that the move Ψ_{2j+1} be distinct from Ψ_1.

(3) The branch Λ has odd length $2j - 1 (j \geq 1)$, its end

$$B_{2j-1} = \text{fin}(B, \Lambda) \in \mathfrak{S}'_T(B, \Psi_1)$$

is terminal in $\mathfrak{S}_T(B)$, and the corresponding position

$$C_{2j-1} = \text{fin}(C, \Lambda) \in \mathfrak{S}''(C, \Psi_1)$$

is non-terminal in the game tree $\mathfrak{S}(C)$. Then B_{2j-1} is a terminal position even in the original game $\mathfrak{A}$, since the move at it belongs to the color opposite to col B. Moreover, if $M(B_{2j-1}) \neq \varnothing$ and $S(B_{2j-1}) = \varnothing$, that color loses. By the axiom on Cases 3 and 4, $L \sim \Lambda$, and so $L \sim \mathfrak{S}'_T(B, \Psi_1)$. If we renounce this axiom and add the condition in item 6, the situation we have in hand is impossible. In fact, since col $\Lambda \neq$ col B, the condition

$$\text{sc}_{\mathfrak{S}_T(B)}(\text{fin}(B, \Lambda)) \underset{\text{col } \Lambda}{\prec} m$$

implies that

$$m \underset{\text{col } B = \text{col } C}{\prec} \text{sc}_{\mathfrak{S}_T(B)}(\text{fin}(B, \Lambda)),$$

and this contradicts the assertion of Condition 5 that the position $B_{2j} = \text{fin}(B, \Lambda)$ is lost.

(4) The branch Λ has even length $2j (j \geq 1)$, its end $C_{2j} = \text{fin}(C, \Lambda) \in \mathfrak{S}''(C, \Psi_1)$ is a terminal position in the game $\mathfrak{S}(C)$ and the corresponding position $B_{2j} = \text{fin}(B, \Lambda) \in \mathfrak{S}'_T(B, \Psi_1)$ is non-terminal in the game $\mathfrak{S}_T(B)$. Then we prove the following facts in exactly the same way as above: the position C_{2j} is terminal in the game $\mathfrak{A}$; if the axiom on Cases 3 and 4 holds we have $L \sim \Lambda$; if Condition 6 holds C_{2j} is a lost position.

(5) Both of the positions $\mathrm{fin}(B, \Lambda)$ and $\mathrm{fin}(C, \Lambda)$ are terminal in the corresponding model games, but both cannot be non-terminal in the original game $\mathfrak{A}$, since then the rule on the definition of scores would imply that both were lost or both were not lost. If one of them is terminal in the model game and the other is terminal in the original game, the axiom on Cases 3 and 4 implies that $L \sim \Lambda$. Finally, if both positions are terminal in the original game $\mathfrak{A}$ the axiom on Case 5 implies that

$$\mathrm{sc}(\mathrm{fin}(C, \Lambda)) \underset{\mathrm{col}\,C}{\prec} \mathrm{sc}(\mathrm{fin}(B, \Lambda))$$

or $L \sim \Lambda$. But Condition 5 implies that the conditions and our postulates for the first possibility are not realized. Thus $L \sim \mathfrak{S}'_T(B, \Psi_1)$. If we renounce the axiom on Case 5 and add the condition of item 6, this situation too will be impossible, since the requirements that the position $\mathrm{fin}(B, \Lambda)$ be lost and $\mathrm{fin}(C, \Lambda)$ not lost and

$$\mathrm{sc}_{\mathfrak{S}(C)}(\mathrm{fin}(C, \Lambda)) \underset{\mathrm{col}\,C}{\preccurlyeq} \mathrm{sc}_{\mathfrak{S}_T(B)}\mathrm{fin}(B, \Lambda)$$

are incompatible.

This completes the proof of the theorem on the transfer of scores. Some of the axioms were not used in the proof, in particular, the axiom on the symmetry of the influence relation. If we define the influence relation constructively in such a way that we use only the above theorem and the following theorem on the transfer of position scores, we do not require symmetry and we construct a relationship that will be more rarely satisfied and so will offer more frequent possibilities for shortening the search.

The following theorem is proved in exactly the same way as the above.

Theorem on the Transfer of Position Scores. *Let the following conditions be satisfied*:

(a) *conditions 1–3 of the theorem on the transfer of scores of moves*;
(b) *the position B has the following score in the game* $\mathfrak{S}_T(B)$

$$\mathrm{sc}_{\mathfrak{S}_T(B)}(B) \underset{\mathrm{col}\,C}{\prec} m,$$

(c) $\mathfrak{S}'_T(B)$ *is a minimal pruned tree of the game* $\mathfrak{S}_T(B)$ *for the color opposite to* $\mathrm{col}\,B = \mathrm{col}\,C$ *and all its terminal positions F have a score*

$$\mathrm{sc}_{\mathfrak{S}_T(B)}(F) \underset{\mathrm{col}\,C}{\prec} m.$$

Then the position C has a score

$$\mathrm{sc}_{\mathfrak{S}(C)}(C) \underset{\mathrm{col}\,C}{\prec} m$$

in the game $\mathfrak{S}(C)$ *else the branch L influences the tree* $\mathfrak{S}'_T(B)$.

As in the theorem on the transfer of scores of moves, we may impose the condition 6 instead of the axiom on Cases 3, 4, and 5.

Let us ask why we discuss the influence of the branch L on the tree $\mathfrak{S}'_T(B)$ or the tree $\mathfrak{S}'_T(B, \Psi_1)$ but not on the pseudocritical branch Λ. When we arrive at the position C in the search of the game tree $\mathfrak{A}$ or one of its models, the search of $\mathfrak{S}'_T(B)$ is normally already completed, so that we already know both it and its subtree $\mathfrak{S}'_T(B, \Psi)$, where $\Psi \in S(B)$, but the C-subtree of $\mathfrak{A}$ or a model of $\mathfrak{A}$ has not yet been searched. Therefore we do not then know the pseudocritical branch Λ but must make a decision about shortening the search. For this reason we must use information about the branch $L(B, \ldots, C)$ and the trees $\mathfrak{S}'_T(B)$ and $\mathfrak{S}'_T(C, \Psi_1)$ for $\Psi \in S(B)$.

The Decomposition of Branches and Application to Shortening the Search

One of the unpleasant constraints in the hypothesis of the theorem on the transfer of move scores is that at the position C, where we wish to shorten the search, we cannot contemplate any move Θ that is inadmissible under the rules of the original game $\mathfrak{A}$ at the base B of the test model tree $\mathfrak{S}_T(B)$. Thus from the point of view of col B the position B should 'strongly majorize' the position C. However, as a rule the moves in the branch L beginning at B and ending at C offer the opponent new possibilities that often bear no relation to the variants of the model game $\mathfrak{S}_T(B)$.

On the other hand, it was assumed in the proof of the theorem that the tree $\mathfrak{S}_T(B)$ was searched without pruning, because some of its branches do not influence others. Thus the search takes more time than may be necessary. It is clearly not easy to investigate the possibility of shortening the search by recursive pruning based on repeated application of the method of analogy. Only the first steps have been made in this direction. They require new ideas, which will probably be of a rather general character. In this section we attempt to present some thoughts about them. We consider certain possibilities for shortening the search when, at the end $C = \text{fin}(B, L)$ of a branch L having even length, certain moves Θ are admissible even though they do not belong to the set $S(B)$ of moves allowed at the base position of the test model game $\mathfrak{S}_T(B)$. Some of these possibilities are connected with the permissibility of studying some of the moves $\Psi \in S(B)$ independently of one another.

Let $\mathfrak{S}$ be a model game with the following properties:
At the positions $\text{fin}(C, \Lambda)$ of its subtrees $\mathfrak{S}(C)$ all virtual moves in the subsets $R(\Lambda)$, and only those, are allowed, i. e.

$$S(\text{fin}(C, \Lambda)) = M(\text{fin}(C, \Lambda)) \cap R(\Lambda) \,|\, \Lambda \diamondsuit \mathfrak{S}(C).$$

The set $R(\varnothing) \subset V$ is also defined, corresponding to the empty branch $\varnothing$:

$$S(C) = M(C) \cap R(\varnothing).$$

The sets decrease as new moves are added to the branch Λ:

$$R(\Lambda_1 + \Lambda_2) \subset R(\Lambda_1) \,|\, \mathrm{col}(\Lambda_1 + \Lambda_2) = \mathrm{col}\,\Lambda_1.$$

In particular, any set $R(\Lambda) \subset R(\varnothing)$.

Many model games satisfy such conditions, for instance, the absolute scheme

$$R(\Lambda) = \begin{cases} V, & \text{if } r(\mathrm{fin}(C, \Lambda)) < n, \\ V_f, & \text{if } r(\mathrm{fin}(C, \Lambda)) \geq n, \end{cases}$$

where $r(D)$ is the rank of the position D in the tree $\mathfrak{A}$, n is the depth of the model, and $V_f \subset V \cup m_\varnothing$ is the set of moves in the forced game. For a quiet game

$$R(\Lambda) = \begin{cases} V, & \text{if } r(\mathrm{fin}(C, \Lambda)) < n \\ & \text{or}\, |V_f \cap \{\Lambda\}| > \rho(\Lambda) - d(C), \\ V_f & \text{in other cases,} \end{cases}$$

where $\{\Lambda\}$ is the set of moves in the branch Λ, $\rho(\Lambda)$ is its length, $|U|$ is the number of elements in the set U, and $d(C)$ is the number of quiet moves allowed in the branch. Finally, the original game $\mathfrak{A}$ itself meets the conditions: we need only assume that $R(\Lambda) = V$ for all Λ.

The test model game $\mathfrak{S}(B)$ that we shall be considering later must satisfy the following conditions:

(1) An arbitrary move $\Psi \in S(B)$ from the base position loses, that is

$$\mathrm{sc}_{\mathfrak{S}_T(B)}(B) \underset{\mathrm{col}\,B}{\prec} m,$$

where m is a given score.

(2) A new move Θ from the position

$$D = \mathrm{fin}(B, \Lambda) \in \mathfrak{S}_1(B)$$

of the color $\mathrm{col}(B)$, i.e. a move not existing at B, is allowed if it belongs to the corresponding set $R(\Lambda)$:

$$M(\mathrm{fin}(B, \Lambda)) \setminus M(B) \cap R(\Lambda) \subset S(\mathrm{fin}(B, \Lambda))$$

for $\mathrm{col}\,\Lambda = \mathrm{col}\,B$.

(3) If a branch Λ of color $\mathrm{col}\,\Lambda = \mathrm{col}\,B$ is incident on the tree of the game $\mathfrak{S}_T(B)$, all the moves $\Psi \in M(\mathrm{fin}(B, \Lambda)) \cap M(B) \cap R(\Lambda)$ that are influenced by Λ are allowed at the position $\mathrm{fin}(B, \Lambda)$:

$$\Psi \in M(\mathrm{fin}(B, \Lambda)) \cap M(B) \cap R(\Lambda) \,\&\, \Lambda \sim \Psi$$

$$\mathrm{col}\,\Lambda = \mathrm{col}\,B \Rightarrow \Psi \in S(\mathrm{fin}(B, \Lambda)).$$

(4) Let
  (a) $\mathfrak{S}'_T(B)$ be a minimal pruned tree of the game $\mathfrak{S}_T(B)$, of color opposite to $\mathrm{col}(B)$;
  (b) the terminal positions $F \in \mathfrak{S}'_T(B)$ have scores $\mathrm{sc}_{\mathfrak{S}(B)}(F) \underset{\mathrm{col}\,B}{\prec} m$;
  (c) $\mathfrak{S}_T(B, \Psi)$ be a subtree of $\mathfrak{S}_T(B)$, consisting of the positions and moves in the branch $\Lambda(\Psi, \dots)\Diamond\mathfrak{S}_T(B)$, of which the first move init Λ is $\Psi \in S(B)$;
  (d) $\mathfrak{S}'_T(B, \Psi) = \mathfrak{S}'_T(B) \cap \mathfrak{S}_T(B, \Psi)$.

Then if the tree $\mathfrak{S}'_T(B, \Psi_1)$ influences the tree $\mathfrak{S}'_T(B, \Psi_2)$ ($\Psi_1, \Psi_2 \in S(B)$), the move Ψ_2 is allowed at all positions of the tree $\mathfrak{S}_T(B, \Psi_1)$ where the side $\mathrm{col}\,B$ has the move and Ψ_2 is admissible under the rules of $\mathfrak{A}$ and belongs to the corresponding set $B(\Lambda)$:

$$\mathfrak{S}'_T(B, \Psi_1) \sim \mathfrak{S}'_T(B, \Psi_2) \& \Lambda\Diamond\mathfrak{S}_T(B, \Psi_1) \&$$

$$\Psi_2 \in M(\mathrm{fin}(B, \Lambda)) \cap R(\Lambda) \Rightarrow \Psi_2 \in S(\mathrm{fin}(B, \Lambda)).$$

In particular, the move Ψ_1 is allowed at all the positions $\mathrm{fin}(B, \Lambda) \in \mathfrak{S}_T(B, \Psi_1)$ where it is admissible and belongs to the corresponding set $R(\Lambda)$.

By the axiom on the symmetry of the influence relation the move Ψ_1 is also allowed at the positions in $\mathfrak{S}_T(B, \Psi_2)$ if it is admissible and belongs to the corresponding sets $R(\Lambda)$.

(5) All the moves allowed at the position $D = \mathrm{fin}(B, \Lambda) \in \mathfrak{S}_T(B)$ where the color opposite to $\mathrm{col}\,B$ has the move, belong to the set $R(\Lambda)$:

$$S(\mathrm{fin}(B, \Lambda)) \subset M(\mathrm{fin}(B, \Lambda)) \cap$$

$$R(\Lambda) \mid \Lambda\Diamond\mathfrak{S}_T(B), \mathrm{col}\,\Lambda \neq \mathrm{col}\,B.$$

Thus we want to study individually only those moves Ψ that are allowed at the base B.

It has been conjectured that at any position one need investigate only new moves Θ and those earlier possible moves whose branches are influenced by the search tree already constructed there. A proof of this conjecture requires an extended concept of influence on branches with different origins, and has not so far been obtained. Its application entails some algorithmic difficulties, since a return may be required to positions from which a backward step has already been taken.

We can construct by iteration a model game $\mathfrak{S}_T(B)$ in which, at positions of color $\mathrm{col}\,B$ that do not coincide with the base B, moves admissible at B are not allowed if possible. Let $G_0 := S(B)$ and let the set $H_0(\Psi)$ be empty for all $\Psi \in S(B)$. Let the sets $G_i \subset S(B)$ and $H_i(\Psi)$ for $\Psi_i \in G_i$ be defined for some $i \geq 0$. Then we define the model game $\mathfrak{S}_i(B, \Psi)$

recursively for all $\Psi \in G_i$:

$$S_\Psi(B) := \{\Psi\};$$

$S_\Psi(\mathrm{fin}(B, \Lambda)) :=$

$$:= \begin{cases} \big((M(\mathrm{fin}(B, \Lambda))\backslash M(B))\cup(M(\mathrm{fin}(B, \Lambda))\cap H_i(\Psi))\big) \\ \quad \cap R(\Lambda)|\mathrm{col}\,\Lambda = \mathrm{col}\,B; \\ M(\mathrm{fin}(B, \Lambda))\cap R(\Lambda)|\mathrm{col}\,\Lambda \ne \mathrm{col}\,B. \end{cases}$$

Let us now see whether the position B is lost in these games, i. e. whether its score $\underset{\mathrm{col}\,B}{\prec} m$. To this end we construct for the losing moves Ψ the corresponding minimal pruned trees $\mathfrak{S}'_i(B, \Psi)$ of the color opposite to col B. Then

$$G_{i+1} := \Big\{\Psi \mid \Psi \in G_i \& \mathrm{sc}_{\mathfrak{S}_i(B,\Psi)}(B) \underset{\mathrm{col}\,B}{\prec} m\Big\};$$

$$H_{i+1}(\Psi) := G_{i+1}\cap\big(H_i(\Psi)\cup\{\Theta \mid \mathfrak{S}'_i(B,\Theta) \sim \mathfrak{S}'_i(B,\Theta)\}\big).$$

In the latter definition we assume that the influence relation is symmetric, else we would need to add the moves Θ for which $\mathfrak{S}'_i(B,\Theta) \sim \mathfrak{S}'_i(B,\Psi)$. Then we may go to the following iteration.

Since $G_0 = S(B)$ is finite and G_{i+1} is imbedded in G_i for all i greater than some N, these sets will stabilize:

$$i > N \Rightarrow G_i = G_N.$$

In the following iterations the sets $H_i(\Psi)$ can only expand, but they remain subsets of the finite set G_N, so that the expansion process must end. So for some $i > N$ we shall have the following conditions:

$$G_{i+1} = G_i = G_N,$$

$$\forall \Psi \in G_i \quad H_{i+1}(\Psi) = H_i(\Psi).$$

Hence the model games $\mathfrak{S}_{i+1}(B, \Psi)$ and $\mathfrak{S}_i(B, \Psi)$ will be identical.

The model game

$$\mathfrak{S}_T(B) = \bigcup_{\Psi \in G_i} \mathfrak{S}_i(B, \Psi)$$

satisfies the conditions we have imposed on a test model, and $\mathfrak{S}_i(B, \Psi)$ and $\mathfrak{S}'_i(B, \Psi)$ $(\Psi \in G_i = S_T(B))$ are the subtrees of $\mathfrak{S}_T(B, \Psi)$ and $\mathfrak{S}'_T(B, \Psi)$, respectively, that we discussed in item 3 of the conditions.

In fact, the following three assertions are valid:

(1) All the moves $\Psi \in G_i = G_{i+1}$ are losing moves.
(2) Let

$$\mathfrak{S}_T(B) = \bigcup_{\Psi \in G_i} \mathfrak{S}_i(B, \Psi) = \bigcup_{\Psi \in G_{i+1}} \mathfrak{S}_{i+1}(B, \Psi),$$

where Λ is a branch incident on the tree $\mathfrak{S}_T(B)$. If $\mathrm{col}\,\Lambda = \mathrm{col}\,B$,

$$(M(\mathrm{fin}(B,\Lambda))\backslash M(B))\cap R(\Lambda) \subset S(\mathrm{fin}(B,\Lambda))$$

since by definition

$$S(\mathrm{fin}(B,\Lambda)) = ((M(\mathrm{fin}(B,\Lambda))\backslash M(B)$$
$$\cup (M(\mathrm{fin}(B,\Lambda))\cap H_{i+1}(\Psi)))\cap R(\Lambda)$$

If however, $\mathrm{col}\,\Lambda \neq \mathrm{col}\,B$ we have

$$M(\mathrm{fin}(B,\Lambda))\cap R(\Lambda) = S(\mathrm{fin}(B,\Lambda)).$$

(3) If $\mathfrak{S}'_T(B,\Psi_1) \sim \mathfrak{S}'_T(B,\Psi_2)$ $(\Psi_1,\Psi_2 \in S(B),\ G_i = G_{i+1})$, then

$$\Psi_1 \in H_i(\Psi_2) = H_{i+1}(\Psi_2),$$
$$\Psi_2 \in H_i(\Psi_1) = H_{i+1}(\Psi_1).$$

Thus if the move Ψ_2 is admissible at the position $\mathrm{fin}(B,\Lambda) \in \mathfrak{S}_T(B,\Psi_1)$ and belongs to the corresponding set, it is allowed. A like assertion is valid also for the move Ψ_1 at positions of color $\mathrm{col}\,B$ in the subtree $\mathfrak{S}_T(B,\Psi_2)$.

Such a process, however, may be too lengthy. In practice, if the second or third iteration alters the sets G_i or $H_i(\Psi)$ we should obviously construct a model with more moves but a simpler rule, as follows:

For every move $\Psi \in M(B)\cap R(\varnothing)$ or belonging to the subset we are interested in, $S(B) \subset M(B)\cap R(\varnothing)$, we define the game $\mathfrak{S}T(B,\Psi)$ recursively:

$$S_\Psi(B) = \{\Psi\};$$
$$S_\Psi(\mathrm{fin}(B,\Lambda)) := M(\mathrm{fin}(B,\Lambda))\cap R(\Lambda)$$
$$|\Lambda \neq \varnothing,\ \Lambda\Diamond S_T(B,\Psi).$$

Next we may adjust m so that

$$\mathrm{sc}_{\mathfrak{S}_T(B,\Psi)}(B) \underset{\mathrm{col}\,B}{\prec} m\,|\,\Psi \in S(B),$$

and look at the earlier defined set $S(B) \subset M(B)$ of moves allowed at the base B, or if m is fixed, adjust the set $S(B)$:

$$S(B) := \left\{\Psi\,|\,\Psi \subset S(B)\,\&\,\mathrm{sc}_{\mathfrak{S}_T(B,\Psi)}(B) \underset{\mathrm{col}\,B}{\prec} m\right\}.$$

In both cases the test model game $\mathfrak{S}_T(B)$ is defined as the union of the games $\mathfrak{S}_T(B,\Psi)$ for $\Psi \in S(B)$:

$$\mathfrak{S}_T(B) := \bigcup_{\Psi \in S(B)} \mathfrak{S}_T(B,\Psi).$$

Theorem on the Transfer of Scores of Groups of Moves. *Let the following conditions be satisfied:*

(1) *In the game $\mathfrak{A}$ White and Black move alternately and there are defined a set V of virtual moves and influence relations of a branch on a move, branch, and subtree, and of a subtree on a subtree, satisfying the axioms and definitions given in this section.*

(2) $\mathfrak{S}(C)$ *is a model game for $\mathfrak{A}$ with base C, and*

$$S(C) = M(C) \cap R(\varnothing),$$
$$S(\mathrm{fin}(C,\Lambda)) = M(\mathrm{fin}(C,\Lambda)) \cap R(\Lambda) \mid \Lambda \diamondsuit \mathfrak{S}(C),$$
$$R(\Lambda_1 + \Lambda_2) \subset R(\Lambda_1) \mid \Lambda \diamondsuit \mathfrak{S}(C), \mathrm{col}\, \Lambda = \mathrm{col}\, \Lambda_1 = \mathrm{col}\, C.$$

(3) *L is a branch of even length with beginning B and end C, $\mathfrak{S}_T(B)$ is a non-empty test model game for the game $\mathfrak{S}(C)$. Thus the following conditions are satisfied:*

$$S(B) \neq \varnothing,$$
$$S(B) \subset S(C),$$
$$\mathrm{sc}_{\mathfrak{S}T(B)}(B) \underset{\mathrm{col}\, B = \mathrm{col}\, C}{\dashv} m,$$

$$\left.\begin{array}{l} S(\mathrm{fin}(B,\Lambda)) \subset M(\mathrm{fin}(B,\Lambda)) \cap R(\Lambda), \\ (M(\mathrm{fin}(B,\Lambda)) \setminus M(B)) \cap R(\Lambda) \subset S(\mathrm{fin}(B,\Lambda)), \\ \Theta \in S(\mathrm{fin}(B,\Lambda)) \mid (\Theta \in M(\mathrm{fin}(B,\Lambda)) \cap S(B) \cap R(\Lambda) \\ \quad \& \mathfrak{S}'_T(B,\Theta) \sim \mathfrak{S}'_T(B,\mathrm{init}\,\Lambda)) \\ \vee (\Theta \in M(\mathrm{fin}(B,\Lambda)) \cap M(B) \cap R(\Lambda) \& \Lambda \sim \Theta), \end{array}\right\} \begin{array}{l} \Lambda \diamondsuit \mathfrak{S}_T(B), \\ \mathrm{col}\, \Lambda = \mathrm{col}\, B \end{array}$$

$$M(\mathrm{fin}(C,\Lambda_1 + \Lambda_2)) \cap S(\mathrm{fin}(B,\Lambda_1))$$
$$\subset S(\mathrm{fin}(C,\Lambda_1 + \Lambda_2)) \mid \Lambda_1 \diamondsuit \mathfrak{S}_T(B), \Lambda_1 + \Lambda_2 \diamondsuit \mathfrak{S}_T(C),$$
$$\mathrm{col}\, \Lambda_1 = \mathrm{col}(\Lambda_1 + \Lambda_2) \neq \mathrm{col}\, B.$$

Here $\mathfrak{S}'_T(B,\Theta) = \mathfrak{S}_T(B,\Theta) \cap \mathfrak{S}'_T(B)$ *and* $\mathfrak{S}'_T(B)$ *is a minimal pruned tree of a color opposite to $\mathrm{col}\, B$, with terminal positions F for which* $\mathrm{sc}_{\mathfrak{S}_T(B)}(F) \underset{\mathrm{col}\, B}{\prec} m$.

$\mathfrak{S}_T(B,\Theta)$ $(\Theta \in S(B))$ *is a subtree of the game tree $\mathfrak{S}_T(B)$ consisting of positions and moves belonging to the branches $\Lambda \diamondsuit \mathfrak{S}_T(B)$ for which* $\mathrm{init}\, \Lambda = \Theta$.

(4) $\mathfrak{S}(C)$ *is a model game with base C where*

$$\tilde{S}(C) = (M(C) \cap R(\varnothing)) \setminus S(B),$$

$$S(\mathrm{fin}(C,\Lambda)) = \left\{\begin{array}{l} (M(\mathrm{fin}(C,\Lambda)) \cap R(\Lambda)) \setminus S(B) \\ \text{for col } \Lambda = \mathrm{col}\, C, \\ M(\mathrm{fin}(C,\Lambda)) \cap R(\Lambda) \\ \text{for col } \Lambda \neq \mathrm{col}\, C, \quad \mathrm{sc}_{\tilde{\mathfrak{S}}(C)}(C) \underset{\mathrm{col}\, C}{\dashv} m. \end{array}\right\} \Lambda \diamondsuit \tilde{\mathfrak{S}}(C).$$

$\tilde{\mathfrak{S}}'(C)$ *is a minimal pruned tree, of color opposite to* $\operatorname{col} C$, *for the game* $\tilde{\mathfrak{S}}(C)$ *with terminal positions F having scores* $\operatorname{sc}_{\tilde{\mathfrak{S}}(C)}(F) \underset{\operatorname{col} C}{\prec} m$.

$\tilde{\mathfrak{S}}'(B) = L \cup \tilde{\mathfrak{S}}'(C)$, *that is, it consists of the positions and moves in the branch L and the tree* $\tilde{\mathfrak{S}}'(C)$.

Then the position C is lost in the game $\mathfrak{S}(C)$ (*i. e.* $\operatorname{sc}_{\mathfrak{S}(C)}(C) \underset{\operatorname{col} B}{\prec} m$) *else the tree* $\mathfrak{S}'_T(B)$ *influences the tree* $\tilde{\mathfrak{S}}'(B)$.

Let us dwell on the conditions of the theorem. We could change them by not requiring that the axioms on Cases 3, 4, and 5 be satisfied, but still making use of the connection between position scores, the color of the turn to move at the positions, the material score, and the blank move we discussed earlier. Then the proof would be substantially more complicated but would involve nothing in principle new. We could also replace the game $\tilde{\mathfrak{S}}(C)$—whose branches issuing from the base C begin with moves $\Theta \in S(C)$ that do not belong to the set $S(B)$—by a test game $\tilde{\mathfrak{S}}_T(C)$ analogous to the game $\mathfrak{S}_T(B)$. At the inner positions in the game $\tilde{\mathfrak{S}}_T(C)$ with the move belonging to the color $\operatorname{col} B$ the moves in the set $S(C) = M(C) \cap R(\varnothing)$ are as far as possible disallowed. However, we shall not consider such generalizations of the theorem. Moreover, for specific games and constructively defined influence relationships we must investigate cases in which other axioms are not satisfied. A complete proof is given in [15] for chess and some chess model games.

The conditions to be satisfied by the set $S(D)$ of moves allowed at the positions D in the test model game $\mathfrak{S}_T(B)$ are in general the same as those prescribed above. The only change is in the condition on the sets $S(D)$ of moves at the positions D where the move belongs to the color opposite to $\operatorname{col} B = \operatorname{col} C$:

$$M\big(\operatorname{fin}(C, \Lambda_1 + \Lambda_2)\big) \cap S\big(\operatorname{fin}(B, \Lambda_1)\big) \subset S\big(\operatorname{fin}(C, \Lambda_1 + \Lambda_2)\big).$$

This means that some moves allowed in the model $\mathfrak{S}_T(B)$ at positions of significantly lower rank are not to be suppressed at high-ranked positions in the model $\mathfrak{S}(C)$. But we are actually playing in the original game $\mathfrak{A}$, not in the model, and we need not fix the latter. If we allow some moves in it that are needed to advance the proof of the theorem, the model does not suffer.

In the first stage of the proof we construct the pseudocritical branch L_0 and its decomposition into the branches $\Lambda_1, \Lambda_2, \ldots, \Lambda_n, \Lambda_{n+1}$, where n is the number of moves $\Psi_1, \Psi_2, \ldots, \Psi_n$ in the set $S(B)$. Each of the branches Λ_i for $i = 1, 2, \ldots, n$ will be incident on the tree $\mathfrak{S}'_T(B, \Psi_i)$, and the branch Λ_{n+1} will be incident on the tree $\tilde{\mathfrak{S}}'(C)$. Some of the branches, however will be empty, and we assume that the empty branch $\varnothing$ is incident on every tree and does not influence any branch or any move. In addition to the branch L we shall consider auxiliary branches $L_1, L_2, \ldots, L_n$. These also all begin at the position B. We shall postulate that the position C is not lost, and prove the second alternative conclusion of the theorem. We denote by

$\mathfrak{S}''(C)$ the minimal col C-pruned tree of the game $\mathfrak{S}(C)$, whose terminal positions (and therefore all other positions) are not lost. Such a tree exists, since C is not lost.

In constructing the pseudocritical branch L_0 we must satisfy the following conditions:

$$\Lambda_i \lozenge \mathfrak{S}'_T(B, \Psi_i), \quad i = 1, 2, \ldots, n,$$

$$\Lambda_{n+1} \lozenge \tilde{\mathfrak{S}}'(C),$$

$$\Lambda_i \nleftrightarrow \Lambda_j, \quad i, j = 1, 2, \ldots, n, \quad i \neq j,$$

$$\mathfrak{S}'_T(B, \Psi_i) \nleftrightarrow \mathfrak{S}'_T(B, \Psi_j), \quad i, j = 1, 2, \ldots, n, i \neq j,$$

$$\Lambda_i, \Lambda_j \neq \varnothing,$$

$$L_{i-1} = L_i + \Lambda_i, \quad i = 1, 2, \ldots, n,$$

$$L_i = L * L_i^c,$$

$$L_i^c = \Lambda_{i+1} + \cdots + \Lambda_{n+1} \lozenge \mathfrak{S}(C),$$

$$\mathrm{fin}(B, L_i) = \mathrm{fin}(C, L_i^c) \in \mathfrak{S}(C), \quad i = 0, 1, \ldots, n,$$

$$\mathrm{fin}(B, L_0) \in \mathfrak{S}''(C),$$

$$\mathrm{fin}(B, L_n) \in \tilde{\mathfrak{S}}'(C).$$

(some of these conditions are immediate consequences of others, and are stated here to facilitate their use later). When the length $\rho(L_0)$ of the constructed portion of the psuedocritical branch L_0 is even, all the branches Λ_i, L_i, L_i^c must have even length as well. If $\rho(L_0)$ is odd, only one of the branches Λ_i, L_h, L_h^c has odd length for $h > i$, and the last moves in these branches are identical.

To start the construction of the pseudocritical branch L_0 we set

$$\Lambda_i := \varnothing \quad (i = 1, 2, \ldots, n+1);$$

$$L_i := L;$$

$$L_i^c := \varnothing \quad (i = 0, 1, \ldots, n).$$

Then all our conditions will be satisfied. Now suppose that they are satisfied when the currently constructed portion of the pseudocritical branch L_0 has even length $\rho(L_0) = \rho(L) + 2j$, where $j \geq 0$. The position $\mathrm{fin}(B, L_0) = \mathrm{fin}(C, L_0^c) \in \mathfrak{S}''(C)$, i. e. it is not lost. Either it is terminal in that game, or there is a unique and non-losing move Ω_{2j+1} from it, belonging to the minimal col C-pruned tree $\mathfrak{S}''(C) \subset \mathfrak{S}(C)$. In the first case we end the construction of L_0 and begin to analyze the reasons for its non-extendability (cf. Part 1 of the analysis, below). In the second case we attempt to add the move Ω_{2j+1} at the end of the branch L_0.

We test the admissibility of the move Ω_{2j+1} at the position $\mathrm{fin}(B, L_i)$ for $i = 1, 2, \ldots, n$. If it is admissible at these positions it is allowed. For, since $\mathrm{fin}(B, L_i) = \mathrm{fin}(C, L_i^c)$, the move belongs to the set $M(\mathrm{fin}(C, L_i^c))$. Furthermore,

$$L_0^c = \underbrace{\Lambda_1 + \cdots + \Lambda_i}_{\tilde{\Lambda}_i} + \underbrace{\Lambda_{i+1} + \cdots + \Lambda_{n+1}}_{L_i^c} = L_i^c + \tilde{\Lambda}_i,$$

and since the branches L_0^c and L_i^c have even length, col $L_0^c = $ col $L_i^c = $ col C. Accordingly,

$$\Omega_{2j+1} \in S(\text{fin}(C, L_0^c))$$
$$= M(\text{fin}(C, L_0^c)) \cap R(L_0^c) \subset R(L_0^c) \subset R(L_i^c),$$
$$\Omega_{2j+1} \in M(\text{fin}(C, L_i^c)) \cap R(L_i^c)$$
$$= S(\text{fin}(C, L_i^c)) = S(\text{fin}(B, L_i)).$$

If for some $i = h$ the move Ω_{2j+1} is admissible at the position $\text{fin}(B, L_{h-1})$ but turns out to be non-admissible at $\text{fin}(B, L_h)$ we test to see if it is allowed at $\text{fin}(B, \Lambda_h)$.

Since $L_{h-1} = L_h + \Lambda_h$, the branch Λ_h is not empty, else the positions $\text{fin}(B, L_{h-1})$ and $\text{fin}(B, L_h)$ would coincide. If $\Omega_{2j+1} \in S(\text{fin}(B, \Lambda_h))$ we set for $i = 1, 2, \ldots, h-1$

$$L_i := L_i * \Omega_{2j+1};$$
$$L_i^c := L_i^c * \Omega_{2j+1};$$
$$\Lambda_h := \Lambda_h * \Omega_{2j+1};$$

and we go to the next step in the construction of the pseudocritical branch L_0 (see Figure 29). If however the move Ω_{2j+1} is not admissible at $\text{fin}(B, \Lambda_h)$ under the rules of the original game $\mathfrak{A}$ or of the test model game $\mathfrak{S}_T(B)$ we analyze the reasons for the non-extendability (Case 2).

If the move Ω_{2j+1} is admissible (and allowed) at all the positions $\text{fin}(B, L_i)$ for $i = 1, 2, \ldots, n$ it is admissible from the position $\text{fin}(C, \Lambda_{an+1})$ also. In fact, $L_n^c = \Lambda_{n+1}$ and $\text{fin}(C, \Lambda_{n+1}) = \text{fin}(C, L_n^c) = \text{fin}(B, L_n)$, since $L_n = L * L_n^c = L + \Lambda_{n+1}$. If

$$\Omega_{2j+1} \in \tilde{S}(\text{fin}(C, \Lambda_{n+1}))$$
$$= M(\text{fin}(C, \Lambda_{n+1})) \cap R(\Lambda_{n+1}) \backslash S(B),$$

we set

$$L_i := L_i * \Omega_{2j+1};$$
$$L_i^c := L_i^c * \Omega_{2j+1};$$
$$\Lambda_{n+1} := \Lambda_{n+1} * \Omega_{2j+1}$$

for $i = 1, 2, \ldots, n$ and pass to the next step in the construction of the pseudocritical branch.

If, however, the move Ω_{2j+1} is admissible but not allowed at the position $\text{fin}(C, \Lambda_{n+1})$, it belongs to the set $S(B)$ and coincides with some move Ψ_i

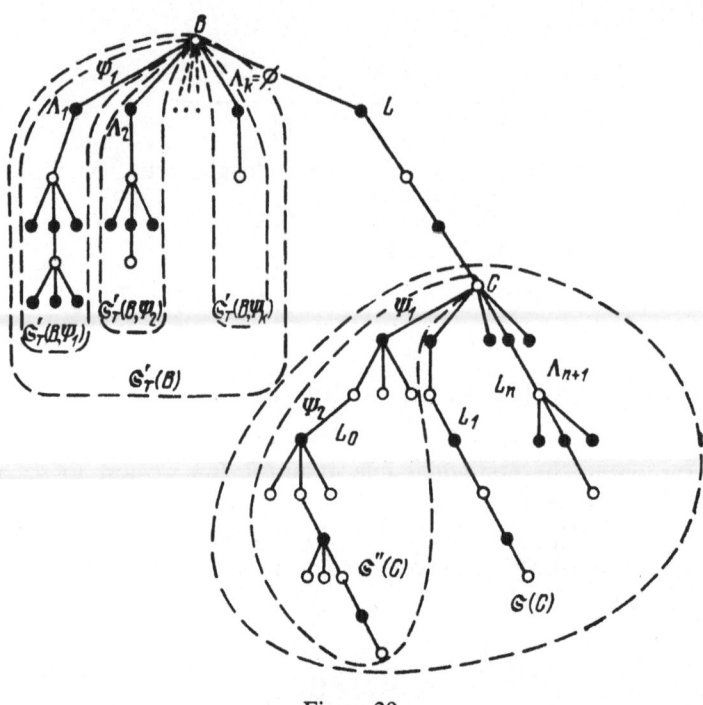

Figure 29

$(1 \le i \le n)$. Then it is allowed at the interior positions of the trees $\mathfrak{S}'_T(B, \Psi_g)$ with the move belonging to $\mathrm{col}\, B = \mathrm{col}\, C$; these trees influence the tree $\mathfrak{S}_T(B, \Psi_i)$, and the move is allowed at all positions of the latter where the turn to move belongs to the same color. (It is allowed at the analogous positions in the corresponding trees $\mathfrak{S}_T(B, \Psi_g)$ and $\mathfrak{S}_T(B, \Psi_i)$, but this has no significance for us.) It is clear that for this move to be allowed it must be admissible under the rules of the original game $\mathfrak{A}$.

Let Λ_h be a non-empty branch incident on the tree $\mathfrak{S}'_T(B, \Psi_h)$ which influences the tree $\mathfrak{S}_T(B, \Psi_i)$ $(1 \le h \le n)$, and let the move $\Omega_{2j+1} = \Psi_i$ be admissible under the rules of $\mathfrak{A}$ at the position $\mathrm{fin}(B, \Lambda_h)$. Since $L_{h-1} = L_h + \Lambda_h$, $L_{h-1} = L * L^c_{h-1}$ and $L_h = L * L^c_h$, the condition $L^c_{h-1} = L^c_h + \Lambda_h$ is satisfied. So $R(L^c_{h-1}) \subset R(\Lambda_h)$. The move $\Psi_i = \Omega_{2j+1}$ is allowed in the game $\mathfrak{S}(C)$ at the position $\mathrm{fin}(B, L_{h-1}) = \mathrm{fin}(C, L^c_{h-1})$. Accordingly

$$\Psi_i = \Omega_{2j+1} \in S\big(\mathrm{fin}\,(C, L^c_{h-1})\big)$$

$$= M\big(\mathrm{fin}\,(C, L^c_{h-1})\big) \cap R(L^c_{h-1}) \subset R(L^c_{h-1}) \subset R(\Lambda_h)$$

and

$$\Psi_i = \Omega_{2j+1} \in M\big(\mathrm{fin}\,(B, \Lambda_h)\big) \cap R(\Lambda_h) \cap S(B).$$

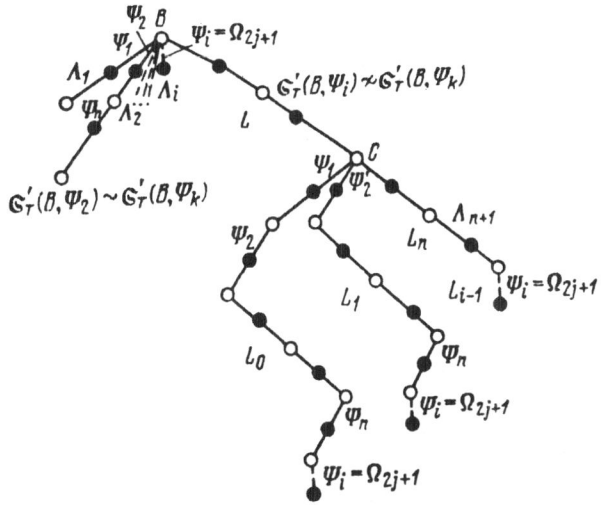

Figure 30

Moreover $\mathfrak{S}'_T(B, \Psi_h) \sim \mathfrak{S}'_T(B, \Psi_i)$. So, because of condition 3 of the theorem

$$\Psi_i = \Omega_{2j+1} \in S\big(\text{fin}(B, \Lambda_h)\big).$$

We add this move at the end of a branch Λ_h such that the number $h \leq n$ is maximal, and at the end of the branches $L_0, L_1, \ldots, L_{h-1}$; then we go on to the next step in constructing the pseudocritical branch L_0.

If the move $\Omega_{2j+1} = \Psi_{h'}$ is not admissible under the rules of the game $\mathfrak{A}$ at the position $\text{fin}(B, \Lambda_n)$ we analyze the reasons for the non-extendability (case 3). Finally, if only empty branches Λ_i are incident on the trees $\mathfrak{S}'_T(B, \Psi_i)$ that influence the tree $\mathfrak{S}'_T(B, \Psi_{h'})$, e. g. for $j = 0$, we extend the branch $\Lambda_{h'}$ for $i = 0, 1, \ldots, h-1$:

$$L_i := L_i * \Omega_{2j+1}$$

$$L_i^c := L_i^c * \Omega_{2j+1}$$

$$\Lambda_{h'} := \Lambda_{h'} * \Omega_{2j+1} = \varnothing * \Omega_{2j+1} = \Lambda\big(\Omega_{2j+1}\big) = \Lambda\big(\Psi_{h'}\big);$$

and we go to the next step (see Figure 30).

After we add the move Ω_{2j+1} of color col B to the pseudocritical branch L_0 all our conditions will be satisfied. In fact, let this move be added also to the branch Λ_h ($1 \leq h \leq n+1$). By their construction, the branches Λ_i are not changed for $h \neq i$ and the move Ω_{2j+1} added at the end of Λ_h belongs to the corresponding tree $\mathfrak{S}'_T(B, \Psi_h)$ for $1 \leq h \leq n$ or to $\tilde{\mathfrak{S}}'(C)$ for $h = n+1$.

Therefore the conditions $\Lambda_i \lozenge \mathfrak{S}'_T(B, \Psi_i)$ will be satisfied for $i = 0, 1, 2, \ldots, n$ and the condition $\Lambda_{n+1} \lozenge \mathfrak{S}'(C)$ will also be satisfied. The equations $L_{i-1} = L_i + \Lambda_j$, $L_i = L * L_i^c$, $L_i^c = \Lambda_{i+1} + \ldots + \Lambda_{n+1}$ will hold for $i = 0, 1, 2, \ldots, n$ since nothing changes on either side of the equations for $h \geq i$; for $h = i$ the move Ω_{2j+1} is added to the branches L_{i-1} and Λ_i, and for $h < i$ it is added to the branches $L_{i-1}, L_{i-1}^c, L_i, L_i^c$. The position $\text{fin}(B, L_i)$ belongs to $\mathfrak{S}(C)$ for $i = 1, 2, \ldots, n$ because it is allowed under the rules of $\mathfrak{S}(C)$ at the ends of the branches L_i, where the move Ω_{2j+1} is added. For a like reason $\text{fin}(B, L_0) \in \mathfrak{S}''(C)$ and $\text{fin}(B, L_n) \in \tilde{\mathfrak{S}}'(C)$.

If the new move Ω_{2j+1} is added at the end of a non-empty branch Λ_h, the non-empty branches Λ_i $(i \leq n)$ remain incident on the mutually non-influencing trees $\mathfrak{S}'_T(B, \Psi_i)$. Then they do not influence each other, and the empty branches Λ_i influence no branch whatsoever. The new non-empty tree $\Lambda_h(\Psi_h)$ is added only when the tree $\mathfrak{S}'_T(B, \Psi_h)$ influences no tree $\mathfrak{S}'_T(B, \Psi_i)$ on which the non-empty branch Λ_i is incident. Therefore the conditions connected with the mutual influences of the branches Λ_i and the trees $\mathfrak{S}'_T(B, \Psi_i)$ continue to hold in this case also.

Finally, the move Ω_{2j+1} was added at the end of some branch Λ_h $(1 \leq h \leq n+1)$ and at the ends of the branches $L_0, L_1, \ldots, L_{h-1}$, $L_0^c, L_1^c, \ldots, L_{h-1}^c$. Their lengths were odd, and the lengths of the branches Λ_i for $i \neq h$, and of L_i and L_i^c for $i \geq h$ remain even.

Now suppose that we have constructed the pseudocritical branch L_0 of odd length $\rho(L_0) = \rho(L) + 2j - 1$, where $j > 0$, that it satisfies all our conditions, and that the last move Ω_{2j-1}, of color $\text{col } C$, has been added at the ends of the branches Λ_h $(1 \leq h \leq n+1)$ and of the branches L_i for $i = 0, 1, \ldots, h-1$. If the position $\text{fin}(B, \Lambda_h)$ is terminal in the corresponding game $\mathfrak{S}_T(B)$ for $h \leq n$ or in $\tilde{\mathfrak{S}}(C)$ for $h = n+1$ we end the construction of L_0 and analyze the reasons for its non-extendability (Case 4). Otherwise we add the unique move Ω_{2j}, which belongs to either $S(\text{fin}(B, \Lambda_h))$ or $\tilde{S}(\text{fin}(C, \Lambda_{n+1}))$, to the corresponding minimal pruned tree $\mathfrak{S}'_T(B)$ or $\tilde{\mathfrak{S}}'(C)$ at the end of the branch Λ_h, of color opposite to $\text{col } B$. This move loses for $\text{col } B$ in the corresponding model game.

Suppose that $\Omega_{2j} \in M(\text{fin}(B, L_i)) = M(\text{fin}(C, L_i^c))$ for $i < h$. Since

$$L_i^c = \Lambda_{i+1} + \ldots + \Lambda_h + \ldots + \Lambda_{n+1}$$

$$= \Lambda_h + \tilde{\Lambda}_{i,h},$$

$$\Omega_{2j} \in M\big(\text{fin}\big(C, \Lambda_h + \tilde{\Lambda}_{i,h}\big)\big)$$

$$\cap S\big(\text{fin}(B, \Lambda_h)\big) \subset S\big(\text{fin}\big(C, \Lambda_h + \tilde{\Lambda}_{i,h}\big)\big) = S\big(\text{fin}(C, L_i^c)\big),$$

i.e. the move Ω_{2j} is allowed in the game $\mathfrak{S}(C)$. For $i = 0$ it leads from the position $\text{fin}(B, L_0)) \in \mathfrak{S}''(C)$ to a position in the tree $\mathfrak{S}''(C)$, which is a $\text{col } C$-pruned tree that also contains all the allowed moves at its positions of color opposed to $\text{col } C$.

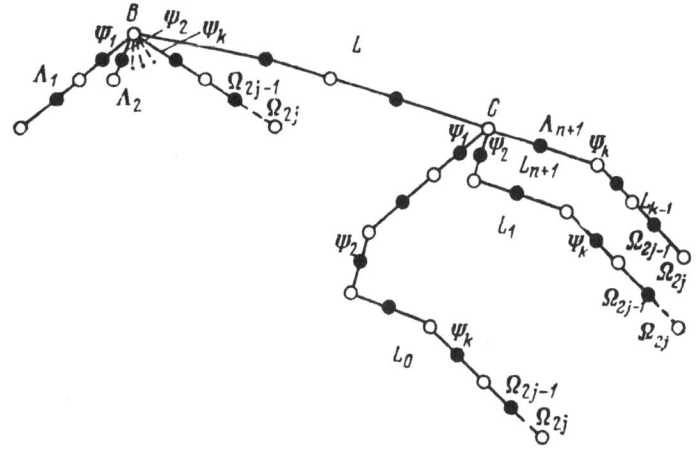

Figure 31

If the move Ω_{2j} is admissible, and therefore allowed, at all the positions $\mathrm{fin}(B, L_i)$ for $i < h$ we set (see Figure 31)

$$L_i := L_i * \Omega_{2j};$$
$$L_i^c := L_i^c * \Omega_{2j} \quad (i = 0, 1, \ldots, h-1)$$
$$\Lambda_h := \Lambda_h * \Omega_{2j}.$$

In the opposite case we end the construction of the pseudocritical branch L_0 and analyze the reasons for its non-extendability. (Case 5).

The branches L_i, L_i^c $(i = 0, 1, \ldots, h-1)$ and Λ_h will have even length after the move Ω_{2j} has been appended at their ends. For $i \geq h$ the lengths of L_i and L_i^c are unchanged and remain even.

The move Ω_{2j} belongs to the subtrees $\mathfrak{S}(C)$ and $\mathfrak{S}''(C)$, and to the corresponding subtree $\mathfrak{S}_T'(B, \Psi_n) = \mathfrak{S}_T'(B) \cap \mathfrak{S}_T(B, \Psi_h)$ for $h \leq n$ or to $\tilde{\mathfrak{S}}'(C)$ for $h = n+1$. For $i < n$ it is added to both sides of the equations $L_{i-1} = L_i + \Lambda_i$, $L_i = L_i * L_i^c$, $L_i^c = \Lambda_{i+1} + \ldots + \Lambda_{n+1}$, and for $i \geq h$ the equations are untouched. This proves that our conditions are satisfied after we append to the pseudocritical branch L_0 the move Ω_{2j} of color opposite to col B just as they are when we append a move of color col B.

Thus we can proceed to the next step in the construction of L_0.

Let us go to the second and third stages of the proof. We consider, in turn, two cases in which L_0 is not extended.

(1) The pseudocritical branch L_0 has even length and its end $\mathrm{fin}(B, L_0) = \mathrm{fin}(C, L_0^c)$ is terminal in the game $\mathfrak{S}(C)$.

Since $S(B) \subset S(C)$ is a non-empty set the position C is non-terminal and some branch Λ_h contains at least two successive moves Ω_{2j-1} and Ω_{2j} of

L_0. It contains these and all the ends $\text{fin}(B, L_i) = \text{fin}(C, L_i^c)$ of the branches L_i and L_i^c for $i < h$. The position $\text{fin}(B, L_0)$ is not lost in the game $\mathfrak{S}(C)$ and the move at it belongs to col C. Therefore it is terminal in the original game $\mathfrak{A}$ (new terminal positions in model games are lost for their own color).

We consider sequentially the positions $\text{fin}(B, L_i)$ for $i = 0, 1, 2, \ldots, h-1$.

If the position $\text{fin}(B, L_{i-1})$ is terminal and $\text{fin}(B, L_i)$ is not, then since $L_{i-1} = L_i + \Lambda_i$, the axiom on Cases 3 and 4 implies that $\Lambda_i \sim L_i$.

If $\text{fin}(B, L_{i-1})$ and $\text{fin}(B, L_i)$ are both terminal, the axiom on Case 5 implies that either $\Lambda_i \sim L_i$ or $\text{sc}(\text{fin}(B, L_{i-1})) \underset{\text{col } B}{\preccurlyeq} \text{sc}(\text{fin}(B, L_i^i))$.

Thus if the branches Λ_i do not influence L_i for $i = 1, 2, \ldots, h-1$, all the positions $\text{fin}(B, L_i)$ are terminal and

$$\text{sc}(\text{fin}(B, L_0)) \underset{\text{col } B}{\preccurlyeq} \text{sc}(\text{fin}(B, L_1))$$

$$\underset{\text{col } B}{\preccurlyeq} \ldots \underset{\text{col } B}{\preccurlyeq} \text{sc}(\text{fin}(B, L_{h-1})),$$

i.e. the position $\text{fin}(B, L_{h-1})$ is also not lost. Since $\text{fin}(B, L_n)$ belongs to $\mathfrak{S}'(C)$ it cannot be terminal and not lost; therefore $h - 1 < n$. Now let us look at the position $\text{fin}(B, \Lambda_h) \in \mathfrak{S}'_T(B, \Psi_h)$. It too cannot be terminal and not lost. Moreover, $L_{h-1} = L_h + \Lambda_h$. Hence the axiom on Cases 3 and 4, or the axiom on Case 5, implies that $\Lambda_h \sim L_h$. Thus we have shown that in the case we are considering there exist the branches Λ_i and L_i ($1 \le i \le n$) that influence one another.

(2) The move Ω_{2j+1} of color col B is admissible at the position $\text{fin}(B, L_{i-1})$ under the rules of the original game $\mathfrak{A}$ and, as we showed during the construction of the pseudocritical branch L_0, it is allowed in the game $\mathfrak{S}(C)$; it is admissible at $\text{fin}(B, L_i)$ and at $\text{fin}(B, \Lambda_i)$ it is either inadmissible or not allowed in the test model game $\mathfrak{S}_T(B)$ for $1 \le i \le n$.

If it is inadmissible at $\text{fin}(B, \Lambda_i)$ the conditions in the first axiom on Case 2 are satisfied, since $L_{i-1} = L_i + \Lambda_i$. Therefore $\Lambda_i \sim L_I$.

Suppose it is admissible but not allowed in the game $\mathfrak{S}_T(B)$, i.e.

$$\Omega_{2j+1} \in M(\text{fin}(B, \Lambda_i)) \setminus S(\text{fin}(B, \Lambda_i)).$$

Since $L_{i-1} = L_i + \Lambda_i$, $L_{i-1} = L * L_{i-1}^c$ and $L_i = L * + L_i^c$, the condition $L_{i-1}^c = L_i^c + \Lambda_i$. The branch L_{i-1}^c has even length, and therefore col $L_{i-1}^c =$ col C. It consists of moves allowed by the rules of $\mathfrak{S}(C)$ at the corresponding positions. So, $L_{i-1}^c \lozenge \mathfrak{S}(C)$. At its end $\text{fin}(C, L_{i-1}^c)$ the move Ω_{2j+1} is allowed, i.e. $\Omega_{2j+1} \in S(\text{fin}(C, L_{i-1}^c))$. The branch Λ_i also has even length, and at its beginning B the move belongs to col $B =$ col C. Then by item 2 in the conditions of the theorem

$$\Omega_{2j+1} \in S(\text{fin}(C, L_{i-1}^c))$$

$$= M(\text{fin}(C, L_{i-1}^c)) \cap R(L_{i-1}^c) \subset R(L_{i-1}^c) \subset R(\Lambda_i).$$

Thus

$$\Omega_{2j+1} \in \left(M(\operatorname{fin}(B, \Lambda_i)) \backslash S(\operatorname{fin}(B, \Lambda_i)) \right) \cap R(\Lambda_i).$$

According to item 3 of the conditions of the theorem

$$(M(\operatorname{fin} B, \Lambda_i)) \backslash M(B)) \cap R(\Lambda_i) \subset S(\operatorname{fin}(B, \Lambda_i)).$$

Therefore

$$\Omega_{2j+1} \in \left(M(\operatorname{fin}(B, \Lambda_i)) \backslash S(\operatorname{fin}(B, \Lambda_i)) \right) \cap R(\Lambda_i)$$
$$= \left(M(\operatorname{fin}(B, \Lambda_i)) \cap R(\Lambda_i) \right) \backslash S(\operatorname{fin}(B, \Lambda_i))$$
$$\subset \left(M(\operatorname{fin}(B, \Lambda_i)) \cap R(\Lambda_i) \right) \backslash \left((M(\operatorname{fin}(B, \Lambda_i)) \backslash M(B)) \right.$$
$$\cap R(\Lambda_i)) = \left(M(\operatorname{fin}(B, \Lambda_i)) \right.$$
$$\backslash (M(\operatorname{fin}(B, \Lambda_i)) \backslash M(B))) \cap R(\Lambda_i) = M(\operatorname{fin}(B, \Lambda_i))$$
$$\cap M(B) \cap R(\Lambda_i) \subset M(B).$$

Thus

$$\Omega_{2j+1} \in M(\operatorname{fin}(B, L^c_{i-1})) \cap M(\operatorname{fin}(B, \Lambda_i)) \cap M(B),$$
$$\Omega_{2j+1} \notin M(\operatorname{fin}(B, L^c_i)),$$
$$L^c_{i-1} = L^c_i + \Lambda_i,$$

i. e. the conditions for the second axiom on Case 2 are satisfied. Then either $\Lambda_i \sim L_i$ or $\Lambda_i \sim \Omega_{2j+1}$. But in the latter alternative the move Ω_{2j+1} would be allowed under the rules of the test model game $\mathfrak{S}_T(B)$ at the position $\operatorname{fin}(B, \Lambda_i)$ (by item 3 of the conditions of the theorem). Therefore the branch Λ_i influences the branch L_i.

(3) The move Ω_{2j+1} of color col B is admissible under the rules of $\mathfrak{A}$ at the positions $\operatorname{fin}(B, L_{h-1})$ and $\operatorname{fin}(B, L_h)$, and belongs to the set $S(B)$, so that it is admissible at B but not at $\operatorname{fin}(B, \Lambda_h)$. $\Omega_{2j+1} = \Psi_{h'}$ for $1 \le h' \le n$ and if $h < i \le n$ and the branch Λ_i is not empty, then $\mathfrak{S}'_T(B, \Psi_i) \twoheadrightarrow \mathfrak{S}'_T(B, \Psi_h)$.

Since $L_{h-1} = L_h + \Lambda_h$, the conditions for the second axiom on Case 2 are fulfilled in the present case. Thus either $L_h \sim \Lambda_h$ or $L_h \sim \Omega_{2j+1} = \Psi_{h'}$. If $L_h \sim \Lambda_h$, the axiom on the symmetry of the influence relation implies that $\Lambda_h \sim L_h$ (from now on we shall not mention this axiom explicitly, and will write $L'' \sim L'$ quite freely in place of $L' \sim L''$). But if $L_h \sim \Psi_{h'}$ we consider the non-empty branch $\tilde{\Lambda}_{h'}(\Psi_{h'}, \ldots)$, which is incident on the tree $\mathfrak{S}'_T(B, \Psi_{h'})$, and has $\Psi_{h'} = \Omega_{2j+1}$ as its first move and necessarily exists, for example the branch $\Lambda(\Psi_{h'})$ consisting of the single move $\Psi_{h'}$. If for some i ($h < i \le n$) the branch Λ_i were to influence the move $\Psi_{h'}$ (it therefore could not be empty), the axiom of influence on a move and a branch would imply that it influenced the branch $\tilde{\Lambda}_{h'}$. Then by the definition of the influence relation of one tree on another, $\mathfrak{S}'_T(B, \Psi_i)$ would influence $\mathfrak{S}'_T(B, \Psi_{h'})$.

Therefore

$$\Lambda_i \leftthreetimes \Psi_{h'} \mid h < i \le n.$$

(4) The end of a branch Λ_h $(1 \le h \le n+1)$ having odd length is a terminal position in the test model game $\mathfrak{S}_T(B)$.

Since $\mathrm{fin}(B, \Lambda_h) \in \mathfrak{S}'_T(B)$ and $\mathrm{col}\,\Lambda_h \neq \mathrm{col}\,B = \mathrm{col}\,C$, this position is terminal in the original game $\mathfrak{A}$ and, moreover, lost (as in the first situation where the pseudocritical branch L_0 could not be extended, a position which is terminal only in the model game must be lost for the color having the move there, i.e. the color leading to its branch—and $\mathrm{col}\,\Lambda_h \neq \mathrm{col}\,C$). Let us consider the positions $\mathrm{fin}(B, L_i)$ for $i = 0, 1, \ldots, h-1$. If there is a non-terminal position among them we choose the position $\mathrm{fin}(B, L_i)$ with the largest index $i < h$ for which $M(\mathrm{fin}(B, L_i)) \neq \varnothing$. Then $L_i = L_{i+1+\Lambda i+1}$ and $M(\mathrm{fin}(B, \Lambda_{i+1})) = \varnothing$, or if $i = h-1$, $M(\mathrm{fin}(B, \Lambda_{i+1})) = \varnothing$. Then by the axiom on Cases 3 and 4, $\Lambda_{i+1} \sim L_{i+1}$.

If, however, $M(\mathrm{fin}(B, L_i)) \neq \varnothing$ for $i = 1, 2, \ldots, h-1$, some of these positions are not lost; at least the position $\mathrm{fin}(B, L_0) \in \mathfrak{S}''(C)$ is not. We choose a non-lost terminal position $\mathrm{fin}(B, L_i)$ with maximal index $i < h$. Then either $\mathrm{fin}(B, L_{i+1})$ or (if $i = h-1$) $\mathrm{fin}(B, \Lambda_{i+1})$ is terminal and lost, i.e. for $i < h-1$

$$\mathrm{sc}\left(\mathrm{fin}(B, L_i)\right) \underset{\mathrm{col}\,B}{\not\leqslant} \mathrm{sc}\left(\mathrm{fin}(B, L_{i+1})\right),$$

and for $i = h-1$

$$\mathrm{sc}\left(\mathrm{fin}(B, L_i)\right) \underset{\mathrm{col}\,B}{\not\leqslant} \mathrm{sc}\left(\mathrm{fin}(B, \Lambda_{i+1})\right).$$

However, since $L_i = L_{i+1} + \Lambda_{i+1}$ the axiom on Case 5 implies that either $\Lambda_{i+1} \sim L_{i+1}$ or for $i < h-1$

$$\mathrm{sc}\left(\mathrm{fin}(B, L_i)\right) \underset{\mathrm{col}\,B}{\leqslant} \mathrm{sc}\left(\mathrm{fin}(B, L_{i+1})\right),$$

and for $i = h-1$

$$\mathrm{sc}\left(\mathrm{fin}(B, L_i)\right) \underset{\mathrm{col}\,B}{\leqslant} \mathrm{sc}\left(\mathrm{fin}(B, \Lambda_{i+1})\right).$$

The second alternative is incompatible with the assumption that $\mathrm{col}\,B$ loses at $\mathrm{fin}(B, L_{i+1})$ or $\mathrm{fin}(B, \Lambda_{i+1})$ and with the absence of a loss at $\mathrm{fin}(B, L_i)$. Thus $\Lambda_{i+1} \sim L_{i+1}$.

(5) The move Ω_{2j}, of color opposite to $\mathrm{col}\,B$, is admissible at the position $\mathrm{fin}(B, \Lambda_h)$ $(1 \le h \le n+1)$ under the rules of the game $\mathfrak{A}$ and inadmissible at $\mathrm{fin}(B, L_i)$ for some $i < h$.

In this case $i < n$ since $\mathrm{fin}(B, L_n) = \mathrm{fin}(C, \Lambda_{n+1})$. Just as in the two preceding cases we choose the position $\mathrm{fin}(B, L_i)$ with the largest value of i for which the move Ω_{2j} is admissible. Since $L_i = L_{i+1} + \Lambda_{i+1}$ the conditions for the axiom on Case 1 are satisfied. Then $L_{i+1} * \Omega_{2j} \sim \Lambda_{i+1}$ for $i < h-1$, and $\Lambda_{i+1} * \Omega_{2j} \sim L_{i+1}$ for $i = h-1$. In the first case $L_{i+1} * \Omega_{2j} \Diamond \tilde{\mathfrak{S}}'(C)$ and in the second case $\Lambda_{i+1} * \Omega_{2j} \Diamond \mathfrak{S}'_T(B, \Psi_{i+1})$.

Thus, when the construction of the pseudocritical path L_0 is completed one of the following three conditions will be satisfied:

(1) Some branch Λ_i influences the corresponding auxiliary branch L_i $(1 \le i \le n)$.
(2) An auxiliary branch L_i influences the move $\Psi_{h'} \in S(B)$ $(1 \le i \le n, 1 \le h' \le n)$, while $L_k \not\sim \Psi_{h'}$ for $k = i+1, \dots, n$.
(3) At the ends of the branches $L_{i+1}, \dots, L_{-1}, \Lambda_h$ $(1 \le i < h \le n)$ we may append a move Ω_{2j} of color opposite to col B:

$$L_k := L_l * \Omega_{2j} \qquad (k = i+1, \dots, h-1);$$
$$\Lambda_h := \Lambda_h * \Omega_{2j}.$$

The conditions

$$L_{k-1} = L_k - \Lambda_k \qquad (k = i+2, \dots, n);$$
$$\Lambda_{i+1} \sim L_{i+1}.$$

are satisfied. Now we show that some branch $\Lambda_k \Diamond \mathfrak{S}'_T(B)$ $(1 \le k \le n)$ influences the branch $L_n \Diamond \mathfrak{S}'(B)$. This will complete the proof of the second alternative of the theorem on the transfer of scores of groups of moves—the influence of the tree $\mathfrak{S}'_T(B)$ on the tree $\tilde{\mathfrak{S}}'(B)$—and therewith complete the proof of the theorem.

Suppose that for $1 \le i \le j \le k$ the branches Λ_i and L_j that we have constructed influence one another, or the branch L_j influences the move Ψ_i $(1 \le i \le n, 1 \le j \le n)$, while $\Lambda_k \not\sim \Psi_i$ for $k = j+1, \dots, n$. We have just shown that for some values of i and j one of these conditions is satisfied (if the first condition, $i = j$). If $j = n$ we have $\mathfrak{S}'_T(B, \Psi_i) \sim \tilde{\mathfrak{S}}'(B)$, i.e. $\mathfrak{S}'_T(B) \sim \tilde{\mathfrak{S}}'(B)$, which is what we were to prove. In fact, either $\Lambda_i \sim L_n$ or, when $L_n \sim \Psi_i$, the axiom on the connection between influence on a move and influence on a branch implies that the branch L_n influences an arbitrary branch $\tilde{\Lambda}_i$ of the non-empty tree $\mathfrak{S}'_T(B, \Psi_i)$, since such a branch contains its first move Ψ_i.

If $i, j < n$ we have $L_j \sim \Psi_i$, and since $L_j = L_{j+1} + \Lambda_{j+1}$ the first axiom on weak composition of branches implies that either $L_{j+1} \sim \Psi_i$ or $\Lambda_{j+1} \sim \Psi_i$ or $\Lambda_{j+1} \sim L_{j+1}$. The case $\Lambda_{j+1} \sim \Psi_i$ is ruled out by the relations we have proved above, and the other two have the form we have already studied, but with a larger index on the auxiliary branch L_{j+1}. If however $\Lambda_i \sim L_j$, the second axiom on the weak composition of branches implies that

$$\Lambda_i \sim L_{j+1} \qquad \text{or} \qquad \Lambda_i \sim \Lambda_{j+1} \qquad \text{or} \qquad \Lambda_{j+1} \sim L_{j+1}.$$

Here the second possibility is ruled out since the several branches Λ_k do not influence one another. The two remaining possibilities have the form we have already studied, but again the index of the branch L_{j+1} increases. So, in the end, we prove one of the relations

$$\Lambda_k \sim L_n \quad (i < k \le n) \quad \text{or} \quad \Lambda_n \sim \Psi_i,$$

as required.

In the 5th section we shall discuss in detail the use of our theorem for shortening the search. Here we shall only offer a few remarks. Suppose that at the position B we are considering the test model $\mathfrak{S}_T(B)$ and the position $C = \text{fin}(B, L)$, where the branch L has even length. If $L \not\rightarrow \mathfrak{S}_T'(B)$, we may begin by inspecting the moves $\Theta \in M(C) \setminus S(B)$ from C. At positions C with the move belonging to $\text{col}\, C = \text{col}\, B$ we need not consider moves belonging to $S(B)$. We define some minimal pruned tree $\tilde{\mathfrak{S}}'(C)$ of color opposite to $\text{col}(B)$. If (a) the score (in the game $\tilde{\mathfrak{S}}(C)$) of all moves from C that belong to the tree $\tilde{\mathfrak{S}}'(B) = L \cup \mathfrak{S}'(C)$ is not better than the score of B in the test model game $\mathfrak{S}_T(B)$, and (b) $\mathfrak{S}_T'(B) \not\rightarrow \tilde{\mathfrak{S}}'(B)$, the moves $\Psi \in S(B)$ from the position C may be neglected: $\text{sc}(C) \preccurlyeq \text{sc}_{\mathfrak{S}_T(B)}(B)$. A single test model $\mathfrak{S}_T(B)$ may serve for several positions C_i $(i = 1, 2, \ldots, l)$, where $C_i = \text{fin}(B, L^{(i)})$ and $\text{col}\, L^{(i)} = \text{col}\, B$.

A Constructive Definition of the Influence Relation for Several Games

Suppose that for some game $\mathfrak{A}$ we are given (a) an equivalence relation among moves from different positions, (b) the predicates $i(L, \Psi)$ and $I(B, L_1, L_2)$ for the influence relationships of the branch L on the move Ψ and of the branch L_1 with origin B on the branch L_2 with the same origin, and (c) the predicates for the influence of L on the subtree $\mathfrak{S}$ and the influence of a subtree $\mathfrak{S}_1$ on a subtree $\mathfrak{S}_2$, expressed in terms of the above elementary predicates:

$$I(B, L, \mathfrak{S}) := \bigvee I(B, L, \Lambda) | \Lambda \Diamond \mathfrak{S};$$

$$I(B, \Sigma_1, \mathfrak{S}_2) := \bigvee I(B, L_1, L_2) | L_1 \Diamond \mathfrak{S}_1, L_2 \Diamond \mathfrak{S}_2;$$

where the position B is the the origin of the branch L and the root of the subtrees $\mathfrak{S}$, $\mathfrak{S}_1$, and $\mathfrak{S}_2$. Suppose further that a theorem like the one we have proved above holds, which allows us to apply the method of analogies in cases where some branch or subtree does does not influence another subtree. Then the same theorem will also hold for the influence relationships defined by the more widely embracing predicate $I'(B, L_1, L_2)$ for which

$$B \in \mathfrak{A} \& L_1 \Diamond \mathfrak{A}(B) \& L_2 \Diamond \mathfrak{A}(B)$$

$$\Rightarrow I(B, L_1, L_2) \rightarrow I'(B, L_1, L_2)$$

where $\mathfrak{A}(B)$ is the B-subtree of the tree $\mathfrak{A}$.

Thus the influence relationships for a game $\mathfrak{A}$ can be defined in various ways. They may be distinguished from one another by the composition and amount of the information about moves and positions in the branches L_1 and L_2 defining the predicate $I(B, L_1, L_2)$, by the complexity of the algorithm for computing it, by the formulation of the theorem on the possibility of applying the method of analogy, and by the concept of equivalence among moves to which the method relates.

For example, if the theorem on the transfer of position scores is to hold, the influence relation need not be symmetric, but if the theorem on the transfer of scores of groups of moves is to hold, the influence relation must be symmetrized (at least partially). The larger the body of information used to define the value of the influence predicate and the more complex the predicates and the methods for using them, then usually the more often will the value of the predicate reduce to 0. Thus we find that we may be able to use the method of analogies effectively in shortening the search.

On the other hand, increasing the volume of information and complicating its structure leads to a growth in the amount of time needed per step in the search of the test model tree, since at each step we must evaluate the predicate. Moreover, the amount of memory needed will increase, since the computed information must be stored for future use.

Complicating the algorithms for evaluating the predicate and the methods for arriving at a decision by analogy—e. g. going over to methods using the notion of resolving a variation into what one might call independent components—leads to a growth in the time required for each step in the search since one must apply these algorithms and arrive at decisions.

We shall consider examples of a constructive definition of the influence relation, for the games of chess and noughts-and-crosses. We shall present predicates of several types, each having its own advantages and weaknesses. To simplify the reasoning we shall substitute for chess a nearly equivalent game that we have mentioned earlier. In it one need not defend against checks, the King may attack, and the game is won by capturing the opponent's King. If stalemate is not counted as a draw, but as a loss for the stalemated side, this game will be equivalent to chess.

Our model allows all moves admissible under the rules of chess, and some others as well. The latter lead to the inclusion of auxiliary positions in the tree of the game itself and of the models used in game programming, and lead also to the arising of 'parasitic' influences from moves that either do not defend one's King from a check, or put him under attack. This effect is clearly more important than the non-equivalence of our game to chess. The latter, in general, is irrelevant when—as is usually the case—we are contemplating a model with limited depth of search, and mate or stalemate are far away.

We shall present (but without supporting it) an influence predicate for chess which takes into account checks to the King and the pinning of pieces that protect it from attacks by long-range pieces. We have mentioned earlier the proofs of theorems on the transfer of scores of positions and groups of moves [15]; these were carried out for the influence relation defined in them. We again assume that there are neither castlings nor captures *en passant* among the moves in the branches we consider. The rules of castling preclude its execution under attacks by enemy pieces on the squares where the King stands before and after castling or on the intermediate squares. We neglect such restrictions, as connected with attacks on the King. A capture *en passant* is the sole chess move in which the capturing piece does not occupy the square on which the captured piece stood. We could to some extent alter the influence predicate to take account of these captures, but it is simpler to treat the branches containing them as influencing arbitrary ones.

To construct concrete influence predicates we study the following scheme. Suppose given the branches $L_1(\Psi_1, \Psi_2, \ldots, \Psi_k)$ and $L_2(\Theta_1, \Theta_2, \ldots, \Theta_l)$ together with their segments $L_1^g(\Psi_1, \Psi_2, \ldots, \Psi_g)$ and $L_2^h(\Theta_1, \Theta_2, \ldots, \Theta_h)$ $(1 \leq g \leq k, 1 \leq h \leq l)$. We consider the composition $L^{g,h} = L_1^g + L_2^h$ of these segments. An influence relation will exist when one of the cases of Type 1 or Type 2 arises:

at the position $\text{fin}(B, L^{2\gamma-1,2\eta})$ the move $\Psi_{2\gamma}$ of color opposite to col B is inadmissible;

at the position $\text{fin}(B, L^{2\gamma,2\eta-1})$ the move $\Theta_{2\eta}$ of the same color is inadmissible; or

at the position $\text{fin}(B, L^{2\gamma,2\eta})$ it is possible to make a move Ω of color col B that is inadmissible at the positions $\text{fin}(B, L_1^{2\gamma})$ and $\text{fin}(B, L_2^{2\eta})$, but in certain circumstances at only one of these (see Figure 32). Information on the branches and subtrees investigated during the search will consist of some set of boards. The definition of a board was given in the preceding chapter, but we shall recall it here. A board is an array of 64 bits, corresponding to the squares on a chessboard; the bits corresponding to squares that are in some way distinguished are set to 1 and the remainder to

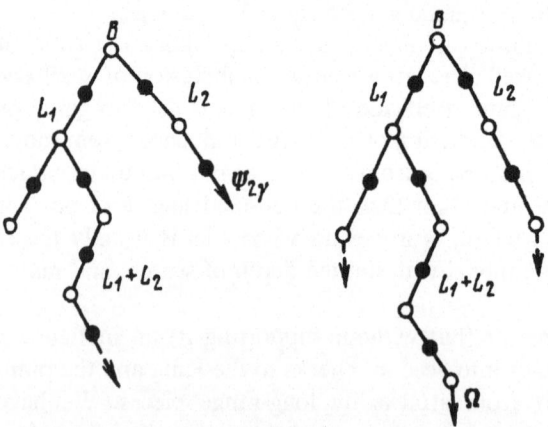

Figure 32

0. The union $T_1 \cup T_2$ of two boards is a board on which a bit is set to 1 if and only if it is set to 1 on one or both of T_1 and T_2; the intersection $T_1 \cap T_1$ is a board in which a bit is set to 1 if and only if it is set to 1 on both T_1 and T_2; in the non-coincidence, or discrepancy board $T_1 \oplus T_2$ a bit is set to 1 if and only if its values differ on the two boards; in the difference board $T_1 \setminus T_2$ when it has the value 1 on T_1 and 0 on T_2; in the inverse board $\neg T$ a bit is set to 1 if it is 0 in T, and to 0 if 1 in T. The predicate $[T]$ has the value 'true' if T is not empty, i. e. if at least one bit in T is set to 1, and 'false' if $T = \varnothing$, i. e. all bits in T are set to 0.

We introduce the following information elements that describe the virtual chess move Ψ:

(1) $E(\Psi)$—a board on which the square where the move originates is distinguished (bit set to 1);
(2) $I(\Psi)$—the board for the square to which the piece moved;
(3) $W(\Psi)$—a board containing the trajectory of the move.

For moves of long-range pieces—Bishops, Rooks, and Queens—the board $W(\Psi)$ contains marks on the squares in diagonals, ranks, and files as appropriate for the move between the squares $E(\Psi)$ and $I(\Psi)$. It is empty for other moves, $W(\Psi) = \varnothing$.

Moves of the color col B will be considered as ours; moves of the opposite color will be moves by the opponent. The boards for these and the other moves in a contemplated branch or subtree will be compounded. Thus

$$E(s, \mathfrak{L}) := \bigcup E(\Psi) | \Psi \in \Lambda, \text{col } \Psi = \text{col } B;$$

$$E(c, \mathfrak{L}) := \bigcup E(\Psi) | \Psi \in \Lambda, \text{col } \Psi \neq \text{col } B;$$

$$I(s, \mathfrak{L}) := \bigcup I(\Psi) | \Psi \in \Lambda, \text{col } \Psi = \text{col } B;$$

$$I(c, \mathfrak{L}) := \bigcup I(\Psi) | \Psi \in \Lambda, \text{col } \Psi \neq \text{col } B;$$

$$W(c, \mathfrak{L}) := \bigcup W(\Psi) | \Psi \in \Lambda, \text{col } \Psi \neq \text{col } B;$$

where $\mathfrak{L}$ is the contemplated branch or subtree. We do not need the board $W(s, \mathfrak{L})$.

Let $P(A, \mu)$ be the board for the *mobility* of the piece μ in the position A. This board distinguishes those squares that are attacked by μ from A. If such a square is not occupied by a piece of the same color, the piece μ can move to it. Pawns are an exception, since they do not move in the same way as they attack. For this reason there are two boards corresponding to a Pawn μ: $p(A, \mu)$ for the squares to which it can move, and $b(A, \mu)$ for the squares it attacks. We denote by $\lambda(A; \Psi_1, \Psi_2)$ the branch beginning at A and consisting of the two moves Ψ_1 and Ψ_2, naturally of opposite color. We shall from time to time contemplate a short branch $\lambda(A; \Psi)$ consisting of one move. The fact that such branches are contained in our contemplated branch L with origin B, or in the subtree $\mathfrak{S}$, is expressed, together with their color, by the following statements:

$$\lambda(A; \Psi_1, \Psi_2) \in L := L = L_1 * \lambda * L_2 \& A = \operatorname{fin}(B, L_1);$$

$$\lambda(A; \Psi) \in L := L = L_1 * \lambda \& A = \operatorname{fin}(B, L_1);$$

$$\lambda \in \mathfrak{S} := \exists L (L \Diamond \mathfrak{S} \& \lambda \in L);$$

$$\operatorname{col} \lambda(A; \Psi_1, \Psi_2) := \operatorname{col} \lambda(A; \Psi) := \operatorname{col} A.$$

In the first of these definitions we do not require that the branch L_2 consist of admissible moves when its origin is the position B. It must be admissible only at the position $\operatorname{fin}(B, L_1 * \lambda)$.

We shall need the so-called 'boards of new mobility',

$$N(\lambda, \mu) := \left\{ \begin{array}{ll} P(\operatorname{fin}(A, \lambda), \mu), & \text{if } \mu = \varphi(\Psi'): \\ P(\operatorname{fin}(A, \lambda), \mu) \backslash P(A, \mu), & \text{if } \mu \neq \varphi(\Psi'), \end{array} \right\} \mu \text{—not a pawn}$$

$$n_{p,b}(\lambda, \mu) := \left\{ \begin{array}{ll} p, b(\operatorname{fin}(A, \lambda), \mu), & \text{if } \mu = \varphi(\Psi'); \\ \varnothing, & \text{if } \mu \neq \varphi(\Psi'), \end{array} \right\} \mu \text{—a pawn}$$

where A is the origin of the branch and Ψ is its first and perhaps only move. We also compound these boards for all pieces and two-move branches of the same color in the contemplated branch or subtree, and include as well the single-move branches that are ends of branches L of odd length:

$$N(s, \mathfrak{L}) := \bigcup N(\lambda, \mu) | \lambda \in \mathfrak{L}, \mu \in \Phi, \operatorname{col} \lambda = \operatorname{col} \mu = \operatorname{col} B;$$

$$n_{p,b}(s, \Lambda) := \bigcup n_{p,b}(\lambda, \mu) | \lambda \in \mathfrak{L}, \mu \in \Pi, \operatorname{col} \lambda = \operatorname{col} \mu = \operatorname{col} B;$$

where Φ is a set of pieces (Kings, Queens, Rooks, Bishops, and Knights) and Π a set of Pawns. We do not need the analogous boards $N(c, \mathfrak{L})$ and $n_{p,b}(c, \mathfrak{L})$.

Let some enemy move $\varphi(\Psi_{2\gamma})$ in the branch L_1 be inadmissible at the position $\operatorname{fin}(B, L_1^{2\gamma-1} + L_2^{2\eta})$. This can happen only in the following cases:

(1) The enemy piece $\varphi(\Psi_{2\gamma})$ does not occupy the square $E(\Psi_{2\gamma})$ at the position $\operatorname{fin}(B, L_1^{2\gamma-1} + L_2^{2\eta})$. Since it occupies this square at the position $\operatorname{fin}(B, L_1^{2\gamma-1})$ it either left it by one of the moves $\Theta_{2\varepsilon} \in L_2$, of color opposite to $\operatorname{col} B$, or it was captured by some move $\Theta_{2\delta+1} \in L_2$ of color $\operatorname{col} B$ ($2\varepsilon \leq 2\eta$ or $2\delta + 1 < 2\eta$). Therefore

$$E(c, L_1) \cap (E(c, L_2) \cup I(s, L_2)) \neq \varnothing.$$

(2) There is no piece $\chi(\Psi_{2\mu})$ on the square $I(\Psi_{2\gamma})$. If $\chi(\Psi_{2\gamma}) \neq \varnothing$, i. e. if the move $\Psi_{2\gamma}$ is a capture, the piece $\chi(\Psi_{2\gamma})$ left this square by one of the moves $\Theta_{2\delta+1} \in L_2$ of color $\operatorname{col} B$ or it was captured by a move $\Theta_{1\varepsilon} \in L_2$ of the opposite color. If however $\chi(\Psi_{2\gamma}) = \varnothing$, the square $I(\Psi_{2\gamma})$ was occupied in one of the moves $\Theta \in L_2$ ($2\delta + 1$ or 2ε or h cannot exceed 2η). Then

$$I(c, L_1) \cap (E(s, L_2) \cup I(s, L_2) \cup I(c, L_2)) \neq \varnothing.$$

(3) The trajectory of the move $W(\Psi_{2\gamma})$ is partitioned by some move $\Theta_h \in L_2$. Then

$$W(c, L_1) \cap (I(s, L_2) \cup I(c, L_2)) \neq \varnothing.$$

For similar reasons a move $\Psi_{2\eta}$ by the opponent may be inadmissible at the position $\mathrm{fin}(B, L_1^{2\gamma} + L_2^{2\eta-1})$.

Let us now consider the cases in which at the position $\mathrm{fin}(B, L_1^{2\gamma} + L_2^{2\eta})$ it is possible to make a move Ω (of color col B) which is inadmissible at both of the positions $\mathrm{fin}(B, L_1^{2\gamma})$ and $\mathrm{fin}(B, L_2^{2\gamma})$. Every piece in the position $\mathrm{fin}(B, L_1^{2\gamma} + L_2^{2\eta})$ is on the same square that it occupies at the end of one or other or both of the branches $L_1^{2\gamma}$ and $L_2^{2\eta}$ issuing from B. Then the piece $\varphi(\Omega)$ is on the square $E(\Omega)$ in one or the other, or both, of the positions $\mathrm{fin}(B, L_1^{2\gamma})$ and $\mathrm{fin}(B, L_2^{2\gamma})$. In the first case, since $\varphi(\Omega)$ is on the same square $E(\Omega)$ in the position $\mathrm{fin}(B, L_1^{2\gamma} + L_2^{2\eta})$ but not in the position $\mathrm{fin}(B, L_2^{2\eta})$ it occupied this position in one of the moves $\Psi_{2\varepsilon+1} \in L_1$ $(\varepsilon < \gamma)$. Thus all the squares it can reach from the position $\mathrm{fin}(B, L_1^{2\gamma})$ belong to one of the positions on the new possibility boards $N(\lambda, \varphi(\Omega))$, $n_p(\lambda, \varphi(\Omega))$ or $n_b(\lambda, \varphi(\Omega))$ (the latter cases hold when $\varphi(\Omega)$ is a Pawn) for one of the two-move branches $\lambda \in L_1^{2\gamma}$ having the color col B. The move Ω is of course not allowed from the position $\mathrm{fin}(B, L_2^{2\eta})$ but may turn out to be so from the position $\mathrm{fin}(B, L_2^{2\gamma})$ for the following reasons.

(1) The square $I(\Omega)$ is not occupied by the same piece that it holds in the position $\mathrm{fin}(B, L_1^{2\gamma} + L_2^{2\eta})$. If $\chi(\Omega) \neq \varnothing$, the piece $\chi(\Omega)$ arrived at the given square in one of the moves $\Theta_{2\varepsilon} \in L_2(\varepsilon \leq \eta)$, i.e. $I(\Omega) = I(\Theta_{2\varepsilon})$. If however $I(\Omega) = \varnothing$, a piece that occupies it in the position $\mathrm{fin}(b, L_1^{2\gamma})$ or arrives on it in some move $\Theta_g \in L_2$, must have left it during one of the moves $\Theta_{2\varepsilon} \in L_2$ $(n \leq 2\eta)$. Accordingly,

$$I(\Omega) \in E(s, L_2) \cup E(c, L_2) \cup I(c, L_2)$$

and, since the square $I(\Omega)$ belongs to one of the boards $N(\lambda, \varphi(\Omega))$, $n_p(\lambda, \varphi(\Omega))$ or $n_b(\lambda, \varphi(\Omega))$ for some two-move branch $\lambda \in L_1^{2\gamma}$ with color col B, the following conditions are satisfied:

$$N(s, L_1) \cap (E(s, L_2) \cup E(c, L_2) \cup I(c, L_2)) \neq \varnothing$$
$$n_p(s, L_1) \cap E(s, L_2) \cup E(c, L_2) \neq \varnothing$$
$$n_b(s, L_1) \cap I(c, L_2) \neq \varnothing$$

(2) The piece $\chi(\Omega)$ occupies the square $I(\Omega)$ in the position $\mathrm{fin}(B, L_1^{2\gamma})$, but the trajectory of the move Ω is interrupted by some other piece. Since this trajectory is free in the position $\mathrm{fin}(B, L_1^{2\gamma} + L_1^{2\eta})$, all of the blocking pieces leave it during one or more moves $\Theta_{ih} \in L_2^{2\gamma}$ $(h \leq 2\gamma)$. The ex-blocking-piece nearest to the square $E(\Omega)$ will be found in the position $\mathrm{fin}(B, L_2^{2\gamma})$ under attack by the long range piece $\varphi(\Omega)$, and the square it occupies belongs to the board $N(s, L_1)$. It may happen that it is not this piece that leaves this square, but rather some capturing piece

in the branch L_2, or a piece capturing the latter, and so on. In any case

$$N(s, L_1) \cap E(s, L_2) \cup E(c, L_2) \neq \varnothing$$

The case in which the piece $\varphi(\Omega)$ occupies the square $E(\Omega)$ in the position $\text{fin}(B, L_2^{2\eta})$ but not in $\text{fin}(B, L_1^{2\gamma})$ is symmetric with the case we have just considered. Therefore we have only the case to consider in which this piece occupies the same square $E(\Omega)$ in all three of the positions $\text{fin}(B, L_2^{2\gamma})$, $\text{fin}(B, L_2^{2\eta})$, and $\text{fin}(B, L_1^{2\gamma} + L_2^{2\eta})$. Then it occupied this same square in the position B. Otherwise it would have arrived at this square in both of the branches $L_1^{2\gamma}$ and $L_2^{2\eta}$, and if it arrived there more than once the corresponding number of arrivals must have been greater by 1 than the number of departures. But then in the branch $L_1^{2\gamma} + L_2^{2\eta}$ the number of arrivals must have been greater by 2 than the number of departures, and this is impossible. Moreover, as we have already noted, in at least one of the positions $\text{fin}(B, L_1^{2\gamma})$ or $\text{fin}(B, L_2^{2\eta})$ the piece $\chi(\Omega)$ (perhaps a null piece) occupies the square $I(\Omega)$. Then two cases may exist.

(1) The move Ω is admissible at both of the positions $\text{fin}(B, L_1^{2\gamma})$ and $\text{fin}(B, L_2)$. This case is of no interest.

(2) The move Ω is admissible at one and only one of the positions $\text{fin}(B, L_1^{2\gamma})$ and $\text{fin}(B, L_2^{2\eta})$. If it is inadmissible at B, the case is of no interest. But it cannot be admissible at B, else the piece $\chi(\Omega)$ would there occupy the square $I(\Omega)$ and the trajectory $W(\Omega)$ would be either empty or free. Suppose that Ω is inadmissible at $\text{fin}(B, L_2^{2\eta})$. Then our King is not taken (it would be taken also at $\text{fin}(B, L_1^{2\gamma} + L_2^{2\eta})$). Thus either the piece $\chi(\Omega)$ is not on the square $I(\Omega)$ or the trajectory $W(\Omega)$ is blocked. In the latter case there was a piece on some square $p \in W(\Omega)$; we shall count the number of arrivals and departures of this piece with respect to the squares $I(\Omega)$ and p in the branches $L_1^{2\gamma}$, $L_2^{2\eta}$, and $L_1^{2\gamma} + L_2^{2\eta}$.

Since the move Ω is admissible at $\text{fin}(B, L_1^{2\gamma})$ the number of exits of the piece $\chi(\Omega)$ at the square $I(\Omega)$ equals the number of its returns (the exit of a null piece $\chi(\Omega) = \varnothing$ will be regarded as an occupation and its arrival as a freeing of the square). In the same way, the number of arrivals of any piece at the square $p \in W(\Omega)$ must equal the number of its departures. At the position $\text{fin}(B, L_1^{2\gamma})$ the move Ω is inadmissible; this means that the piece $\chi(\Omega)$ has made an uncompensated exit from the square $I(\Omega)$ or that some piece has made an uncompensated entrance at the square p. The number of arrivals at a square and exits from it in the branch $L_1^{2\gamma} + L_2^{2\eta}$ equals the sum of the corresponding numbers of arrivals and exits in the constituent branches $L_1^{2\gamma}$ and $L_2^{2\eta}$. Thus at the position $\text{fin}(B, L_1^{2\gamma} + L_2^{2\eta})$ the piece $\chi(\Omega)$ is not on the square $I(\Omega)$, or there is a piece on the square p: the move Ω is inadmissible. We treat in the same way the case in which Ω is admissible only at the position $\text{fin}(B, L_2^{2\eta})$. So, the conditions for applying the second axiom on Situation 2 are not satisfied.

(3) The move Ω is inadmissible at $\operatorname{fin}(B, L_1^{2\gamma})$ because the piece $\chi(\Omega)$ is not on the square $I(\Omega)$, and is inadmissible at $\operatorname{fin}(B, L_2^{2\eta})$ because the trajectory $W(\Omega)$ is blocked. It cannot have been free at B, else it would also be free at $\operatorname{fin}(B, L_2^{2\eta})$ or blocked at $\operatorname{fin}(B, L_1^{2\gamma})$ {a piece on this trajectory in the branch $L_2^{2\eta}$ cannot leave it in the branch $L_1^{2\gamma}$}. Thus in the position $\operatorname{fin}(B, L_1^{2\gamma})$ the square $I(\Omega)$ is attacked by the piece $\varphi(\Omega)$ as a result of the move $\Psi \in L_1$ ($h \le 2\gamma$), and $I(\Omega) \in N(s, L_1)$. On the other hand, the piece $\chi(\Omega)$ enters the square $I(\Omega)$ in the branch $L_2^{2\eta}$ (If $\chi(\Omega) = \varnothing$, the square $I(\Omega)$ becomes free in the branch $L_2^{2\eta}$.) Therefore

$$N(s, L_1) \cap (E(sL_2) \cup E(c, L_2) \cup I(c, L_2)) \ne \varnothing.$$

(4) The trajectory $W(\Omega)$ is freed of friendly pieces by moves in the branches L_1 and L_2. Among the squares thus freed, let the square that is nearest to $E(\Omega)$ be freed by a move in L_1, and among the squares freed by a move in L_2 let p be that one nearest to $E(\Omega)$. Then in the position B the square p is not under attack by the piece $\varphi(\Omega)$, but is under attack by it in the position $\operatorname{fin}(B, L_1^{2\gamma})$. Therefore $p \in N(s, L)$. At the same time,

$$p \in E(s, L_2) \cup E(c, L_2)) \ne \varnothing.$$

(5) The move Ω is the initial jump of a Pawn from the second rank to the fourth, or from the seventh to the fifth, while one of the squares blocking it is freed by a move in $L_1^{2\gamma}$ and another by a move in $L_2^{2\eta}$. Let hor_ν be a board on which the squares in the ν-th rank are marked, let $T^\uparrow$ represent a displacement of the squares on the board T by one rank upward, and $T^\downarrow$ by one rank downward. Let $\Pi(s, B)$ be a board on which our own Pawns stand in the position B. If White has the move at B

$$G(B, L_1, L_2) :=$$

$$= \Pi(s, B)^\uparrow \cap \operatorname{hor}_3 \cap \Big(\big((E(s, L_1) \cup E(c, L_1))^\downarrow$$

$$\cap (E(s, L_2 \cup E(c, L_2))) \big)$$

$$\cup \big((E(s, L_1) \cup E(c, L_1)) \cap (E(s, L_2) \cup E(c, L_2))^\downarrow \big) \Big)$$

$$\ne \varnothing;$$

If Black has the move,

$$G(B, L_1, L_2) :=$$

$$= \Pi(s, B)^\downarrow \cap \operatorname{hor}_6 \cap \Big(\big((E(s, L_1) \cup E(c, L_1))^\uparrow$$

$$\cap (E(s, L_2) \cup E(c, L_2))) \big)$$

$$\cup \big((E(s, L_1) \cup E(c, L_1)) \cap (E(s, L_2) \cup E(c, L_2))^\uparrow \big) \Big)$$

$$\ne \varnothing.$$

For some models this case need not be taken into account.

The influence predicate $\mathrm{Inf}((B, L_1, L_2)$ takes on the value *True* when one of the above conditions, or its symmetric counterpart, is satisfied. Thus

$$\mathrm{Inf}(B, L_1, L_2) :=$$

$$= \big[(E(c, L_1) \cap E(c, L_2)) \cup ((I(c, L_1) \cup W(c, L_1))$$
$$\cap (I(s, L_2) \cup I(c, L_2))) \cup ((I(s, L_1) \cup I(c, L_1))$$
$$\cap (I(c, L_2) \cup W(c, L_2)))$$
$$\cup ((I(c, L_1) \cup N(s, L_1) \cup n_p(s, L_1)) \cap (E(s, L_2) \cup E(c, L_2)))$$
$$\cup ((E(s, L_1) \cup E(c, L_1)) \cap (I(c, L_2) \cup N(s, L_2) \cup n_p(s, L_2)))$$
$$\cup ((N(s, L_1) \cup n_b(s, L_1)) \cap I(c, L_2)) \cup (I(c, L_1)$$
$$\cap (N(s, L_2) \cup n_b(s, L_2))) \cup G(B, L_1, L_2)\big].$$

This formula is more convenient for calculation, since during the traversal of branches or subtrees we need to form a limited number of boards, and in testing the influence of branches on subtrees or of subtrees on subtrees we need to form a strongly limited number of unions and intersections. However, to study the properties of the influence predicate it is better to use a different formulation of it, which we now proceed to define.

Every board on which it depends is the union of two well-defined boards corresponding to the two-move branches $\lambda(A; \Psi', \Psi'')$ belonging to the branches L under consideration, with origin B, and the single-move branches $\lambda'(A; \Psi)$ with origins at the ends of branches of odd length (the latter may be formally considered to be two-move branches, with a null second move and empty boards corresponding to it). All these short branches have the color col B. Since the operations of union and intersection are distributive, the formula given above for the influence predicate can be written in the form

$$\mathrm{Inf}(B, L_1, L_2) = \left[\bigcup_{\lambda_1 \in L_1} \bigcup_{\lambda_2 \in L_2} \bigcup_{\zeta \in Z} \left(T_{\sigma_1(\zeta)}(\lambda_1) \cap T_{\sigma_2(\zeta)}(\lambda_2) \right) \right]$$

where Z is the completely determined set of intersections of (a) the boards $E(\Psi)$, $I(\Psi)$, $W(\Psi)$, $N(\lambda, \mu)$, $n_p(\lambda_1, \mu)$ and $n_b(\lambda_1, \mu)$ corresponding either to the branch $\lambda_1 \in L_1$ or to its moves Ψ (we denote these boards by the symbol $T_{\sigma_1(\zeta)}(\lambda_1)$) and (b) the analogous boards $T_{\sigma_2(\zeta)}(\lambda_2)$ corresponding to the branch $\lambda_2 \in L_2$.

Let us dwell on the question of how we prove theorems on the transfer of scores of positions and the scores of groups of moves for the influence predicate we have just constructed. We do not need the axioms on the relationship between influence on moves and on a branch, nor the first axiom on the composition of branches, since we do not need to consider the influence of branches on moves. The axiom of symmetry is satisfied since the influence predicate $\text{Inf}(B, L_1, L_2)$ is symmetric with respect to the boards corresponding to the branches L_1 and L_2. We use the second axiom on the composition of branches when the branch $\Lambda_1 + \Lambda_2$ is composed of two-move branches λ of color col B belonging entirely either to Λ_1 or to Λ_2 (with perhaps single-move branches λ' of color col B, lying at the end of one of the branches Λ_1, Λ_2 and at the end of the branch $\Lambda_1 + \Lambda_2$).

The influence predicate $\text{Inf}(B, \Lambda_1 + \Lambda_2, \Lambda_3)$ can be written in the form

$$\text{Inf}(B, \Lambda_1 + \Lambda_2, \Lambda_3)$$

$$= \left[\bigcup_{\lambda_1 \in \Lambda_1 + \Lambda_2} \bigcup_{\lambda_2 \in \Lambda_2} \bigcup_{\varsigma \in Z} \left(T_{\sigma_1(\varsigma)}(\lambda_1) \cap T_{\sigma_2(\varsigma)}(\lambda_2) \right) \right]$$

$$= \left[\bigcup_{\lambda_1 \in \Lambda_1} \bigcup_{\lambda_2 \in \Lambda_2} \bigcup_{\varsigma \in Z} \left(T_{\sigma_1(\varsigma)}(\lambda_1) \cap T_{\sigma_2(\varsigma)}(\lambda_2) \right) \right]$$

$$\cup \left[\bigcup_{\lambda_1 \in \Lambda_2} \bigcup_{\lambda_2 \in \Lambda_3} \bigcup_{\varsigma \in Z} \left(T_{\sigma_1(\varsigma)}(\lambda_1) \cap T_{\sigma_2(\varsigma)}(\lambda_2) \right) \right]$$

$$= \widetilde{\text{Inf}}(B, \Lambda_1, \Lambda_3) \vee \widetilde{\text{Inf}}(B, \Lambda_2, \Lambda_3),$$

where the predicates $\widetilde{\text{Inf}}(B, \Lambda_1, \Lambda_3)$ and $\widetilde{\text{Inf}}(B, \Lambda_2, \Lambda_3)$ differ from the predicates $\text{Inf}(B, \Lambda_1, \Lambda_3)$ and $\text{Inf}(B, \Lambda_2, \Lambda_3)$ only in the fact that the boards $T_\sigma(\lambda)$ corresponding to the two-move branches λ and belonging to the branches Λ_1 and Λ_2 are intersected in $\widetilde{\text{Inf}}$ by the boards $T_\sigma(\lambda')$ corresponding to the branches λ' composed of the same moves but belonging to the composite branch $\Lambda_1 + \Lambda_2$. But, all of the boards $T_\sigma(\lambda)$, except $Nf(\lambda, \mu)$ for the long-range pieces (Queens, Rooks, Bishops), are defined by the moves Ψ' and Ψ'' in the branch λ and do not depend on its origin A. This means that for these boards $T_\sigma(\lambda) = T_\sigma(\lambda')$.

Let $\text{Inf}(B, \Lambda_1 + \Lambda_2, \Lambda_3) = 1$, $\text{Inf}(B, \Lambda_1, \Lambda_3) = 0$, and $\text{Inf}(B, \Lambda_2, \Lambda_3) = 0$ (as always we write 1 for *True* and 0 for *false*), i.e. $\Lambda_1 \nrightarrow \Lambda_3$, and $\Lambda_2 \nrightarrow \Lambda_3$. Then at least one of the predicates $\widetilde{\text{Inf}}(B, \Lambda_1, \Lambda_3)$ and $\widetilde{\text{Inf}}(B, \Lambda_2, \Lambda_3)$ is true, say the first, and one of the intersections of the boards $T_{\sigma_1}(\lambda_1') \cap T_{\sigma_2}(\lambda_2)$ that define it is not empty. The corresponding intersection $T_{\sigma_1}(\lambda_1) \cap T_{\sigma_2}(\lambda_2)$, which enters the expression defining the influence predicate $\text{Inf}(B, \Lambda_1, \Lambda_3)$, is empty. Therefore $T_{\sigma_1}(\lambda') = N(\lambda_1', \mu)$ for some long range piece μ of color col B, and $T_{\sigma_2}(\lambda_2)$ is the board $E(\Theta)$ for some move $\Theta \in \Lambda_3$ or the board $I(\Theta)$ for a move $\Theta \in \Lambda_3$ of color opposite to col B. In either case the board marks only a single square $p \in N(\lambda_1', \mu)$, $p \notin N(\lambda_1, \mu)$.

Since $\lambda_1' \in \Lambda_1 + \Lambda_2$, the branch $\Lambda_1 + \Lambda_2$ is a strict composition of the branches Λ', λ_1', and Λ'', i. e.

$$\Lambda = \Lambda' * \lambda_1' * \Lambda''.$$

The branches Λ' and Λ'' consist of moves in the branches Λ_1 and Λ_2; the Λ_1 moves precede the Λ_2 moves in the corresponding branches Λ' and Λ'', while the moves belonging to λ'_1 belong to the branch Λ_1 and occur in it between moves belonging to Λ' and Λ''. Thus

$$\Lambda' = \Lambda_1' + \Lambda_2',$$

$$\Lambda' * \lambda_1' = \Lambda_1' * \lambda_1 + \Lambda_2',$$

where Λ_1' and $\Lambda_1' * \lambda_1$ are initial portions of the branch Λ_1, and Λ_2' is an initial portion of Λ_2.

We consider first the case in which the first move Ψ' of the two-move branches $\lambda_1(A; \Psi', \Psi'') \in \Lambda_1$ and $\lambda'(A'; \Psi', \Psi'') \in \Lambda_1 + \Lambda_2$ is a move by the piece μ, i. e. $\varphi(\Psi') = \mu$. Then

$$p \in N(\lambda', \mu) = P\big(\mathrm{fin}(B, \Lambda' * \lambda_1'), \mu\big)$$

$$p \notin N(\lambda_1, \mu) = P\big(\mathrm{fin}(B, \Lambda_1' * \lambda_1), \mu\big)$$

and the piece μ occupies one and the same square (Ψ') in the positions $\mathrm{fin}(B, \Lambda' * \lambda_1')$ and $\mathrm{fin}(B, \Lambda_1' * \lambda_1)$. This means that in the second position the trajectory from the square (Ψ') to the square p is blocked by pieces. The blocking piece nearest to $I(\Psi')$ is on some square $p' \in P(\mathrm{fin}(B, \Lambda_1' * \lambda_1), \mu) = N(\lambda_1, \mu)$, which is freed by some move $\Theta_n \in \Lambda_1'$. Therefore $E(\Theta_n) = p'$ and the square p' is marked in the intersection $N(\lambda_1, \mu) \cap E(\Theta_n)$. This intersection is therefore not empty, so that $\Lambda_1 \sim \Lambda_2$.

Now let $\varphi(\Psi') \neq \mu$. Then the piece μ occupies one and the same square q in the positions $\mathrm{fin}(B, \Lambda')$ and $\mathrm{fin}(B, \Lambda' * \lambda_1')$, while

$$p \in N(\lambda_1', \mu) = P\big(\mathrm{fin}(B, \Lambda' * \lambda_1'), \mu\big) \setminus P\big(\mathrm{fin}(B, \Lambda'), \mu\big).$$

That is to say

$$p \in P\big(\mathrm{fin}(B, \Lambda' * \lambda_1'), \mu\big),$$

i.e. the position $\mathrm{fin}(B, \Lambda')$ contains at least one piece on the trajectory between the squares q and p. The nearest of these (possibly there is only one) left its square by the move Ψ' or Ψ''. The piece μ is on the square q either in one of the positions $\mathrm{fin}(B, \Lambda_1')$ or $\mathrm{fin}(B, \Lambda_2')$ or in both. Let us examine the two cases.

If the piece μ is on the square q in only one of these positions, it arrived there by one of the moves Θ_{2e} in the corresponding branch Λ_1' or Λ_2' (the proof of this assertion is based, as above, on a count of the number of the arrivals and departures of μ at q in the branch $\Lambda_1' + \Lambda_2'$). Since the piece does not move in the branch λ_1 ($\varphi(\Psi') \neq \mu$), it occupies the same square in the positions $\mathrm{fin}(B, \Lambda_1' * \lambda_1)$ and $\mathrm{fin}(B, \Lambda_1')$ (it cannot have been captured in the move Ψ'', else it would not appear in $\mathrm{fin}(B, \Lambda' * \lambda_1)$). Therefore, in that one of the positions $\mathrm{fin}(B, \Lambda_1' * \lambda_1)$, $\mathrm{fin}(B, \Lambda_2')$ in which it occupies the square q, all the squares attacked by it belong to one of the new-possibilities boards $N(\lambda, \mu)$, where the two-move branch λ belongs to $\Lambda_1' * \lambda_1$ or Λ_2', respectively. Since $\Lambda_1 \nsim \Lambda_3$ and $\Lambda_2 \nsim \Lambda_3$, the square p appears on none of these boards.

Therefore the square p is not attacked by the piece μ in that one of the positions $\mathrm{fin}(B, \Lambda_1' * \lambda_1)$, $\mathrm{fin}(B, \Lambda_2')$ where μ is on the square q. The trajectory between the squares p and q is blocked in that position. Let p' be the square nearest q occupied by a blocking piece. It is attacked by the piece μ and, as we showed earlier, belongs to one of the boards $N(\lambda, \mu)$, where the branch λ belongs to one of the branches $\Lambda_1' * \lambda_1$, Λ_2'. The square p' is freed by a move in the other branch (Λ_2' or $\Lambda_1' * \lambda_1$). Therefore $N(\lambda, \mu) \cap E(\Theta_n) \neq \varnothing$, and so $\Lambda_1 \frown \Lambda_2$.

Finally, let μ be on the square q in both the positions $\mathrm{fin}(B, \Lambda_1')$ and $\mathrm{fin}(B, \Lambda_2')$, and therefore in the position $\mathrm{fin}(B, \Lambda_1' * \lambda_1)$. At the move Ψ' or Ψ'' in the branch $\Lambda_1' * \lambda_1$ some piece leaves the trajectory between the squares q and p. But $p' \in N(\lambda_1, \mu)$, so that the trajectory remains blocked. The blocking pieces set it free by moves $\Theta_n \in \Lambda_2'$, since it is free in the position $\mathrm{fin}(B, \Lambda' * \lambda_1')$. That is to say, our trajectory is liberated by the 'combined forces' of moves in the branches $\Lambda_1' * \lambda_1$ and Λ_2'. During the construction of the influence predicate it was shown that in this case $N(s, \Lambda_1' * \lambda_1) \cap (E(s, \Lambda_2') \cup E(c, \Lambda_2')) \neq \varnothing$ or $N(s, \Lambda_2') \cap (E(s, \Lambda_1 * \lambda_1) \cup E(c, \Lambda_1' * \lambda_1)) \neq \varnothing$. Thus $\Lambda_1' * \lambda_1 \frown \Lambda_2$, whence it follows that in our current case we have $\Lambda_1 \frown \Lambda_2$.

An immediate consequence of the construction of the influence predicate $I(B, L_1, L_2)$ is that the axiom on Situation 1 and the first axiom on Situation 2 are satisfied. We showed in the construction that the conditions of the second axiom on Situation 2 never hold (hence the axiom is formally satisfied). Therefore in the proofs of the theorems on the transfer of scores we may omit the cases in which it is applied, and so dispense with the influence of branches on a move and the corresponding axioms. We may however introduce another concept of equivalent moves, which helps in reducing the number of different virtual moves. The different moves of a long range piece that all take place in a single rank, file, or diagonal and all lead to the same square may be taken as equivalent. Then the satisfaction of the second axiom on Situation 2 becomes possible, and we may then so define the predicate for the influence of branches on a move that the second axiom on Situation 2 and the axioms on the influence of branches on a move are satisfied.

The axiom on Situations 3 and 4 is needed only when (a) a branch Λ_2 whose end is a terminal position is of odd length and the position $\mathrm{fin}(B, \Lambda_2)$ is lost for col B, or (b) the end of the branch $\Lambda_1 + \Lambda_2$, of even length, is terminal and not lost. We may suppose that in our game such positions do not exist. In fact, a complete loss for either side occurs in a position where the losing side is to move after the opponent has captured the King on the preceding move (if a loss occurs in obtaining a mate, it can occur only in positions where one has the move). Since our game does not admit of a stalemate, a terminal position with a drawn score is a conditional concept. We may therefore suppose that our opponent will agree to a draw only in positions where the move belongs to the side that considers a draw a loss for itself. Then the terminal positions $\mathrm{fin}(B, L_1)$ and $\mathrm{fin}(B, L_1 + L_2)$ are either both lost or both not lost. Accordingly, we need not apply the axiom on Situation 5.

Thus in our game both the theorems proved earlier, on the transfer of scores, are valid. If, however, we are to consider all the rules of chess, we must include in our construction of an influence predicate boards connected with checks. Then some of the influence axioms will fail to be satisfied in some cases and we must supplement the theorems on the transfer of scores by an investigation of such special cases. We introduce without proof the construction of the influence predicate when checks are taken into account. We define the following supplementary boards:

(1) $C(\Psi)$ is a trajectory in which the move Ψ gives check (if it gives check at all), including the square occupied by the checking piece and the square occupied by the checked King;

(2) $C'_{w,b}(\Psi)$ is a trajectory in which a pin arises (if it arises at all) including the squares containing the pinning and pinned pieces and the King (White or Black) of the color of the pinning piece;

(3) $\overline{W}(\Psi) := W(\Psi) \cup C(\Psi) \cup C'_{\mathrm{col}\,\Psi}(\Psi)$;

(4) $N_c(\lambda)$ represents the new possibilities of giving check to the King of color opposed to $\mathrm{col}\,\lambda$, i.e. the squares from which King can be attached by an opposing piece not yet removed from the board after moves in the branch λ and not attacked before those moves

$$S(\lambda(A; \Psi', \Psi''), \mu) := P(A, \mu) \backslash (P(\mathrm{fin}(A, \lambda), \mu)$$

(5) $R(\Psi)$ are the squares occupied by opposing pieces around the opponent's King, if the move Ψ is the check;

(6) $Q(\Psi)$ are the squares around the King checked by the move Ψ that are attacked by pieces of color $\mathrm{col}\,\Psi$.

The new boards corresponding to moves and two-move branches belonging to the branch under investigation or to the subtree $\mathfrak{L}$ will also be integrated along with the pieces of the given color in positions belonging to $\mathfrak{L}$. We shall need the following boards:

$$\overline{W}(c, \mathfrak{L}) := \bigcup_{\Psi \in \mathfrak{L}, \mathrm{col}\,\Psi \ne \mathrm{col}\,B} \overline{W}(\Psi);$$

$$C(c, \mathfrak{L}) := \bigcup_{\Psi \in \mathfrak{L}, \mathrm{col}\,\Psi \ne \mathrm{col}\,B} C(\Psi);$$

$$N_c(s, \mathfrak{L}) := \bigcup_{\lambda \in \mathfrak{L}, \mathrm{col}\,\Psi = \mathrm{col}\,B} N_c(\lambda);$$

$$S(c, \mathfrak{L}) := \bigcup_{\lambda \in \mathfrak{L}, \mu \in \phi \cup \Pi, \mathrm{col}\,\lambda = \mathrm{col}\,\mu \ne \mathrm{col}\,B} S(\lambda, \mu);$$

$$R(s, \mathfrak{L}) := \bigcup_{\Psi \in \mathfrak{L}, \mathrm{col}\,\Psi \ne \mathrm{col}\,B} R(\Psi);$$

$$Q(s, \mathfrak{L}) := \bigcup_{\Psi \in \mathfrak{L}, \mathrm{col}\,\Psi \ne \mathrm{col}\,B} Q(\Psi);$$

$$\overline{N}(s, \mathfrak{L}) := N(s, \mathfrak{L}) \cup n_p(s, \mathfrak{L}) \cup n_b(s, \mathfrak{L}).$$

The influence predicate is given by the formula

$\mathrm{Inf}(B, L_1, L_2) :=$

$$= (((E(s, L_1) \cup E(c, L_1) \cup I(c, L_1)) \cap (E(c, L_2) \cup I(s, L_2)$$
$$\cup I(c, L_2) \cup \overline{W}(c, L_2) \cup \overline{N}(s, L_2))) \cup (I(s, L_1) \cap \overline{W}(c, L_2))$$
$$\cup ((E(s, L_1) \cup E(c, L_1)) \cap (N_c(s, L_2) \cup R(s, L_2))) \cup (N_c(s, L_1)$$
$$\cap (E(s, L_2) \cup E(c, L_2) \cup \overline{N}(s, L_2)))$$
$$\cup ((I(s, L_1) \cup I(c, L_1) \cup R(s, L_1)) \cap (E(s, L_2) \cup E(c, L_2)))$$
$$\cup (\overline{W}(c, L_1) \cap (E(s, L_2) \cup E(c, L_2) \cup I(s, L_2) \cup I(c, L_2)))$$
$$\cup (\overline{N}(s, L_1) \cap (E(s, L_2) \cup E(c, L_2)$$
$$\cup I(c, L_2) \cup C(c, L_2) \cup N_c(c, L_2))) \cup (C(c, L_1) \cap \overline{N}(s, L_2))$$
$$\cup (S(c, L_1) \cap Q(s, L_2)) \cup (Q(s, L_2)$$
$$\cap S(c, L_1)) \cup G(B, L_1, L_2).$$

Let us now pass to the construction of the influence predicate for the game of noughts-and-crosses. We note first that situations of type 2 cannot occur in this game. For, let Λ_1 and Λ_2 be branches of even length, and let $\Lambda = \Lambda_1 + \Lambda_2$. In the position $\mathrm{fin}(B, \Lambda_1 + \Lambda_2)$ the side corresponding to col B can place its mark (say a nought) on any free square and only on such a square. But the free squares in this position are also free in the positions $\mathrm{fin}(B, \Lambda_1)$ and $\mathrm{fin}(B, \Lambda_2)$. At the end of the composite branch $\Lambda_1 + \Lambda_2$ only those moves are admissible that were admissible at the ends of the constituent branches Λ_1 and Λ_2. Thus there remains for consideration Situation 1.

Let L_1 be a branch with odd length, L_2 a branch with even length. Suppose that in the position $\mathrm{fin}(B, L_1)$ a cross may be marked in the square p, but cannot be marked there in the position $\mathrm{fin}(B, L_1 + L_2)$. This may happen for one of the following reasons:

(1) The square p contains a nought, which must have been placed there in a move $\Theta \in L_2$.
(2) The square p, in which the cross is to be marked, already contains a cross, which must have been put there in some move $\Theta_{2\varepsilon} = \Psi_p \in L_2$. In this case we do not suppose that $L_1 \sim L_2$. In fact, in constructing a pseudocritical branch we make the move Ω_q to a free square q from the position $\mathrm{fin}(B, L_1 + L_2)$ with move belonging to the opponent of col B. By the inductive assertion this position is not lost; the beginning of the induction is the assertion that the position $C = \mathrm{fin}(B, L_2)$ is not lost. The position $\mathrm{fin}(B, (L_1 + L_2) * \Omega_q)$ is not lost since no move from $\mathrm{fin}(B, \Lambda_1 + \Lambda_2)$ wins. This position coincides with the position $\mathrm{fin}(B, L_1 * \Xi_q + L_2')$, where L_2' is the branch in which the move $\Theta_{2\varepsilon} = \Psi_p$ is replaced by the move $\Theta_{2\varepsilon'} = \Omega_q$ (see Figure 33). Such a replacement is permissible since the square q, which is free in the position $\mathrm{fin}(B, L_1 + L_2)$ is free in all the preceding positions in the branch $L_1 + L_2$.

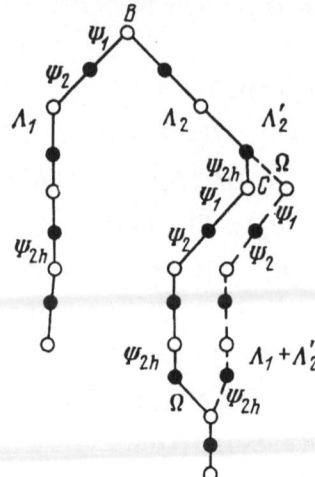

Figure 33

(3) No move can be made from the position $\mathrm{fin}(B, L_1 + L_2)$ since it is a terminal position, that is, the side of color col B completed a quintuple in the preceding move. The quintuple consists of three sets of noughts: those that were in place in the position B and those that were placed by moves in the branches L_1 and L_2, these latter sets being non-empty. The branch L_2 can be distinguished from the initial branch $L_2^{(0)}$ with which we began the construction of the pseudocritical branch, but we do not change the moves of color col B. That is, the quintuple consists of noughts from the branches L_1 and $L_2^{(0)}$ and from the position B.

We arrive at the following definition of the influence predicate $\mathrm{Inf}(B, L_1, L_2)$): Let $T^0(B)$ be a board on which the noughts are in the same places as in B; let $T^0(L_1)$ and $T^0(L_2)$ be boards for the noughts placed by moves in the branches L_1 and L_2: and let $T^x(L_1)$ be the board for crosses placed by moves in the branch L_1. Then

$$\mathrm{Inf}(B, L_1, L_2) := \left[T^x(L_1) \cap T^0(L_2) \right]$$
$$\cup \mathrm{Pent}\left(T^0(B) \cup T^0(L_1) \cup T^0(L_2) \right);$$

where $\mathrm{Pent}(T)$ is a predicate with the value 1 if the board T contains a quintuple of squares (in a horizontal, vertical, or diagonal line), and with the value 0 otherwise. The influence predicate so defined yields a natural definition of the influence of a branch on a branch or subtree.

We have not symmetrized this influence predicate, and have not added the requirement that the second axiom on the composition of branches should be satisfied. (The first axiom is not needed since we have avoided the influence of a branch on a move. For an arbitrary branch L of even length, only those moves are admissible from the position $\mathrm{fin}(B, L)$ that were admissible from the position B.) We can therefore manage without (a) the theorem on the transfer of scores of groups of moves and (b) the resolution of a pseudocritical path into its components. The axioms in question are not used in the proof of the theorem on the transfer of a position score (nor in the similar proof of the theorem on the transfer of the score of a move).

Influence-Based Algorithms

All of the algorithms described below have been implemented in the chess program KAISSA and in this section we shall therefore discuss only the game of chess.

The first algorithm prunes subtrees in the forced game. We may think of the forced game as an auxiliary search from a terminal position in some model game with the aim of producing a dynamic score for the position since a static score may be unreliable.

For instance, a position F in which the Queen of color opposite col F is under attack, cannot be evaluated without taking into account the consequences of the capture. Therefore, to obtain a reliable score we make an auxiliary search in which we recognize only captures, checks, and replies to a check. In the preceding chapter we called such a model a forced game.

Forced variations are searched in all terminal positions and almost always in nearby positions they are identical. The theorem on the transfer of position scores holds for a forced game and for the constructively defined influence relation developed in the preceding section (in essence, this was proved in Section 2 of this chapter, where situations connected with checks were not considered). Suppose that we have produced a forced game from the position B and it turns out that it yields a loss. Then from the position $C = \text{fin}(B, L)$ with col B having the move, the forced game also loses if the following conditions are satisfied:

in the forced game the branch L does not influence the minimally pruned subtree of color opposite to col B from the position B;
no new forcing moves can be made from the position $C = \text{fin}(B, L)$.

Thus in many positions we need not run a forced variation. The effectiveness of the algorithm using the result of the theorem on the transfer of position scores depends in large measure on a successful choice of the position B. In a non-terminal position B a trial forced variation is run with the sole aim of applying the results of the search to the terminal positions in the B-subtree.

In choosing B we encounter conflicting requirements. On the one hand, it should be near the root of the tree so that its subtree will contain many terminal positions. On the other hand, we need a small value for the probability that the branches stemming from B and leading to the terminal position C will influence the trial forced game. Therefore the position B should be near the final positions, i. e. should be rather far from the root of the tree. Moreover, if it turns out that the trial forced game from B does not lose, the time spent in running it will have been wasted.

In the implemented algorithm for a search to a fixed depth n the trial forced games are run from positions of depth $n-2$ for which the static score is unsatisfactory. This criterion increases the probability of a loss for the forced variations from terminal positions of depth n.

The depth $n-2$ is maximal for the test models $\mathfrak{S}_T(B)$. The branches L from B to C, where their results are applied, have length $d(L)=2$. However, an increase in the number of trial model games to be investigated is often compensated by the possibility of pruning branches in the base model. If the position B is won for its own color $\mathrm{col}(B)$ in the game $\mathfrak{S}_T(B)$, it is won in the base model. Then we may step back from B without investigating the quiet moves in the base-model B-subtree.

In fact, if quiet moves are admissible in the base model at a position D which arises after a winning move $\Psi=(B,D)$ in the forced game, then a blank move $m_\varnothing$ from D is admissible in the game $\mathfrak{S}_T(B)$. Since it loses, the material balance at D favors $\mathrm{col}\,B$. After a quiet move $\Theta=(D,F)$ from the position D the position F has the same material balance and is terminal in the base model. At F we consider only moves in the forced game, while the blank move $m_\varnothing$ is admissible (else $\Theta=(D,F)$ would not be a quiet move) and it wins. Moves from D in the forced game were investigated in the trial model game $\Sigma_T(B)$; they also lead to positions in the base model from which only moves in the forced game are considered. These have been investigated in the trial model $\Sigma_T(B)$, and some of them win. Therefore B is a won position for $\mathrm{col}\,B$ in the base model.

The fundamental defect in this algorithm is its instability: if a new forced move appears at C, the forced variation must be repeated. The algorithm does not recognize situations in which the new moves bear no relation to the trial forced game.

On close examination the forced games turn out not to be identical, but instead similar. More exactly, there exists a set of fragments—subtrees of the base tree such that in almost every position of the forced game we find a union of a subset of these fragments. However this subset may be unique for each position.

It is natural to examine influence with respect to fragments. There are two different algorithms using such methods for shortening the search. The second is based on theory; we shall turn to it later.

We begin with an algorithm equipped with a table of poor moves. At every point in the search the table of poor moves contains fragments assigned to the various positions in the current branch. In fact the table contains the first move in the fragment, the opponent's best reply, the amount of material loss sustained in the fragment, and the boards belonging to the fragment.

We define the material loss for a poor move as the difference between the material score at the position A as seen by the side with color $\mathrm{col}\,A$ and the material score at a position reached from A as a result of that move. Thus if the move Ψ loses a minor piece, its material loss is equal to the weight h_{mp}. If the move Ψ loses no more than a Pawn, the material loss is equal to h_p. Nevertheless the move Ψ may be poor if there is reason to try for greater gain. All the moves in the fragment, except possibly the first, are forced. The fact that the fragment is assigned to a position in the current branch implies that in this position the fragment would also run its course. It is not at all necessary that the fragment should in fact have been generated in this position. Moreover, as we shall see, it is not necessary that the fragment should have been generated in any position whatever. It may have been put together from smaller fragments that were generated in various positions. We know only that it is assembled from the subtree of the position to which it is assigned.

Fragments with blank first moves occupy a special place. They characterize winning forced variations; the need for them arises in attempts to put together a strong fragment from weak ones. They have, however, a meaning of their own. Assigned to a node of opposing color they point to the possibility of a forced win for the side having the move: it is for this reason that they are called threats.

Fragments that define threats or, what amounts to the same thing, fragments with a blank first move, will be called k^0-threats and the remaining fragments will be called l^0-poor moves. The color of an l^0-poor move is the color of the side that opens the fragment. The color of a k^0-threat is defined quite naturally as the color of the side making the blank move, e. g. a 'White k^0-threat' is a threat to White.

The formula developed in the preceding section, for the value of the influence predicate $I(B, L_1, L_2)$, can be used to determine the mutual influence of branches and subtrees stemming from different nodes. However, our observations on the usability of such influence predicates have only a heuristic character.

The position B to which a fragment is assigned is regarded as the root of the corresponding subtree of the game. In practice, however, the fragment may be formed during the traversal of a subtree having its root at another node. It is transferred to the position B along some branch on which it has no influence. Figure 34 depicts a portion of a game tree. Certain subtrees in this portion are distinguished (by the fact that they lead to a change in the material balance). They represent k^0-threats and l^0-poor moves assigned to the position D. After a traversal of the D-subtree of the game $\mathfrak{A}$ they are all assigned to the position D. After a backward step from D some of them in a transformed version will be assigned to the preceding position B.

The traversal algorithm consists of forward and backward steps; we first consider the elaboration of fragments during a backward step.

Some of the fragments assigned to D are assembled to form fragments that may be transferred to B. Other fragments are merely lifted from the table. Suppose the move $\Psi = (B, D)$ leads from B to D. We map it into

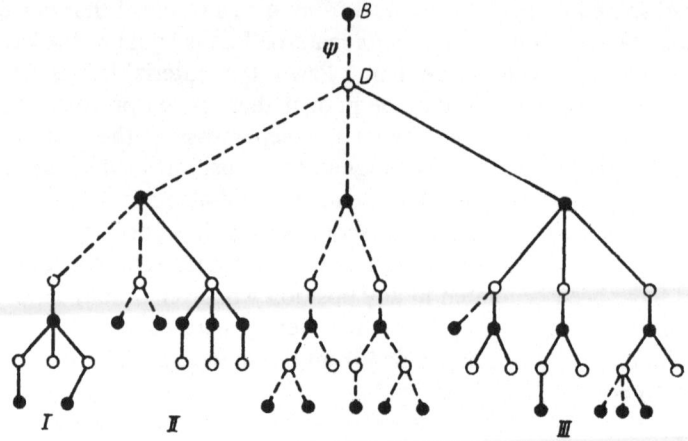

D-search-tree and selected fragments

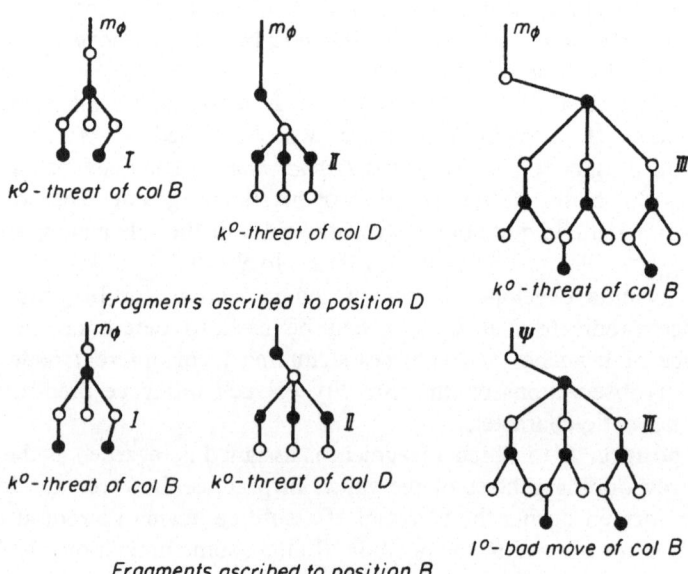

Figure 34

two (formally) two-move branches $\lambda_1(B; \Psi, m_\varnothing)$ and $\lambda_2(B; m_\varnothing, \Psi)$. The origin of the first is the position B; the origin of the second has the same configuration of pieces as B, but the move belongs to the opposite side.

We test each k^0-threat for its influence on the branches λ_1 and λ_2 of its own color: if col k^0 = col Ψ, the influence on λ_1, else on λ_2. If there is no influence we test again to see whether the threat was first obtained from the minimax portion of the search tree from D (the union of the minimal

White- and Black-pruned subtrees). We get a negative answer in two cases: $\mathrm{col}\,k^0 = \mathrm{col}\,\Psi$ and Ψ is a poor move, or $\mathrm{col}\,k^0 \neq \mathrm{col}\,\Psi$ and Ψ is closing. In both cases it must be remembered that the threat in question did not arise in the minimax portion of the game tree. We call it a marked threat; once marked, a threat remains marked until it is excluded from the table.

If a k^0-threat of color $\mathrm{col}\,k^0 = \mathrm{col}\,\Psi$ does not influence the branch λ_1, or if a k^0-threat of color $\mathrm{col}\,k^0 \neq \mathrm{col}\,\Psi$ does not influence the branch λ_2, it is assigned to the position B. If it does influence the corresponding branch, however, we proceed differently: k^0-threats of color $\mathrm{col}\,k^0 \neq \mathrm{col}\,\Psi$ and marked threats k^0 of color $\mathrm{col}\,k^0 = \mathrm{col}\,\Psi$ are excluded from the table, and instead of unmarked k^0-threats of color $\mathrm{col}\,\Psi$ we form l^0-poor moves and assign them to B. The first move $\Theta_0(l^0)$ of the new l^0-poor move, the opponent's best reply $\Theta_1(l^0)$, the material loss $\mathrm{Mat}(l^0)$ and the ensemble of boards $\{T_\gamma(l^0)\}$ are defined in the following way:

$$\Theta_0(l^0) := \Psi;$$

$$\Theta_1(l^0) := \Theta_1(k^0);$$

$$\mathrm{Mat}(l^0) := \mathrm{Mat}(k^0) - h(\Psi);$$

$$\{T_\gamma(l^0)\} := \{T_\gamma(k^0)\} \cup \{T_\gamma(\lambda_1)\}.$$

Here $h(\Psi)$ is the weight of the piece captured in the move Ψ.

Let us see how we process l^0-poor moves. At their outset we form an ensemble of empty boards of the so-called summator, in which we later form the logical union of the boards of some of the contemplated l^0-poor moves assigned to the position D. For each of the l^0-poor moves we test to see whether it influences the branch λ_1 or λ_2 of its own color; if it has no influence, it is assigned to B and is marked in the same way as a k^0-threat.

If an l^0-poor move of color $\mathrm{col}\,B$ influences the branch λ_1, it is excluded from the table. We also exclude l^0-poor moves of color $\mathrm{col}\,l^0 \neq \mathrm{col}\,\Psi$ that influence the branch λ_2 when Ψ is a poor move. The boards for the remaining l^0-poor moves that influence the branch λ_2 are combined, i. e. every set of such boards is in turn combined with the set of boards in the summator. In addition, we determine the minimum material loss for all such l^0-poor moves.

We now form the fragments assigned to the position B. If the move Ψ is poor and some k^0-threats of color $\mathrm{col}\,B$ appear on the branch λ_1, we form a new l^0-poor move:

$$\Theta_0(l^0) := \Psi;$$

$$\Theta_1(l^0) := m_\varnothing;$$

$$\mathrm{Mat}(l^0) := -h(\Psi);$$

$$\{T_\gamma(l^0)\} := \{T_\gamma(\lambda_1)\}.$$

We have no further information about the new l^0-poor move.

If the move Ψ is a refutation, we construct the corresponding k^0-threat, of color opposite to $\operatorname{col}\Psi = \operatorname{col}B$:

$$\Theta_0(k^0) := m_{\varnothing};$$

$$\Theta_1(k^0) := \Psi;$$

$$\operatorname{Mat}(k^0) := \min_{\operatorname{col}l^0 \neq \operatorname{col}\Psi,\, l^0 \sim \lambda_2} \operatorname{Mat}(\lambda_0) + h(\Psi);$$

$$\{T_\gamma(k^0)\} := \{T_\Sigma\}.$$

Here $\{T_\gamma(k^0)\}$ is the set of boards for the new k^0-threat, and $\{T_\Sigma\}$ is the set of boards in the summator. The l^0-poor moves that have influence are excluded from the table.

If Ψ is an improving move the rules that we have stated will correctly construct the k^0-threats and the l^0-poor moves. There may be more than one of the latter, since there may be no k^0-threat that influences the branch λ_1.

After the completion of a forward step the table so constructed is used to shorten the search from the position D whence the step was made. It is used, however, only when nothing but moves in the forced game are considered from the position D. For every k^0-threat of color $\operatorname{col}k^0 = \operatorname{col}D$ we ask whether:

it influences the branch leading to D from the position B to which it is assigned;
it satisfies the inequality

$$f_m(D) \underset{\operatorname{col}D}{\leqslant} \operatorname{Mat}(k^0) + m_m,$$

where $f_m(D)$ is the value of the material evaluation function for the position D and m_m is the material portion of the bound for the corresponding color ($m_m = \underline{\lim}_m$ if White has the move at D, else $m_m = \overline{\lim}_m$).

If there is a k^0-threat of color $\operatorname{col}B$ with sufficient material gain and no influence on the corresponding branch, and if at the same time there is no k^0-threat of the opposing color, the search is not prolonged beyond D; instead a backward step is made at once, on the grounds that the threat is automatically realized. Thus there is a winning move at the given position.

The table is also used to shorten the search by pruning various moves. After we have generated a move Ψ at D and the controlling algorithm proposes to make it, we test to see whether the table contains l^0-poor moves yielding information about the move Ψ. For every fragment with first move $\Theta_0(l^0) = \Psi$ we ask whether the material loss $\operatorname{Mat}(l^0)$ is sufficient to cause pruning, i. e. whether

$$f_m(D) \underset{\operatorname{col}D}{\leqslant} \operatorname{Mat}(l^0) + m_m.$$

If the material loss is great enough we ask whether the l^0-poor move influences the branch $L(B, \ldots, D)$ leading from the position B to which the move is assigned, to the position D under scrutiny. If we find at least one l^0-poor move with first move Ψ and with sufficient material loss, not influencing the corresponding branch, the investigation of the move Ψ is postponed. At the same time, a note is made of the l^0-poor moves that caused the postponement.

If the investigation of Ψ is not postponed, the fragment with first move Ψ is entered into the table and assigned to the position D. We do not construct any boards, but we note that the fragment is in service. Given a move Ψ at a position in the D-subtree, we first ask if such a fragment exists. If it does we do not test the influence of fragments with first move Ψ assigned to higher positions. Thus we waste no time on tests whose results we already know.

After completing the search in a given position for urgent moves, we test each postponed move to see whether the fragments giving rise to the postponement influence the minimax subtree of the given position. Moves for which all fragments influence that subtree are investigated; the remainder are again postponed. At the end of the search of the admitted moves we again test the postponed moves to see whether their fragments influence the new minimax subtree, and repeat the process as necessary.

This procedure comes to an end in one of two cases. Either we have investigated all the moves, or at some point we have found for each admitted move a fragment that caused its postponement and that does not influence the minimax subtree at the given position. In the latter case all uninvestigated moves are pruned.

Above, we described in principle an outline of an algorithm that has been implemented in a chess-playing program. Several important details were omitted, and for ease of understanding others were presented in a form somewhat different from that which was implemented in the program. For this reason the account that was given above must not be taken as a technical description of the algorithm, the more so because the algorithm is continually being revised and optimized. In particular, in the first version there was no notion of the k^0-threat, which was introduced as an aid in the construction of l^0-poor moves and in the course of time became one of the basic concepts in the algorithm.

The table of l^0-poor moves not only aids in shortening the search; it is also used to improve the order in which moves are investigated and to detect some meaningful circumstances related to the position under scrutiny.

The basic shortcoming in the algorithm considered above is the frequent occurrence of situations in which information is lacking.

A characteristic case occurs when a k^0-threat is not connected with the opponent's preceding move but is connected with the move of one's own that preceded the latter. Then the corresponding k^0-threat cannot be constructed by the method we have just described, since the execution of such a threat requires two successive moves by a single side and is complicated by the opponent's intervening move. On the other hand, an old threat cannot be carried to the top level.

As an example of such a situation we choose a Knight fork against Queen and Rook. Any reply results in the loss of Queen or Rook. These captures are k^0-threats and the opponent's move preceding the capture is not connected with it. What information can be carried back to the level of the position at which the forking move was made? Clearly no k^0-threat existed there, since the forking move is not forced and is not connected with the subsequent move by the opponent. On the other hand, in the pure chess sense, this move is a threat and we have all the information we need about it. Such a threat is simply not comprehended in the scheme of our algorithm.

Let us now bring in the notion of the k^1-threat. A move Ψ is a k^1-*threat* when it gives rise to a set of k^0-threats. A k^1-threat is defined constructively as a move immediately connected to at least one k^0-threat. Then any attack on a piece is a k^1-threat.

Nevertheless, a k^1-threat may be a losing move if it allows the opponent to make a k^0-threat of his own, connected with it. If so, the k^1-threat will be included in the table of l^0-poor moves, on general grounds.

The structure of the information concerning a k^1-threat is as follows: the header contains, in addition to all items conserved for a k^0-threat, the

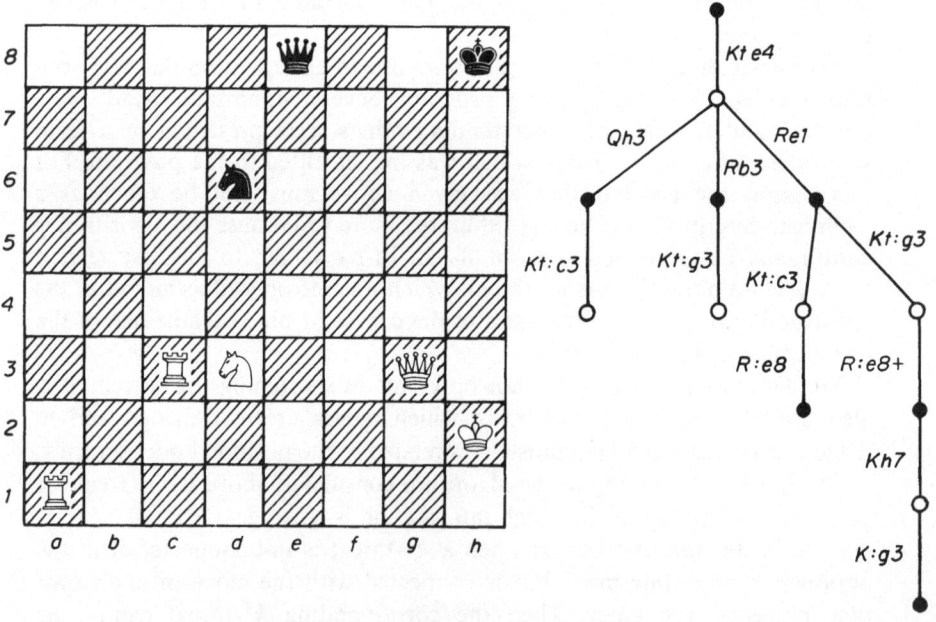

Figure 35

headers for two lists—the list of its own admissible k^0-threats, transferred from the base table of k^0-threats to a special table, and the list of refutation moves by the opponent that are his best replies to the k^1-threat.

The refutations need not be k^0-threats, since they are not forced. They must however influence the k^1-threat. The information on every refutation contains a list of l^0-poor moves of the same color as the k^1-threat and a list of k^0-threats of the opposite color.

Let us look at the information connected with a move that creates a fork. We shall suppose the Knight pinned by an opposing Rook (see Figure 35).

The move $Ne4$ is a k^1-threat because the search for k^0-threats that follows it brings up the moves $N:c3$ and $N:g3$. The information about these k^0-threats is contained in a special field, inaccessible to the basic algorithm, and retained in a list whose header is contained in the information element describing the move $Ne4$ as a k^1-threat.

The refutation list in this case consists of the single move $Re1$. The list of l^0-poor moves assigned to this refutation consists of the moves $N:c3$ and $N:g3$ (this list is also contained in the reference table). In this case the list of k^1-threats is empty.

Information about k^1-threats is used in the following way during the search: If a move Ψ contained in the table of k^1-threats is made, all the k^0-threats assigned to it are moved from the reference table to the k^0-threat table, where they will be treated from then on as a common basis. However, after the backward step corresponding to the move Ψ these k^0-threats will be removed from the table without examination.

If in the search a k^1-threat is immediately followed by one of the refutation moves assigned to it, the corresponding l^0-poor moves and k^0-threats belonging to the opponent are transferred to the general table of l^0-poor moves and k^0-threats.

Thus the program does not compel a new search and new construction of these fragments. Not only do we save the time that would have been spent in reconstructing some already constructed (but lost) elements of the table, but we also increase our ability to shorten the search. Thus, in the example of the Knight fork, if the algorithm did not use the information about k^1-threats it would construct the whole tree shown in Figure 35 after the move $Ne4$ and after a backward step corresponding to the move $Ne4$ would lose the information developed during the search of the tree just constructed.

Of course, during the search of the tree it is necessary to verify the correctness of the prior information on the k^1-threat. This is done on general grounds with the aid of the notion of influence. A separate examination is made for the influence of the k^1-threat itself, each of the refutation moves, and each l^0-poor move and each k^0-threat in the reference reserve. In this way the several elements in the list of l^0-poor moves and k^0-threats may be deleted individually. We can also delete refutations together with all of their lists. A k^1-threat itself is eliminated not only when the current move influences it but also when the list of k^0-threats has been emptied by deletions.

When several individual k^1-threats arise at the same level, their lists are edited and merged.

The structure of k^1-threats falls naturally far short of prescribing all the situations in which a precise treatment with k^0-threats is necessary. In our opinion, however, it does reflect certain concepts connected with the notion of 'tempo' or 'time' in several games. On a careful scrutiny of the algorithms we have obtained, it becomes clear that threats are templates for the searchless estimates of trees at level 2, as defined by Botvinnik. In a model of the active game, k^1-threats and l^1-*losing moves* can be used to reduce the tree of the forced game in the test of moves for activity and safety.

One might try to construct algorithms that would work with k^2-threats and others of higher order. It is very easy to define such algorithms formally, but in such games as chess they would require very great amounts of memory for the corresponding reserve tables, while for simpler games it appears that the k^1-threats suffice.

A second algorithm using fragments to shorten the search is based on the notion of a test-correlated variation (TCV). The TCV is constructed recursively; the first step is to construct it at the base position of the search. Here we list all the forced moves and for each move construct a fragment, whose boards initially consist of the board for that move alone. At any move in the search we allow only those of the original list that influence the branch leading from the base position to the move in question. We also list those moves that were originally excluded because the necessary influence was lacking. After inspection and consideration of a move in the search, its fragment is augmented by the boards of the minimax tree in our search. After considering all moves, we determine which ones need no further examination. These are the moves whose fragments influence moves that were excluded from our initial inspection on the grounds that their influence indicated the possibility of an essential change in the results via combination. Moves that win at some stage of the iteration may still be excluded from further inspection. As a result of the successive steps in the iteration the fragments of losing moves will expand, so that multiple inspections of a single move may be required. It can be shown that the number of such iterations does not exceed the number of forced moves at the position. Such a constraint, however, does not forbid the construction of more moves than would be constructed during a search without pruning. Of course, many positions in the tree of an iterative search, which are intentionally much fewer in number, are re-inspected many times.

Nevertheless, even the first application of the iterative search algorithm, in which the iterative search is merely replaced by a customary search of forced variations, will reduce the running time of the program by 10%.

The most recent development of this algorithm involved the creation of a table of fragments in the TCV which, like the k^0-threats and the l^0-poor moves, was carried from one position to another. The boards of fragments from the table were used instead of the boards from an initial move to form the boards of a fragment at the blank iteration.

If a branch leading to a given position from another position where the corresponding TCV was produced does not influence a fragment of the TCV, the latter is treated as the result of the first iteration rather than the result of the blank iteration, i.e. is re-inspected only if some other move influences its fragment.

This focussing of the algorithm led to a rapid convergence of the iteration, and reduced the number of positions in the tree by a factor of 4 or 5.

CHAPTER 4

Algorithms for Games and Probability Theory

Probabilistic Models for Two-Person Games

In Chapter 2 we considered a probabilistic model of a two-person game in which the probability of a correct decision as to the score of the base position is an increasing function of the depth of the search. The results we obtained had a qualitative flavor, inasmuch as their proofs depended on the hypothesis that the values of the evaluation function $f(A)$ for different positions A are independent. It is not clear that this hypothesis is valid; one of the postulates does not hold: The value of the evaluation function $f(B)$ at the position B, following the move $\Psi = (A, B)$ from the position A, depends on the value of $f(A)$. Nevertheless, the method we used (recursive determination of the probability that the model score is correct, as a function of the value of the evaluation function $f(D)$ for positions D of rank $r(D) = n$) can be applied under a much more general set of postulates.

We suppose that the probability characteristics of a position A in our given game $\mathfrak{A}$ are defined by a set of parameters

$$\chi(A) = \chi(t_1(A), t_2(A), \ldots, t_k(A)).$$

The parameters $t_1, t_2, \ldots, t_h$ $(h \leq k)$ depend on the score of the position A, the scores of the positions where branches leading to B originate, the scores of positions reached in branches originating at B, the values of the evaluation function $f(B)$ at the above listed positions, and the color of the move at A. The parameters $t_{h+1}, \ldots, t_k$ have a character independent of the color of the move. These include, for example, the expected number of moves from A, their precision as related to the divergence between the score at A and the true score, etc. The probability characteristics of positions with White or Black to move are symmetric in the sense that if we replace t_i by $1 - t_i$ in the set of parameters to which the color of the move pertains, the probability characteristics do not change:

$$\chi(1 - t_1, 1 - t_2, \ldots, 1 - t_h, t_{h+1}, \ldots, t_k) =$$
$$= \chi(t_1, t_2, \ldots, t_h, t_{h+1}, \ldots, t_k).$$

Among the parameters on which these characteristics depend we single out the turn to move—col A— and we write col $A = 0$ for a White position A, col $A = 1$ for a Black position. We also single out the true score sc(A) and the value of the evaluation function $f(A)$. The other parameters $t_4, \ldots, t_h$ that change their values in the symmetry formula just as col A does, form a vector q; the $t_{h+1}, \ldots, t_k$ form the vector r. The symmetric vector with coordinates $1 - t_4, \ldots, 1 - t_h$ will be denoted by $1 - q$. Using these symmetries we exclude the color of the turn to move from the explicit parameters and we assume from now on that all the characteristics we deal with relate to a White position. As usual we consider a game $\mathfrak{A}$ in which White and Black move alternately.

We assume that we have a Shannon model or other model whose definition depends on the value of the depth parameter n. In these models we denote the score of the base position A by msc(A). If A is a White position with sc(A) = x, $f(A) = y$, and the remaining probability characteristics having the values q and r, we introduce the notation $\mathbf{P}_-(x, y, z, q, r, n)$ and $\mathbf{P}_+(x, y, z, q, r, n)$ for the respective probabilities of the events $\mathrm{msc}_n(A) < z$ and $\mathrm{msc}_n(A) \leq z$ which may be written out as

$$\mathbf{P}_-(x, y, z, q, r, n) := P(\mathrm{msc}_n(A) < z | \mathrm{col}\, A$$
$$= 0, \mathrm{sc}(A) = x, f(A) = y, q(A) = q, r(A) = r);$$
$$\mathbf{P}_+(x, y, z, q, r, n) := P(\mathrm{msc}_n(A) \leq z | \mathrm{col}\, A$$
$$= 0, \mathrm{sc}(A) = x, f(A) = y, q(A) = q, r(A) = r);$$

The inequalities $\mathrm{msc}(A) < z (\leq z)$ hold if and only if the respective inequalities $\mathrm{msc}(B_i) < z (\leq z)$ hold for all positions B_i in the corresponding models of depth $n - 1$ that are reached by moves $\Psi_i = (A, B_i)$ from A ($i = 1, 2, \ldots, m$). If the inequalities $\mathrm{msc}_n(A) < z$ or $\mathrm{msc}_n(A) \leq z$ hold, so do many similar relationships. From now on we shall treat these all alike, using the symbols $\lessgtr$, for 'less than' or 'not exceeding'.

Since the B_i are Black positions their characteristics are determined by setting the corresponding parameters to the value $1 - z$ (we suppose that the minimum values of the scores and the evaluation functions are equal to zero and the maximum values to one).

We now postulate that for the given depth n the functions $\mathbf{P}_-(x, y, z, q, r, n)$ and $\mathbf{P}_+(x, y, z, q, r, n)$ satisfy the symmetry conditions

$$\mathbf{P}_-(x, y, z, q, r, n) = P(\mathrm{msc}_n(A) > 1 - z | \mathrm{col}\, A = 1,$$
$$\mathrm{sc}(A) = 1 - x, f(A) = 1 - y, q(A) = 1 - q, r(A) = r),$$
$$\mathbf{P}_+(x, y, z, q, r, n) = P(\mathrm{msc}_n(A) \geq 1 - z | \mathrm{col}\, A = 1,$$
$$\mathrm{sc}(A) = 1 - x, f(A) = 1 - y, q(A) = 1 - q, r(A) = r).$$

These conditions are satisfied for the Shannon model of depth $n = 0$, since in that case $\mathrm{msc}(A) = f(A)$. Then the probabilities that the model score $\mathrm{msc}_0(A)$ will have a given value are independent of all the parameters except $f(A)$ and we have

$$P\left(\mathrm{msc}_0(A) \underset{<}{\leq} z | f(A) = y\right) = \begin{cases} 1, & \text{if } y \underset{<}{\leq} z, \\ 0, & \text{if } y \underset{\geq}{>} z, \end{cases}$$

$$P\left(\mathrm{msc}_0(A) \underset{>}{\geq} 1 - z | f(A) = 1 - y\right) = \begin{cases} 1, & \text{if } 1 - y \underset{>}{\geq} 1 - z, \\ & \text{i.e. } y \underset{\leq}{<} z, \\ 0, & \text{if } 1 - y \underset{\leq}{<} 1 - z, \\ & \text{i.e. } y \underset{\geq}{>} z. \end{cases}$$

This means that the model scores $\mathrm{msc}_n(B_i)$ for the positions arising after the moves $\Psi_i = (A, B_i)$ from A $(i = 1, 2, \ldots, m)$ are mutually independent random variables depending only on the parameters $\mathrm{sc}(B_i) = \xi_i$, $f(B_i) = \eta_i$, $q(B_i) = \gamma_i$, $r(B_i) = \rho_i$. Then the probabilities of the compound events

$$P\left(\left\{\mathrm{msc}_n(B) \underset{>}{\geq} \zeta_i\right\}_1^m | \mathrm{col}\, A = 0, M(A) = \{\Psi_i = (A, B_i)\right.$$

$$\left. \mathrm{sc}(B_i) = \xi_i, f(B_i) = \mu_i, q(B_i) = \gamma_i, r(B_i) = \rho_i\}_1^m\right)$$

are equal to the product of the component probabilities

$$\prod_{i=1}^{m} P_{\pm}\left(1 - \xi_i, 1 - \eta_i, 1 - \zeta_i, 1 - \gamma_i, \rho_i, n\right)$$

(we assume that the B_i are Black positions, and we use the symmetry relationships).

With these assumptions and the easily proved relationships

$$P\left(\mathrm{msc}_n(B) \underset{<}{\leq} \zeta | \mathrm{col}\, B = 1, \mathrm{sc}(B) = \xi, f(B) = \eta, q(B) = \gamma,\right.$$

$$\left. r(B) = \rho, n\right) = 1 - P_{\pm}\left(1 - \xi, 1 - \eta, 1 - \zeta, 1 - \gamma, \rho, n\right),$$

we derive the recursive equation

$$P_{\pm}(x, y, z, q, r, n + 1)$$

$$= \pi_{\pm}(x, y, z, q, r) + \sum_{m=1}^{M} p(x, y, z, q, r, m)$$

$$\times \int_{\mathfrak{B}_m} \prod_{i=1}^{m} \left(1 - p_{\pm}\left(1 - \xi_i, 1 - \eta_i, 1 - z, 1 - \gamma_i, \rho_i, n\right)\right)$$

$$\times d\Sigma\left(x, y, q, r, m; \{\xi_i, \eta_i, \gamma_i, \rho_i\}_1^m\right)$$

Here the $\pi_{\pm}(x, y, z, \mathbf{q}, \mathbf{r})$ are the probabilities that a White position A will turn out to be terminal with score not exceeding z (less than z) if $\mathrm{sc}(A) = x$, $f(A) = y$, $\mathbf{q}(A) = \mathbf{q}$, $\mathbf{r}(A) = \mathbf{r}$:

$$
\pi_{\pm}(x, y, z, \mathbf{q}, \mathbf{r}) = \begin{cases} \pi(x, y, \mathbf{q}, \mathbf{r}), & \text{if } x \underset{<}{\leq} z, \\ \\ 0, & \text{if } x \underset{\geq}{>} z; \end{cases}
$$

$p(x, y, \mathbf{q}, \mathbf{r}, m)$ are the probabilities that m moves $\Psi_i = (A, B_i)$ can be made from a White position A if $\mathrm{sc}(A) = x$, $f(A) = y$, $\mathbf{q}(A) = \mathbf{q}$, $\mathbf{r}(A) = \mathbf{r}$; M is the maximum number of moves that can be made from positions in the given game $\mathfrak{A}$. $\mathfrak{P}_m$ is the $k \times m$-dimensional parameter space, the direct product of m k-dimensional parameter spaces defining the probability characteristics of a position:

$$
\mathfrak{P}_m = \overset{m}{\underset{i=1}{\bigotimes}} (X \otimes Y \otimes \mathbf{Q} \otimes \mathbf{R}),
$$

where X is the set of possible values of $\mathrm{sc}(A)$, Y is the set of values of $f(A)$, and $\mathbf{Q}$ and $\mathbf{R}$ are the sets of values of $\mathbf{q}$ and $\mathbf{r}$. We write $\Sigma(x, y, \mathbf{q}, \mathbf{r}, m; \Omega)$ for the completely additive set function $\Omega \subset \mathfrak{P}_m$ equal to the conditional probability that the set of parameters $\{\mathrm{sc}(B_i) = \xi_i, f(B_i) = \eta_i, \mathbf{q}(B_i) = \gamma_i, \mathbf{r}(B_i) = \rho_i\}_1^m$ defines a point in the space $\mathfrak{P}_m$ lying in the set Ω when $\mathrm{sc}(A) = x$, $f(A) = y$, $\mathbf{q}(A) = \mathbf{q}$, $\mathbf{r}(A) = \mathbf{r}$, and m is the number of moves $\Psi_i = (A, B_i)$ admissible under the rules of the game $\mathfrak{A}$.

Our recursion equation implies that the symmetry property is hereditary, in the sense that if the probabilities $P_{\pm}(x, y, z, \mathbf{q}, \mathbf{r}, n)$ and the characteristics $\pi(x, y, \mathbf{q}, \mathbf{r})$, $p(x, y, \mathbf{q}, \mathbf{r}, m)$ and $\Sigma(x, y, \mathbf{q}, \mathbf{r}, m; \Omega)$ satisfy the symmetry conditions, then so do the probabilities $P_{\pm}(x, y, z, \mathbf{q}, \mathbf{r}, n+1)$. We shall be considering some more or less simple probability models that provide concrete definitions of the probability characteristics and we discuss the relationship these models bear to real games and the evaluation functions that can be constructed for them. We shall suppose that the game $\mathfrak{A}$ is completely uniform (i.e. the number m of moves at a non-terminal position A is the same for all such A) and that the terminal positions have a very high rank $N \gg n$ (consequently $\pi(x, y, \mathbf{q}, \mathbf{r}) = 0$).

We shall normally assume that the scores in any given position have only two values, 1 and 0. When there are intermediate values, we can reduce the situation to this case. For instance, if we wish to know whether White can obtain the score z, we may replace the score $\mathrm{sc}(A)$ in the game $\mathfrak{A}$ by the threshold function

$$
\mathrm{sc}_z(A) := \begin{cases} 1, & \text{if } \mathrm{sc}(A) \geq z; \\ 0, & \text{if } \mathrm{sc}(A) < z. \end{cases}
$$

When the evaluation function $f(A)$ can take any value between 0 and 1, and the functions $P_+(x, y, z, \mathbf{q}, \mathbf{r}, n)$ and $P_-(x, y, z, \mathbf{q}, \mathbf{r}, n)$ coincide, we may use the notation $P(x, y, z, \mathbf{q}, \mathbf{r}, n)$:

$$P(x, y, z, q, r, n) = P(\mathrm{msc}_n(A) \le z | \mathrm{col}\, A = 0, \mathrm{sc}(A) = x,$$
$$f(A) = y, q(A) = q, r(A) = r)$$
$$= P(\mathrm{msc}_n(A) < z | \mathrm{col}\, A = 0, \mathrm{sc}(A) = x, f(A)$$
$$\le y, q(A) = q, r(A) = r).$$

Let us return to the model used in Chapter 2. The evaluation function $f(A)$ and the true score $\mathrm{sc}(A)$ take on the values 0 and 1, and the probability characteristics for the position A depend only on the turn to move, $\mathrm{col}\, A$, and the true score $\mathrm{sc}(A)$. If our given game $\mathfrak{A}$ is completely uniform, with m moves at each non-terminal position, and if at each non-terminal position won for the side having the move there are s winning moves, our general recursion equation reduces to the one considered in Chapter 2:

$$P_+(1, 0, n+1) = 1 - P_{n+1} = (1 - P_+(0, 0, n))^s (1 - P_+(1, 0, n))^{m-s}$$
$$= Q_n^s P_n^{m-s},$$
$$P_+(0, 0, n+1) = 1 - Q_{n+1} = (1 - P_+(1, 0, n))^m = P_n^m.$$

These equations yield the probability P_n that the model score $\mathrm{msc}_n(A)$ is correct for a position A which is won for its own color, and the probability Q_n of an error in the model score for a position A lost for its own color. We need to know only the probabilities for the values of the evaluation function $f(A)$:

$$P_0 = p = P(f(A) = 1 | \mathrm{col}\, A = 0, \mathrm{sc}(A) = 1),$$
$$Q_0 = q = P(f(A) = 1 | \mathrm{col}\, A = 0, \mathrm{sc}(A) = 0).$$

According to our fundamental assumptions these probability characteristics satisfy the symmetry conditions

$$p = P(f(A) = 0 | \mathrm{col}\, A = 1, \mathrm{sc}(A) = 0),$$
$$q = P(f(A) = 0 | \mathrm{col}\, A = 1, \mathrm{sc}(A) = 1).$$

In the space of possible values of P_n and Q_n ($0 \le P_n \le 1$, $0 \le Q_n \le 1$) there are three stationary points, the solutions of the system of equations

$$1 - P = Q^s P^{m-s},$$
$$1 - Q = P^m.$$

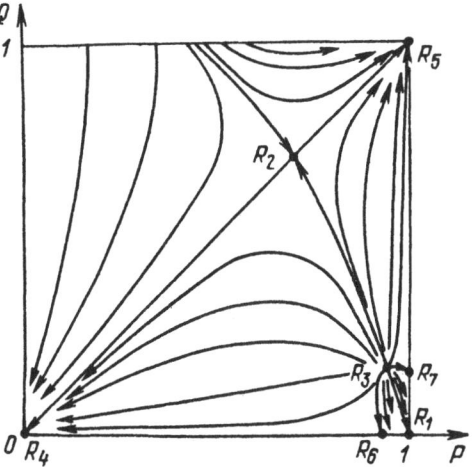

Figure 36

One of these, $R_1(1,0)$ corresponds to the case in which the value of the evaluation function $f(A)$ coincides with the true score $sc(A)$. Another, $R_2(P_2, P_2)$, lies on the diagonal $P = Q$. Its coordinates P_2 are the solutions of the equation $1 - P = P^m$, and lie in the interval $(0,1)$. They are approximately equal to $1 - \ln m / m$. The coordinates of the third stationary point $R_3(P_3, Q_3)$ are approximately equal to $P_3 \approx 1 - \left(\frac{1}{m}\right)^{\frac{s}{s-1}}$, $Q_3 \approx \left(\frac{1}{m}\right)^{\frac{1}{s-1}}$. There are two pairs of points which exchange places: $R_4(0,0)$, $R_5(1,1)$ and $R_6(P_6,0)$, $R_7(1,Q_7)$, where $P_6 \approx 1 - \left(\frac{1}{m}\right)^{\frac{s}{s-1}}$ and $Q_7 \approx \left(\frac{1}{m}\right)^{\frac{1}{s-1}}$.

A curve through the points R_1, R_3, R_2 divides the PQ unit square (see Figure 36) into a lower left and an upper right portion (in fact we are interested in the more restricted area $0 \leq P \leq 1$, $0 \leq Q \leq P$). The point $S_n(P_n, Q_n)$ in one portion goes into the point $S_{n+1}(P_{n+1} = 1 - Q_n^s P_n^{m-s}$, $Q_{n+1} = 1 - P_n^m)$ in the other. Therefore the arrows in Figure 36 show the direction in which the point $S_n(P_n, Q_n)$ moves as the depth n is increased by 2. The stationary point R_1 is stable. The nearby values of the probability characteristics p and q are satisfactory. For such values $P_n \to 1$ and $Q_n \to 0$ as the depth n increases. The stationary points R_4 and R_5 are also stable. The points $S_n(P_n, Q_n)$ approach them when the initial values $P_0 = p$, $Q_0 = q$ are not in the satisfactory zone. The remaining stationary points and pairs are unstable.

Our reason for going into such detail in this simple probability model is not that it is interesting in itself. Other models that better represent the real situation in respect of algorithms for games involve more parameters and are more complex. In some cases, however, the behaviors of the characteristics of positions A with different parameters y, q, and r tend to resemble each other as the depth n increases, and the characteristics begin to vary as they do in our current simple model, in which they are independent of $y(A)$, $q(A)$ and $r(A)$.

Now suppose that the evaluation function $f(A)$ can take on values between 0 and 1, but the relevant probabilities depend only on the turn to move col A and the true score sc(A):

$$p_\pm(z) := P\left(f(A) \lessgtr z | \text{col } A = 0, \text{sc}(A) = 1\right)$$

$$= P\left(f(A) \gtreqless 1 - z | \text{col } A = 1, \text{sc}(A) = 0\right);$$

$$q_\pm(z) := P\left(f(A) \lessgtr z | \text{col } A = 0, \text{sc}(A) = 0\right)$$

$$= P\left(f(A) \gtreqless 1 - z | \text{col } A = 1, \text{sc}(A) = 1\right).$$

Then in essence little has changed. If the game $\mathfrak{A}$ is completely uniform, with m moves at non-terminal positions, and if there are exactly s winning moves at each non-terminal position won for its own color, the general recursion equations have the form

$$P_\pm(1, z, n+1)$$

$$= \prod_{i=1}^{s}\left(1 - P_\mp(0, 1-z, n)\right) \prod_{i=s+1}^{m}\left(1 - P_\mp(1, 1-z, n)\right)$$

$$= \left(1 - P_\mp(0, 1-z, n)\right)^{s}\left(1 - P_\mp(1, 1-z, n)\right)^{m-s},$$

$$P_\mp(0, z, n+1)$$

$$= \prod_{i=1}^{m}\left(1 - P_\mp(1, 1-z, n)\right) = \left(1 - P_\mp(1, 1-z, n)\right)^{m}.$$

We introduce the notation:

$$P_n(z) := \begin{cases} 1 - P_\pm(1, z, n), & n \text{ even}; \\ 1 - P_\mp(1, 1-z, n), & n \text{ odd}; \end{cases}$$

$$Q_n(z) := \begin{cases} 1 - P_\pm(0, z, n), & n \text{ even}; \\ 1 - P_\mp(0, 1-z, n), & n \text{ odd}. \end{cases}$$

In this notation the recursion equations we have just written coincide with the recursion equations for our earlier model in which the evaluation

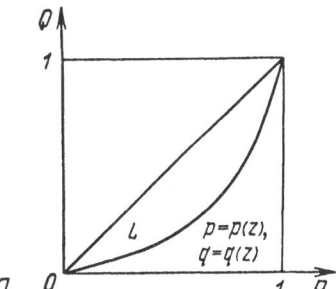

Figure 37

function had only the values 0 and 1:

$$1 - P_{n+1}(z) = Q_n^s(z) P_n^{m-s}(z),$$
$$1 - Q_{n+1}(z) = P_n^m(z),$$

with

$$P_0(z) = 1 - p_\pm(z),$$
$$Q_0(z) = 1 - q_\pm(z).$$

Suppose that the functions $p_\pm(z)$ and $q_\pm(z)$ are strictly monotone, continuous, and differentiable. Such functions can approximate arbitrary functions $p_\pm(z) = P(f(A) \lessgtr z \mid \text{col } A = 0,\ \text{sc}(A) = 1)$ and $q_\pm(z) = P(f(A) \lessgtr z \mid \text{col } A = 0,\ \text{sc}(A) = 0)$. Then $p_+(z) = p_-(z)$, $q_+(z) = q_-(z)$; in the interval $[0,1]$ these functions vary from 1 to 0. If the evaluation function $f(A)$ is relevant, the inequality $P_0(z) > Q_0(z)$ holds over the range $0 < z < 1$. The functions $P_0(z)$, $Q_0(z)$ define a segment of a curve, parameterized by z, lying in the unit square $0 \le P \le 1$, $0 \le Q \le 1$ of the PQ plane (see Figure 37). More exactly, the segment lies in the triangle $0 \le P \le 1$, $0 \le Q \le P$, joins the points $O(0,0)$ and $I(1,1)$, and is the graph of a monotone increasing continuous and differentiable function $Q = \varphi(P)$ (its derivative may equal $+\infty$ at some points).

The curves parameterized by the functions $P_n(z)$ and $Q_n(z)$ for $n = 1, 2, \dots$ have similar properties. In fact, if $P_{n-1}(z)$ and $Q_{n-1}(z)$ are continuous, differentiable, and monotone decreasing (increasing), varying from 1 to 0 (0 to 1) in the interval $0 \le z \le 1$, the functions $P_n(z) = 1 - Q_{n-1}^s(z) P_{n-1}^{m-s}(z)$ and $Q_n(z) = 1 - P_{n-1}^m(z)$ are also continuous and differentiable; they increase monotonely if $P_{n-1}(z)$ and $Q_{n-1}(z)$ decrease monotonely: they decrease monotonely if the latter functions increase monotonely. They vary from 0 to 1 or 1 to 0, as the case may be. Furthermore, for $0 < z < 1$

$$P_n(z) = 1 - Q_{n-1}^s(z) P_{n-1}^{m-s}(z)$$
$$> 1 - P_{n-1}^s(z) P_{n-1}^{m-s}(z) = 1 - P_{n-1}^m(z) = Q_n(z).$$

Now, for a model of depth n, and using Zermelo's formula for selecting a best move, we define the probability of selecting a winning move at a White position A whose score is 1, or at a Black position whose score is 0. Suppose that at the position A there are m moves $\Psi_i = (A, B_i)$ $(i = 1, 2, \ldots, m)$, of which s win and the rest lose. With probability $|Q'_{n-1}(z)|dz$ the position B_i arising after the move $\Psi_i = (A, B_i)$ will have the model score $\mathrm{msc}_n(B_i) = z$ (this position is lost for its own color col B_i). With probability $Q_{n-1}^{s-1}(z)P_{n-1}^{m-s}(z)$ the score z of the position B_i is the maximum of the scores of all the positions B_j $(j = 1, 2, \ldots, m)$. (In all the positions B_j the move belongs to the color opposite to col A, so that the scores msc_{n-1} (B_j) cannot be less favorable to the color col $B_j =$ col B_i than $\mathrm{msc}_{n-1}(B_i) = z$.) Thus the probability of choosing one of the s winning moves is equal to

$$\pi_n = s\int_0^1 Q_{n-1}^{s-1}(z)P_{n-1}^m(z)|Q'_{n-1}(z)|dz.$$

The functions $P_{n-1}(z)$ and $Q_{n-1}(z)$ define the curve $L_{n-1}\{P_{n-1}(z), Q_{n-1}(z)\}_0^1$, and

$$\pi_n = s\int_{L_{n-1}} Q_{n-1}^{s-1}P_{n-1}^{m-s}\,dQ = s(m-s)\int_\Omega Q_{n-1}^{s-1}P_{n-1}^{m-s-1}\,dP\,dQ,$$

where the curve L_{n-1} and the region Ω are as shown in Figure 38. If there is a point R_0 on the curve L_0 and lying in the regions of stability shown in Figures 36 and 38, there are points R_n on the curves L_n $(n = 1, 2, \ldots)$ which tend to the point $R_\infty(1, 0)$. The domains of integration Ω_n approach the unit square $0 \le P \le 1$, $0 \le Q \le 1$. Then

$$\lim_{n \to \infty} \pi_n = s(m-s)\int_0^1\int_0^1 P^{m-s-1}Q^{s-1}dP\,dQ = 1.$$

If, however, the curve L_0 lies wholly outside the region of stability, the curves L_n converge to the diagonal of the unit square (a rigorous proof can be developed by using the theory of dynamic systems in a plane region, and is omitted here). Then

$$\lim_{n \to \infty} \pi_n = s(m-s)\int_0^1\int_0^1 P^{m-s-1}Q^{s-1}dP\,dQ = s/m.$$

The values of $P_n(z)$ and $Q_n(z)$ are shown in Table 4 for several values of z together with the probabilities of selecting a winning move when $m = 10$, $s = 2$, $n = 0.2.4.6.8$, and the functions $P_0(z)$ and $Q_0(z)$ are equal respectively to $1 - z^4$ and $1 - z\sqrt{z}$, $1 - z^4$ and $1 - \sqrt{z}$, $1 - z^8$ and $\sqrt[8]{z}$. [V. P. Akimov studied the probability of choosing a winning move in the current model and carried out the computations present in this section.]

We now consider models in which the value of the evaluation function $f(B)$ [in the position B after the move $\Psi = (A, B)$ from the position A] depends on the value of $f(A)$ at the origin A of the move Ψ. For simplicity we shall suppose for the present that we have a completely uniform game $\mathfrak{A}_{m,N}$ with s winning moves for the side having color col A at the non-terminal position A, and with a search depth N much greater than the depth n of the models. It is comparatively easy to derive similar formulae

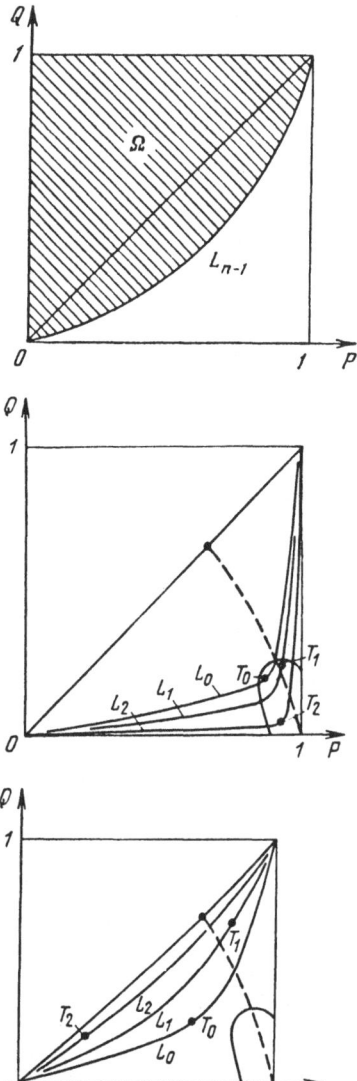

Figure 38

when we are given the distribution functions for the values of m and s, i.e. for the total number of moves and the number of winning moves at the positions of interest.

If the evaluation function $f(B)$ and the true score $sc(B)$ take on only the values 0 and 1, and their probability characteristics depend only on 1) the color col B, 2) the score $sc(B)$, 3) the score $sc(A)$ of the antecedent position A from which the move $\Psi = (A. B)$ was made, and 4) the value of $f(A)$, and if they satisfy the symmetry conditions, they are determined by six

Table 4. Probabilities that Model Scores will be Correct for Shannon Models with Evaluation Functions Taking Intermediate Values

$$m = 10, \qquad s = 2$$

$$P_0(x) = x^4, \qquad Q_0(x) = x\sqrt{x}$$

Search depth n	$P_n(0,57)$	$Q_n(0,57)$	$P_n(0,58)$	$Q_n(0,58)$	R_n
0	0.1056	0.4303	0.1132	0.4414	0.4173
2	0.1444	0.2402	0.1770	0.2809	0.2618
4	0.1462	0.1631	0.2923	0.3158	0.2264
6	0.1082	0.1104	0.7385	0.4716	0.2259
8	0.0221	0.0222	0.9999	0.9999	0.2063

$$P_0(x) = x^4 \qquad Q_0(x) = \sqrt{x}$$

Search depth n	$P_n(0,47)$	$Q_n(0,47)$	$P_n(0,48)$	$Q_n(0,48)$	R_n
0	0.0488	0.6856	0.0531	0.6928	0.7867
2	0.0895	0.5038	0.1068	0.5330	0.5137
4	0.1378	0.2906	0.2186	0.3966	0.2820
6	0.1572	0.1885	0.5529	0.5950	0.2144
8	0.1548	0.1594	0.9973	0.9974	0.2010

$$P_0(x) = x^8 \qquad Q_0(x) = \sqrt[8]{x}$$

Search depth n	$P_n(0,42)$	$Q_n(0,42)$	$P_n(0,43)$	$Q_n(0,43)$	R_n
0	0.0000096	0.8972	0.00967	0.9301	0.9941
2	0.000008	0.9000	0.00827	0.9557	0.9982
4	0.00000007	0.9045	0.00626	0.9818	0.9998
6	0	0.9124	0.00370	0.9969	0.9999
8	0	0.9259	0.00132	0.9999	1.0000

Notation: m—Number of moves from any position, s—number of winning moves from a won non-terminal position, n—depth of the model, $P_n(x)$—probability that the model score $msc_n(A) \le x$ for positions W in the original game, $Q_n(x)$—probability that the model score $msc_n(A) \le x$ for position lost in the original game, R_n—probability of correctly choosing the best move.

parameters:

$$p(x_0, x, y) :=$$
$$P(f(B) = |1 - \mathrm{col}\ A||\mathrm{sc}(A) = |x_0 - \mathrm{col}\ A|,$$
$$\mathrm{sc}(B) = |x_1 - \mathrm{col}\ A|, f(A) = |y - \mathrm{col}\ A|);$$

Thus

$$p(1,1,1) := P(f(B) = 1|\mathrm{col}\ A = 0, \mathrm{sc}(A) = 1, \mathrm{sc}(B) = 1, f(A) = 1)$$
$$= P(f(B) = 0|\mathrm{col}\ A = 1, \mathrm{sc}(A) = 0, \mathrm{sc}(B) = 0, f(A) = 0);$$
$$p(1,1,0) := P(f(B) = 1|\mathrm{col}\ A = 0, \mathrm{sc}(A) = 1, \mathrm{sc}(B) = 1, f(A) = 0)$$
$$= P(f(B) = 0|\mathrm{col}\ A = 1, \mathrm{sc}(A) = 0, \mathrm{sc}(B) = 0, f(A) = 1);$$
$$p(1,0,1) = P(f(B) = 1|\mathrm{col}\ A = 0, \mathrm{sc}(A) = 1, \mathrm{sc}(B) = 0, f(A) = 1)$$
$$= P(f(B) = 0|\mathrm{col}\ A = 1, \mathrm{sc}(A) = 0, \mathrm{sc}(B) = 1, f(A) = 0);$$
$$p(1,0,0) = P(f(B) = 1|\mathrm{col}\ A = 0, \mathrm{sc}(A) = 1, \mathrm{sc}(B) = 0, f(A) = 0)$$
$$= P(f(B) = 0|\mathrm{col}\ A = 1, \mathrm{sc}(A) = 0, \mathrm{sc}(B) = 1, f(A) = 1);$$
$$p(0,0,1) = P(f(B) = 1|\mathrm{col}\ A = 0, \mathrm{sc}(A) = 0, \mathrm{sc}(B) = 0, f(A) = 1)$$
$$= P(f(B) = 0|\mathrm{col}\ A = 1, \mathrm{sc}(A) = 1, \mathrm{sc}(B) = 1, f(A) = 0);$$
$$p(0,0,0) = P(f(B) = 1|\mathrm{col}\ A = 0, \mathrm{sc}(A) = 0, \mathrm{sc}(B) = 0, f(A) = 0)$$
$$= P(f(B) = 0|\mathrm{col}\ A = 1, \mathrm{sc}(A) = 1, \mathrm{sc}(B) = 1, f(B) = 1).$$

The parameters $p(0,1, y)$ do not exist, since from a position that loses for its own color all moves are losing moves.

Let $P(x, y, n)$ be the probabilities of obtaining the model score $|1 - \mathrm{col}\ A|$, given 1) the score $|x - \mathrm{col}\ A|$ for the base position A, 2) the value $f(A) = |y - \mathrm{col}\ A|$ for the evaluation function, and 3) the depth n of the model:

$$P(x, y, n) := P(\mathrm{msc}_n(A)$$
$$= |1 - \mathrm{col}\ A||\mathrm{sc}(A) = |x - \mathrm{col}\ A|, f(A) = |y - \mathrm{col}\ A|)$$

i.e.

$$P(1, 1, n) := P(\mathrm{msc}_n(A) = 1|\mathrm{col}\ A = 0, \mathrm{sc}(A) = 1, f(A) = 1)$$
$$= P(\mathrm{msc}_n(A) = 0|\mathrm{col}\ A = 1, \mathrm{sc}(A) = 0, f(A) = 0);$$
$$P(1, 0, n) := P(\mathrm{msc}_n(A) = 1|\mathrm{col}\ A = 0, \mathrm{sc}(A) = 1, f(A) = 0)$$
$$= P(\mathrm{msc}_n(A) = 0|\mathrm{col}\ A = 1, \mathrm{sc}(A) = 0, f(A) = 1);$$
$$P(0, 1, n) := P(\mathrm{msc}_n(A) = 1|\mathrm{col}\ A = 0, \mathrm{sc}(A) = 0, f(A) = 1)$$
$$= P(\mathrm{msc}_n(A) = 0|\mathrm{col}\ A = 1, \mathrm{sc}(A) = 1, f(A) = 0);$$
$$P(0, 0, n) := P(\mathrm{msc}_n(A) = 1|\mathrm{col}\ A = 0, \mathrm{sc}(A) = 0, f(A) = 0)$$
$$= P(\mathrm{msc}_n(A) = 0|\mathrm{col}\ A = 1, \mathrm{sc}(A) = 1, f(A) = 1).$$

It is easily seen that

$$P(x,1,0) = 1,$$

$$P(x,0,0) = 0 \qquad (x = 0,1).$$

The general recursion equations for the values of the characteristics $P(x, y, n+1)$ can be reduced to the form

$$1 - P(1, y, n+1)$$

$$= (p(1,1, y)P(0,0, n) + (1 - p(1,1, y))P(0,1, n))^s$$

$$\times (p(1,0, y)P(1,0, n) + (1 - p(1,0, y))P(1,1, n))^{n-s},$$

$$1 - P(0, y, n+1)$$

$$= (p(0,0, y)P(1,0, n) + (1 - p(0,0, y))P(1,1, n))^m \quad (y = 0,1).$$

A study of the way in which the probabilities $P(x, y, n)$ vary with an increase in n, as a function of the probability charactistics $p(x, y, n)$, is complicated by the large number (i.e. 6) of these. Therefore we do not have even a qualitative knowledge of the general situation. Table 5 lists the values of $P(x, y, n)$ for $n = 0,1,2,3,4,5,6$ and for a few sets of values of $p(x_0, x_1, y)$. The computations were made for $m = 10$, $s = 2$.

If for some depth n_0 the value of $P(1,0, n_0)$ is close to that of $P(1,1, n_0)$ and the value of $P(0,1, n_0)$ is close to that of $P(0,0, n_0)$, then for $n > n_0$ our recursion equations imply that $P(1,1, n)$ and $P(1,0, n)$ will be close to P_n, and that $P(0,1, n)$ and $P(0,0, n)$ will be close to Q_n for models in which the values of the evaluation functions $f(B)$ at the positions B after the moves $\Psi = (A.B)$ are uncorrelated with the values of $f(A)$ at those preceding positions A for which $P_{n_0} \approx P(1,1, n_0) \approx P(1,0, n_0)$ and $Q_{n_0} \approx P(0,1, n_0) \approx P(1,1, n_{n_0})$. The probabilities $P(1,1, n)$ and $P(1,0, n)$ will tend to 1; $P(0,1, n)$ and $P(0,0, n)$ will tend to 0, whenever for some depth n_0 of the Shannon model tree the point in the PQ-plane with coordinates $P = P(1,0, n_0)$, $Q = P(0,1, n_0)$ lies in the region of stability depicted in Figure 36.

They will converge together, alternately to 1 and 0, if the point with coordinates $P = P(1,1, n_0)$, $Q = P(0,0, n_0)$ lies at some distance outside that region.

Let us pause to estimate some values of of the parameters $p(x_0, x_1, y)$ that specify the increase in reliability of the model scores with an increase in the depth of the model tree. Let $p(1,1,1) = p(1,0,1) = p(0,0,1) = 1$; $p(1,1,0) = 0.7$; $p(1,0,0) = p(0,0,0) = 0.01$. The fact that the first three parameters are equal to 1 means that the evaluation function $f(A)$ is based on a highly optimistic estimate of the positions B resulting from arbitrary moves from A, at which the evaluation function favors the side having the move. An example would be a purely material-based evaluation function since one does not lose material on one's own move. The values $p(1,1,0) = 0.7$ and $p(0,0,0) = 0.1$ indicate that in approximately 90% of one's own positions that are evaluated as bad, the true opinion is found after a search of depth $n = 1$. However, our current model corresponds only very roughly to the situations that occur in real games, with the evaluation functions that are actually used.

We may consider an analogous model in which the evaluation function $f(A)$ may take on values intermediate between 0 and 1. We introduce the following notation:

$$P_{\pm}(x, y, z, n) :=$$

$$= P\left(\text{msc}_n(A) \underset{>}{\geq} z | \text{col } A = 0, \text{sc}(A) = x, f(A) = y\right)$$

$$= P\left(\text{msc}_n(A) \underset{\leq}{\leq} 1 - z | \text{col } A = 1, \text{sc}(A) = 1 - x, f(A) = 1 - y\right);$$

$$p(x_0, x_1, y_0, y_1) :=$$

$$= P(f(B) \leq y_1 | \text{col } A = 0, \Psi = (A, B) \in \mathfrak{A}, \text{sc}(A) = x_0,$$

$$\text{sc}(B) = x, f(A) = y_0)$$

$$= P(f(B) \geq 1 - y_1 | \text{col } A = 1, \Psi = (A, B) \in \mathfrak{A}, \text{sc}(A) = 1 - x_0,$$

$$\text{sc}(B) = 1 - x_1, f(A) = 1 - y_2);$$

In such a model the general recursion formula has the form:

$$1 - P_{\pm}(1, y, z, n+1)$$

$$= \left(\int_0^1 P_{\mp}(0, 1 - \mu, 1 - z, n) \, dp(1, 1, y, \mu)\right)^s$$

$$\times \left(\int_0^1 P_{\mp}(1, 1 - \mu, 1 - z, n) \, dp(1, 0, y, \mu)\right)^{m-s}.$$

$$1 - P_{\pm}(0, y, z, n+1)$$

$$= \left(\int_0^1 P_{\mp}(1, 1 - \mu, 1 - z, n) \, dp(0, 0, y, \mu)\right)^m$$

We use Stieltjes integration for the integrals entering the formula; then the formula is the same in the cases when the evaluation function $f(A)$ may take on all values in the interval $[0,1]$ or only a finite set of them.

Table 5. Probability of Correct Scores for a Shannon Model when the Value of the Evaluation Function at Points B Arising After Moves $\Psi = (A,B)$ Depends on the Value of the Evaluation Function at A Before the Move Ψ

	$p(1,1,0) = 0.8$	$p(1,0,0) = p(0,0,0) = 0.01$		
n	$P(1,1,n)$	$P(1,0,n)$	$P(0,1,n)$	$P(0,0,n)$
0	1.0000	0.0000	1.0000	0.0000
1	1.0000	0.9631	1.0000	0.0956
2	0.9932	0.9238	0.3135	0.0037
3	1.0000	0.9959	0.5474	0.0732
4	0.9950	0.9720	0.0398	0.0005
5	1.0000	0.9999	0.2469	0.0515
6	0.9974	0.9918	0.0007	0.0000
7	1.0000	1.0000	0.0790	0.0267
8	0.9993	0.9986	0.0000	0.0000
9	1.0000	1.0000	0.0137	0.0072
10	0.9999	0.9999	0.0000	0.0000

	$p(1,1,0) = 0.6$ $p(1,0,0) = p(0,0,0) = 0.01$				$P(1,1,0) = 0.5.$ $p(1,0,0) = p(0,0,0) = 0.01$			
n	$P(1,1,n)$	$P(1,0,n)$	$P(0,1,n)$	$P(0,0,n)$	$P(1,1,n)$	$P(1,0,n)$	$P(0,1,n)$	$P(0,0,n)$
0	1.0000	0.0000	1.0000	0.0000	1.0000	0.0000	1.0000	0.0000
1	1.0000	0.8524	1.0000	0.0956	1.0000	0.7693	1.0000	0.0956
2	0.9975	0.7933	0.7976	0.0147	0.9989	0.7054	0.9274	0.0228
3	1.0000	0.8964	0.9013	0.0450	1.0000	0.7815	0.9695	0.0398
4	0.9992	0.8511	0.6650	0.0106	0.9998	0.7498	0.9150	0.0219
5	1.0000	0.9535	0.6912	0.0230	1.0000	0.7853	0.9439	0.0269
6	0.9997	0.8891	0.5305	0.0076	0.9999	0.7685	0.9109	0.0217
7	1.0000	0.9764	0.5694	0.0139	0.9999	0.7867	0.9281	0.0239
8	0.9999	0.9192	0.3786	0.0049	0.9999	0.7773	0.9092	0.0217

	$p(1,1,0) = 0.6$ $p(1,0,0) = p(0,0,0) = 0.01$				$p(1,1,0) = 0.5$ $p(1,0,0) = p(0,0,0) = 0.01$			
n	$P(1,1,n)$	$P(1,0,n)$	$P(0,1,n)$	$P0,0,n)$	$P(1,1,N)$	$P(1,0,n)$	$P(0,1,n)$	$P(0,0,n)$
9	1.0000	0.9925	0.4281	0.0061	0.9999	0.7874	0.9194	0.0229
10	0.9999	0.9457	0.2128	0.0025	0.9999	0.7819	0.9084	0.0217
16	1.0000	0.9979	0.0001	0.0000				
20					0.9999	0.7872	0.9081	0.0217

	$P(1,1,0) = 0.8$ $P(1,0,0) = P(0,0,0) = 0.015$				$P(1,1,0) = 0.6,$ $p(1,0,0) = P(0,0,0) = 0.015$			
n	$P(1,1,n)$	$P(1,0,n)$	$P(0,1,n)$	$P(0,0,n)$	$P(1,1,n)$	$P(1,0,n)$	$P(0,1,n)$	$P(0,0,n)$
0	1.0000	0.0000	1.0000	0.0000	1.0000	0.0000	1.0000	0.0000
1	1.0000	0.9646	1.0000	0.1403	1.0000	0.8582	1.0000	0.1403
2	0.9853	0.9029	0.3029	0.0053	0.9942	0.7695	0.7832	0.0211
3	1.0000	0.9963	0.6398	0.1488	0.9999	0.9013	0.9271	0.0879
4	0.9785	0.9390	0.0363	0.0007	0.9966	0.8227	0.6462	0.0152
5	1.0000	0.9999	0.4668	0.2000	1.0000	0.9317	0.8579	0.0582
6	0.9600	0.9358	0.0005	0.0000	0.9981	0.8583	0.5070	0.0107
7	1.0000	1.0000	0.4851	0.3377	1.0000	0.9576	0.7831	0.0395
8	0.8859	0.8652	0.0000	0.0000	0.9989	0.8871	0.3514	0.0067
9	1.0000	1.0000	0.7650	0.7032	1.0000	0.9796	0.6982	0.0274
10	0.5056	0.4880	0.0000	0.0000	0.9994	0.9128	0.1866	0.0032
18					1.0000	0.9949	0.0000	0.0000

Notation: m —number of moves from non-terminal positions, s —number of winning moves from won non-terminal positions, n —depth of the model, $p(x, y, z)$ —probability that $f(B) = |1 - \text{col } A|$ provided that $\text{sc}(A) = |x - \text{col } A|$, $\text{sc}(B) = |y - \text{col } A|$, $F(A) = |z - \text{col } A|$, $P(x, z, u)$ —probability that $\text{msc}(A) = |1 - \text{col } A|$ provided that $\text{msc}(A) = |x - \text{col } A|$, $f(A) = |z - \text{col } A|$. In all the examples $m = 10$, $s = 2$, $p(1,1,1) = p(1,0,1) = p(0,0,1) = 1$.

The calculations in the above formulae have a definite meaning, but for qualitative studies we may use models in which the evaluation function takes on only the two values 0 and 1. In such models we can reflect various relationships between the values of the evaluation function and the properties of positions. The characteristics of positions may depend not only on the value of the evaluation function but also on other parameters such as, for example, the so-called sharpness of the position. Later we shall offer three examples in order to show how to construct models of this kind; we shall examine in detail the influence of various parameters. We can construct combined models by similar methods, in which we take into account several parameters.

Let the number of winning moves at various positions depend on the value of the evaluation function $f(A)$. In the simplest case we suppose that in a winning position where $f(A) = 1$ there are s_1 winning moves, and in a winning position where $f(A) = 0$ there are only $s_2 < s_1$ such moves. If the value of $f(B)$ depends probabilistically only on the true scores of the positions A and B (where A is the position from which the move $\Psi = (A, B)$ leading to B was made) and the value of the evaluation function $f(A)$ at A, and if the parameters $p(x_0, x_1, y)$ have the same meaning as in the second model considered in this section, then the general recursion formula has the form:

$$1 - P(1,1,n+1)$$
$$= (p(1,1,1) P(0,0,n)$$
$$+ (1 - p(1,1,1)) P(0,1,n))^{s_1} (p(1,0,1) P(1,0,n)$$
$$+ (1 - p(1,0,1)) P(1,1,n))^{m - s_1},$$
$$1 - P(1,0,n+1)$$
$$= (p(1,1,0) P(0,0,n)$$
$$+ (1 - p(1,1,0)) P(0,1,n))^{s_2} (p(1,0,0) P(1,0,n)$$
$$+ (1 - p(1,0,0)) P(1,1,n))^{m - s_2},$$
$$1 - P(0,1,n+1)$$
$$= (p(0,0,1) P(1,0,n) + (1 - p(0,0,1)) P(1,1,n))^{m},$$
$$1 - P(0,0,n+1)$$
$$= (p(0,0,0) P(1,0,n) + (1 - p(0,0,0)) P(1,1,n))^{m}.$$

Suppose that the positions are either quiet or acute, and that to every position A there correspond three parameters on which the probability characteristics depend: the color $\mathrm{col}(A)$ of the side having the turn to move; the true score $\mathrm{sc}(A)$; and the acuteness $\varphi(A)$ of the position (for an acute position $\varphi(A) = 1$, for a quiet position $\varphi(A) = 0$). In an acute won position, for example, there will be only one winning move, which will lead with probability r_1 to an acute position and with probability $1 - r_1$ to a quiet position. Both the acute and the quiet positions, naturally, are lost for the opponent. The remaining $m - 1$ moves lead to positions won for the opponent, acute with probability r_2 and quiet with probability $1 - r_2$. All moves from an acute lost position lead to positions won for the opponent, acute with probability r_3 and quiet with probability $1 - r_3$. For simplicity we shall suppose that all moves from a quiet position lead to quiet positions and that from a won quiet position there are $s > 1$ winning moves.

We denote by $P(1, n)$ the probability of correctly scoring an acute won position, and by $P(0, n)$ the probability of correctly scoring a quiet won position; $Q(1, n)$ and $Q(0, n)$ are the corresponding probabilities that acute and quiet lost positions will be evaluated as winning positions. In all the above definitions the parameter n represents the depth of a Shannon model for which the score of the base position is defined by the use of Zermelo's formula. The general recursion formula for such a model has the form:

$$1 - P(0, n+1) = Q^{s}(0, n) P^{m-s}(0, n),$$
$$1 - Q(0, n+1) = P^{m}(0, n),$$
$$1 - P(1, n+1) = (r_1 Q(1, n) + (1 - r_1) Q(0, n))$$
$$\times (r_2 P(1, n) + (1 - r_2) P(0, n))^{m-1},$$
$$1 - Q(1, n+1) = (r_3 P(1, n) + (1 - r_3) P(0, n))^{m}.$$

We must also prescribe the initial probabilities $P(0,0) = p_0, P(1,0) = p_1, Q(0,0) = q_0, Q(1,0) = q_1$.

Lastly we contemplate a model in which the total number of moves from a position, and the number of winning moves from a won position, may vary from position to position. For simplicity we shall again limit ourselves to the minimal number of parameters for the probability characteristics. Suppose these parameters are the color col A of the side having the move at A and the true score sc(A). We consider the following characteristics: $\pi(m, s)$–the probability that from a won position we may make m moves, of which s win; $\rho(m)$–the probability that from a lost position A we may make m moves—all, naturally, losing. Thus $\pi(0,0)$ and $\rho(0)$ are the corresponding probabilities that won and lost positions will turn out to be terminal. The general recursion formula for our model has the form:

$$1 - P_{n+1} = \pi(0,0) + \sum_{m=1}^{M} \sum_{s=1}^{m} \pi(m,s) Q_n^s P_n^{m-s},$$

$$1 - Q_{n+1} = \sum_{m=1}^{M} \rho(m) P_n^m,$$

where M is the maximum number of moves that can be made from any position in our game, while P_n and Q_n have the same meaning that we assigned them in our first model. As before, we must prescribe the initial values $P_0 = p$ and $Q_0 = q$.

In conclusion we again recall that by similar methods we can construct models in which the probability characteristics depend on all the parameters we have discussed, including the number of admissible moves at the positions being studied, and on parameters not discussed in this section.

Construction of Models and Calculation of Model Scores on a Probabilistic Basis

In this section we consider some examples (due to N. E. Kosatcheva, who also carried out the necessary computations) for the case in which the probability of correctly assessing the score at the base position can be increased if we replace the classical Shannon model by one in which the branches in the search tree have varying lengths, or if we change the method for computing the scores at positions in the model. The basic methods for cutting down the number of positions to be examined in the search are founded on the following considerations:

(1) The search depth from a position B in the model game tree may depend on the value of the evaluation function $f(B)$ and perhaps on other parameters defined there.

(2) In determining the scores of positions in the model we may bypass Zermelo's formula and instead take account of the fact that several moves from a won position B should lead to a position with a value of $f(B)$ favorable to col B.

(3) The probability that a move from a given position will turn out to be the best one may be used not only to define the order in which the moves should be examined but also to determine the search depth.

All these opportunities will be illustrated for some probabilistic assumptions that are as simple as possible, in order to see what results they lead to. Simple as these assumptions are, they are applicable in more complex cases and the methods we illustrate may be combined.

Let B be the position arising after the move $\Psi = (A, B)$ and suppose that $f(B)$ is correlated with the true scores of A and B and with the value of $f(A)$. Then we are given the values of the probability characteristics $p_0(x_0, x_1, y)$ defined in the preceding section. With our standard assumptions (m moves from any non-terminal position, and s winning moves from a won position) for a Shannon model in which all nodes are inspected to a prescribed depth n, we have the recursion equations

$$1 - P(1, y, n+1)$$
$$= \left(p(1,1, y)P(0,0, n) + (1 - p(1,1, y))P(0,1, n) \right)^s \left(p(1,0, y) \right.$$
$$\times P(1,0, n)$$
$$+ (1 - p(1,0, y))P(1,1, n) \right)^{m-2}.$$
$$1 - P(0, y, n+1)$$
$$= \left(p(0,0, y)P(1,0, n) + (1 - p(0,0, y))P(1,1, n) \right)^m$$

where $P(x,0,1) = 1$ and $P(x,0,0) = 0$ $(x = 0,1)$.

As an example we list the values of $P(x, y, n)$ in Table 6 for $n = 0.6$ and for $p(1,1,1) = p(1,0,1) = p(0,1,1) = 1$; $p(1,1,0 = 0.8$; $p(1,0,0) = p(0,0,0) = 0.02$; $m = 10$, $s = 2$. Inspection of the table shows that even for values of

Table 6. Probabilities of $P(x, y, n)$ as a Function of the Search Depth n

	Shannon Model $\mathfrak{A}_{SH,n}$				Model $\mathfrak{A}_{NK,n}$			
n	$P(1,1,n)$	$P(1,0,n)$	$P(0,1,n)$	$P(0,0,n)$	$P(1,1,n)$	$P(1,0,n)$	$P(0,1,n)$	$P(0,0,n)$
0	1.0000	1.0000	1.0000	0.0000				
1	1.0000	0.9660	1.0000	0.1829				
2	0.9746	0.8807	0.2927	0.0068	1.0000	0.8807	1.0000	0.0068
3	1.0000	0.9967	0.7193	0.2414	1.0000	0.9586	0.7193	0.0236
4	0.9433	0.8865	0.0323	0.0008	0.9996	0.9737	0.3488	0.0084
5	1.0000	1.0000	0.7001	0.4491	0.9999	0.9943	0.2340	0.0091
6	0.7983	0.7507	0.0003	0.0000	0.9999	0.9971	0.0554	0.0017

n_0 as low as 2 the point $(P(1,1,n_0), P(0,0,n_0))$ lies outside of the region of stability depicted in Figure 36. Therefore the probabilities $P(x, y, n)$ do not tend to desirable limits but instead tend alternately to the limits 0 and 1. Note that the applicability of the above recursion equations is not limited to the classical Shannon model.

Suppose that we have constructed a model in which the search depths may vary from node to node and that we have specified the probabilities $P(x, y)$ (at the base position A with $\mathrm{sc}(A) = |x - \mathrm{col}\, A|$) that the value of the model evaluation function $f(A) = |y - \mathrm{col}\, A|$ will be $\mathrm{msc}(A) = |1 - \mathrm{col}\, A|$. (We assume that in our model the natural symmetry conditions for probabilities are satisfied.) We now set up a mapping between our model game $\mathfrak{A}_\mu$ and a value n_0 of the formal search depth n; e.g. the maximum depth of a branch in the search tree $\mathfrak{A}_{\mu,n_0} = \mathfrak{A}_\mu$. Then for $n > n_0$ the model games $\mathfrak{A}_{\mu,n}$ are defined as follows: their trees consist of all the nodes A_k in the original tree that have rank $r(A_k) = k \le n - n_0$, the moves leading from them, and the subtrees $\mathfrak{A}_{\mu,n_0}(A_{n-n_0})$. The base positions of these trees have rank $r(A_{n-n_0}) = n - n_0$ and the subtrees themselves, by construction, are trees in the model game $\mathfrak{A}_\mu$. It is easily seen that for $n \ge n_0$ we have the following recursion equations, similar to those developed above:

$$1 - P_\mu(1, y, n+1)$$
$$= \left(p(1,1,y) P_\mu(0,0,n) + (1 - p(1,1,y)) P_\mu(0,1,n) \right)^s$$
$$\times \left(p(1,0,y) P_\mu(1,0,n) + (1 - p(1,0,y)) P_\mu(1,1,n) \right)^{m-s},$$
$$1 - P_\mu(0, y, n+1)$$
$$= \left(p(0,0,y) P_\mu(1,0,n) + (1 - p(0,0,y)) P_\mu(1,1,n) \right)^m.$$

Let us consider a model game $\mathfrak{A}_{NK,2}$, with the following convention: if at the base position A the evaluation function has the value $f(A) = |1 - \mathrm{col}\, A|$ we take A as terminal with the model score $\mathrm{msc}_{NK,2} = f(A)$, else $\mathfrak{A}_{NK,2}(A)$ is taken as the tree of a Shannon model game $\mathfrak{A}_{SH,2}(A)$ of depth 2. This tree contains the position A, all moves from it, all positions B_i ($i = 1, 2, \ldots, m$) to which these moves lead, and all moves in the game $\mathfrak{A}$ from the B_i. The positions reached by these latter moves are taken as terminal. Then

$$P_{NK}(1,1,2) = P_{NK}(0,1,2) = 1,$$
$$P_{NK}(x,0,2) = P_{SH}(x,0,2) \ (x = 0,1).$$

The values of the $P_{NK}(x, y, n)$ are shown in Table 6 for the displayed values of the parameters $m, s, p(x_0, x_1, y)$ and for $n = 2, 3, 4, 5, 6$. For $n = 6$ the point $(P_{NK}(1,0,n), P_{NK}(0,1,n))$ lies in the region of stability illustrated in Figure 36. This means that with further increase in the search depth n the probabilities $P_{NK}(1,1,n)$ and $P_{NK}(1,0,n)$ will tend to 1 while $P_{NK}(0,1,n)$ and $P_{NK}(0,0,n)$ will tend to 0.

The rules for constructing the model $\mathfrak{A}_{NK,n}$ may be so formulated that they appear as well-defined considerations with a sensible meaning. If the search depth n, from the base position A_0, is odd we must compute the evaluation function $f(A_{n-2})$ for the positions A_{n-2} of rank $n-2$. If the value is bad for us (the side with color col A_0), we should believe it and not consider moves from the positions A_{n-2}. If the depth is even we must make an auxiliary search of depth 2 in order to sharpen our earlier good opinion of the position A_{n-2}. Such a reasonable argument, however, fails to explain why for even n we should trust our earlier good opinion of the positions of rank $n-2$ but sharpen a bad opinion.

Let us now consider some possibilities arising from our rejection of Zermelo's formula as the means of defining the model scores of non-terminal positions. Some authors (e.g. [27]) propose that the score of a position A should be defined as the mean value of the most favorable (for col A) of the model scores at the positions B_i that can be reached in one move from A:

$$
\mathrm{msc}(A) := \begin{cases} \displaystyle \max_{i_1, i_2, \ldots, i_k} \frac{1}{k} \sum_{h=1}^{k} \mathrm{msc}(B_{i_h}), \\[4pt] \text{White to move} \\[8pt] \displaystyle \min_{i_1, i_2, \ldots, i_l} \frac{1}{l} \sum_{h=1}^{l} \mathrm{msc}(B_{i_h}), \\[4pt] \text{Black to move} \end{cases}
$$

When the numbers k and l are given, this method is known as the k,l-heuristic. However, the values of k and l may vary for positions of different ranks in the tree of the model game, and may also depend on other circumstances.

We can broaden the class of such formulae by considering weighted averages, writing

$$
\mathrm{msc}_n(A) := \begin{cases} \displaystyle \varkappa_{w,0} f(A) + \sum_{i=1}^{m} \varkappa_{wi} \mathrm{msc}(B_i), \\[4pt] \text{White to move}; \\[8pt] \displaystyle \varkappa_{b,0} f(A) + \sum_{i=1}^{m} \varkappa_{bi} \mathrm{msc}(B_i), \\[4pt] \text{Black to move}; \end{cases}
$$

where the positions B_i that arise after the moves $\Psi_i = (A, B_i)$ ($i = 1, 2, \ldots, m$) are ordered by decreasing values of the model score $\mathrm{msc}(B_i)$ when A is a White position and in the reverse order when A is a Black position. The formulae we have proposed form a particular case. We have for them

$$x_{w,0} = x_{b,0} = 0, \quad x_{w,1} = x_{w,2} = \cdots = x_{w,k} = \frac{1}{k},$$

$$x_{w,k+1} = \cdots = x_{w,m} = 0, \quad x_{b,1} = x_{b,2} = \cdots = x_{b,l} = \frac{1}{l},$$

$$x_{b,l+1} = \cdots = x_{b,m} = 0.$$

In applying such formulae we should use a pruning rule weaker than the α, β-heuristic, since we cannot conclude that we have obtained at a position A a model score favorable to col A merely because we have found a score unfavorable to col B at a position B reached by a single move from A. The consequent increase in the number of positions to be inspected may be offset by a decrease in the search depth n that suffices to obtain values near 1 for the probabilities P_n and $1 - Q_n$ of obtaining correct model scores for the positions won and lost, respectively, for col A. The probability characteristics in the game $\mathfrak{A}$ may be such that the use of Zermelo's formula does not guarantee the convergence of P_n to 1 and of Q_n to 0, but the use of a more general formula does so.

The model scores defined by our formulae take on values in the open interval $(0, 1)$ even in the case when the evaluation function $f(A)$ and the true score $\mathrm{sc}(A)$ take on only the values 0 and 1. Therefore the probabilities of various values of the model scores defined by our formulae are computed by equations containing integrals. We shall present such formulae for a completely uniform game $A_{m,N}$ with s winning moves from every position won for its own side, and values of the evaluation function $f(A)$ that are independently distributed random variables depending probabilistically only on $\mathrm{sc}(A)$ and col A.

Suppose that we have a Shannon model with depth $n \ll N$ and that for positions of rank k we have the probabilities

$$P_{\pm n,k}(z) := P\Big(|\mathrm{col}\, A - \mathrm{msc}(A)| \gtrless z \,|\, \mathrm{sc}(A) = |1 - \mathrm{col}\, A|, r(A) = k\Big);$$

$$Q_{\pm n,k}(z) := P\Big(|\mathrm{col}\, A - \mathrm{msc}(A)| \gtrless z \,|\, \mathrm{sc}(A) = \mathrm{col}\, A, r(A) = k\Big);$$

The values of the $P_{\pm n,n}(z)$ amd $Q_{\pm n,n}(z)$ are determined by the probability characteristics of the evaluation function $f(A)$:

$$P_{\pm n,n}(z) :=$$
$$P\Big(|\mathrm{col}\, A - f(A)| \gtrless z \,|\, \mathrm{sc}(A) = |1 - \mathrm{col}\, A|, r(A) = n\Big);$$

$$Q_{\pm n,n}(z) :=$$
$$P\Big(|\mathrm{col}\, A - f(A)| \gtrless z \,|\, \mathrm{sc}(A) = \mathrm{col}\, A, r(A) = n\Big).$$

The recursion equations for these probabilities have the form

$$1 - P_{\pm n, k-1}$$

$$= \int_{\Omega_{\pm}(z)} \left| \prod_{i=1}^{s} dQ_{n,k}(\zeta_i) \prod_{i=s+1}^{m} dP_{n,k}(\zeta_i) dp_{k-1}(\zeta_0) \right|,$$

$$1 - Q_{\pm n, k-1}$$

$$= \int_{\Omega_{\pm}(z)} \left| \prod_{i=1}^{m} dP_{n,k}(\zeta_i) dq_{k-1}(\zeta_0) \right|,$$

where $\Omega_{\pm}(z)$ is the region in the $(m+1)$-dimensional unit cube $(0 \leq \zeta_i \leq 1)_0^m$ whose points satisfy the condition

$$\varkappa_0 \zeta_0 + \sum_{h=1}^{m} \varkappa_h (1 - \zeta_{i_h}) \overset{<}{\underset{\leq}{}} z$$

for an arbitrary permutation $\begin{pmatrix} 1_2 \ldots m \\ i_1 i_2 \ldots i_m \end{pmatrix}$ (it suffices that the condition be satisfied for those permutations that carry the sequence of coordinates $(\zeta_1, \zeta_2, \ldots, \zeta_m)$ into a monotone non-increasing sequence). The set functions $p_k(z)$ and $q_k(z)$ in the recursion equations represent the probabilities of the values of the evaluation function $f(A)$:

$$p_k(z) := P(|\text{col } A - f(A)| \geq z | \text{sc}(A) = |1 - \text{col } A|, \ r(A) = k):$$

$$q_k(z) := P(|\text{col } A - f(A)| \geq z | \text{sc}(A) = \text{col } A, \ r(A) = k).$$

Thus the probability characteristics of the evaluation function $f(A)$ may depend on the rank $r(A)$ of the position A.

The above formulae are complex, even under simple probabilistic postulates, and no investigation of them has been made. Some simpler methods were considered (for study, not for application) in the case of non-minimax search, i.e. a search in which Zermelo's method is not used in finding the scores of non-terminal positions. This investigation showed that in principle it is possible to obtain results better than those yielded by a minimax search.

Suppose that the model score msc(A) of a position A which is non-terminal in the model has the value $|1 - \text{col } A|$ when l or more of the moves $\Psi = (A, B)$ leading from it lead to positions B with msc(B) = $|1 - \text{col } A|$, and the remaining moves lead to positions B with msc B = col A. Then for a position A of rank $r(A) = k - 1$, the respective probabilities $P_{n, k-1}$ and $Q_{n, k-1}$ that msc(A) = $|1 - \text{col } A|$ when A is won or lost for col A satisfy the recursion equations

$$1 - P_{n, k-1}$$

$$= \sum_{g=0}^{l} \sum_{h=0}^{\min(s, l-g)} C_s^h (1 - Q_{n,k})^h Q_{n,k}^{s-h} C_{m-s}^g (1 - P_{n,k})^g P_{n,k}^{m-s-g}$$

$$1 - Q_{n, k-1} = \sum_{g=0}^{l} C_m^g (1 - P_{n,k})^g P_{n,k}^{m-g}$$

(we assume $l \leq m - s$). The model scores at positions of rank k are equal to the corresponding values of the evaluation function. Accordingly,

$$P_{n,n} = p := P(f(A) = |1 - \operatorname{col} A| | \operatorname{sc}(A) = |1 - \operatorname{col} A|);$$
$$Q_{n,n} = q := P(f(A) = |1 - \operatorname{col} A| | \operatorname{sc}(A) = \operatorname{col} A);$$

Later we shall consider the possibility of varying the threshold l for positions (l is a number of moves, not rank) in the tree of a Shannon model having several ranks. Let $P_n(l)$ and $Q_n(l)$ be the probabilities (for won and lost positions, respectively) that $\operatorname{msc}(A) = |1 - \operatorname{col} A|$. When l is constant, it follows from our formulae that these probabilities satisfy the recursion equations

$$1 - P_{n+1}(l)$$
$$= \sum_{g=0}^{l} \sum_{h=0}^{\min(s, l-g)} C_s^h (1 - Q_n(l))^h Q_n^{s-h}(l) C_{m-s}^g (1 - P_n(l))^g P_n^{m-s-g}(l),$$
$$1 - Q_{n+1}(l) = \sum_{g=0}^{l} C_m^g (1 - P_n(l))^g P_n^{m-g}(l).$$

If we set $l = 0$ we obtain our earlier recursion equations for a minimax search. The values of $P_0(l) = p$ and $Q_0(l) = q$ are found by the methods developed earlier. They do not depend on l.

Let $P_n(l) = 1 - \varepsilon$, $Q_n(l) = \delta$. Then

$$\varepsilon' = 1 - P_{n-1}$$
$$= \sum_{g=0}^{l} C_{m-s}^g \varepsilon^g (1 - \varepsilon)^{m-s-g} \left(\sum_{h=0}^{\min(s, l-g)} C_s^h (1 - \delta)^h \delta^{s-h} \right)$$
$$= C_s^{\min(l,s)} \Pi'_{m,s,l}(\varepsilon, \delta) \delta^{\max(0, s-l)},$$

$$\delta' = Q_{n+1}$$
$$= 1 - \sum_{g=0}^{l} C_m^g \varepsilon^g (1 - \varepsilon)^{m-g} = \sum_{g=l+1}^{m} C_m^g \varepsilon^g (1 - \varepsilon)^{m-g} = C_m^{l+1} \Pi''_{m,l}(\varepsilon) \varepsilon^{l+1},$$

$$\varepsilon'' = 1 - P_{n+2}$$
$$= C_s^{\min(l,s)} \Pi'_{m,s,l}(\varepsilon', \delta') \delta'^{\max(0, s-l)} = C_s^{\min(l,s)} \left(C_m^{l+1} \right)^{\max(0, s-l)}$$
$$\times \Pi'_{m,s,l}(\varepsilon', \delta') \Pi''^{\max(0, s-l)}_{m,l}(\varepsilon) \varepsilon^{(l+1)\max(0, s-l)},$$

$$\delta'' = Q_{n+2}$$
$$= C_m^{l+1} \Pi''_{m,l}(\varepsilon') \varepsilon'^{l+1} = C_m^{l+1} \left(C_s^{\min(l,s)} \right)^{l+1} \Pi''_{m,l}(\varepsilon')$$
$$\times \Pi'^{l+1}_{m,s,l}(\varepsilon, \delta) \delta^{(l+1)\max(0, s-l)},$$

where the polynomial $\Pi'_{m,s,l}(\varepsilon, \delta)$ is equal to 1 for $\varepsilon = \delta = 0$, $\Pi''_{m,l}(\varepsilon) = 1$ for $\varepsilon = 0$, and

$$0 \leq \Pi'_{m,s,l}(\varepsilon, \delta) < 1,$$
$$0 \leq \Pi''_{m,l}(\varepsilon) < 1 \quad (0 < \varepsilon \leq 1, 0 < \delta \leq 1).$$

When $l < s$ these formulae yield

$$\varepsilon'' = C_s^l \left(C_m^{l+1} \right)^{s-l} \Pi'_{m,s,l}(\varepsilon', \delta') \Pi''^{s-l}_{m,l}(\varepsilon) \varepsilon^{(l+1)(s-l)},$$

$$\delta'' = C_m^{l+1} \left(C_s^l \right)^{s+1} \Pi''_{m,l}(\varepsilon') \Pi'^{l+1}_{m,s,l}(\varepsilon, \delta) \delta^{(l+1)(s-l)},$$

while for $l \geq s$

$$\varepsilon'' = \Pi'_{m,s,l}(\varepsilon', \delta'),$$

$$\delta'' = C_{ml+1} \Pi''_{m,l}(\varepsilon') \Pi'^{(l+1)}_{m,s,l}(\varepsilon, \delta).$$

Thus if $l \geq s$, P_n does not tend to 1, nor Q_n to 0, for any value of p or q. In real games the number s of winning moves at won terminal positions is not constant; therefore special significance attaches to positions from which there is only one winning move. In our model a non-minimax search using the same formulae at all positions in the search tree is of doubtful effectiveness. Also, all the scores in this section are based on deliberately simplified probabilistic assumptions about the games considered. We are interested in new approaches to the construction of models of programmable games, definition of model scores, and selection of moves. The application of such approaches requires supplementary work on more realistic probabilistic hypotheses and on experimental verifications.

We return to the question of which values of l promote rapid convergence of the $P_n(l)$ and $Q_n(l)$ to the values we desire. From the formulae derived above it follows that when ε and δ are small enough the rate of convergence is determined by their exponent $(l+1)(s-l)$, which attains its maximum when $l = \lceil (s-1)/2 \rceil$ (for even s the exponents have the same value for $l = \lceil (s-1)/2 \rceil$ and $l = \lfloor (s-1)/2 \rfloor$, but the coefficients of the $\delta^{(l-1)(s-l)}$ increase as l increases). The rate of convergence must be defined for other probabilistic assumptions also. Moreover, one must take into account the fact that the number of positions in the search tree increases with increasing l. Thus it may be useless to decrease the depth of the search tree by going to a non- minimax search method.

Some values of $P_0(l) = p$ and $Q_0(l) = q$ that lie outside the region of stability depicted in Figure 36 will lie within it for a non-minimax search and some value of $l > 0$. For example, let $m = 10$, $s = 3$, $p = 0.95$, and $q = 0.1$; Table 7 shows the results of the computations for $l = 0,1$, $n = 0,1,2,3,4$. If $P_{n+2}(l) < P_n(l)$ and $Q_{n+2}(l) < Q_n(l)$, our recursion equations imply the inequalities $P_{n+3}(l) > P_{n+1}(l)$, $Q_{n+3}(l) > Q_{n+1}(l)$. Conversely, if $P_{n+2}(l) > P_n(l)$ and $Q_{n+2}(l) > Q_n(l)$, it follows that $P_{n+3}(l) < P_{n+1}(l)$ and $Q_{n+3}(l) < Q_{n+1}(l)$.

Accordingly, whenever for some $n = n_0$ the probabilities $P_{n_0+2}(l)$ and $Q_{n_0+2}(l)$ differ from $P_{n_0}(l)$ and $Q_{n_0}(l)$ in the same direction, $P_n(l)$ and $Q_n(l)$ will not tend to the desired limits. For instance, if $P_{n_0+2}(l) > P_{n_0}(l)$ and $Q_{n_0+2}(l) > Q_{n_0}(l)$, for $n = n_0 + 2k$ these probabilities will increase monotonely, and for $n = n_0 + 2k + 1$ they will decrease monotonely. The

Table 7. Various Methods for Computing Model Scores

		$m = 10,$	$s = 3$	
	$l = 0$		$l = 1$	
n	P_n	Q_n	P_n	Q_n
0	0.9500	0.1000	0.9500	0.1000
1	0.9993	0.4013	0.9802	0.0861
2	0.9357	0.0070	0.9817	0.0159
3	1.0000	0.4855	0.9993	0.0137
4	0.8856	0.0000	0.9994	0.0000

Notation: m — number of moves from non-terminal position, s — number of winning moves from a won non-terminal position, l — number of moves from a position A to a position B with $msc(B) = |1 - \text{col } A|$ when $msc(A) = |1 - \text{col} A|$, n — search depth, P_n — probability that $msc_n(A) = |1 - \text{col } A|$ when $sc(A) = |1 - \text{col } A|$, Q_n — probability that $msc_n(A) = |1 - \text{col } A|$ when $sc(A) = \text{col } A$.

probabilities $P_n(0)$ and $Q_n(0)$ behave in this way in our example, as shown in Table 7.

The equations determining the distances of $P_{n+2}(l)$ and $Q_{n+2}(l)$ from the desired limits imply the inequalities

$$\varepsilon'' \le 6075\varepsilon^4,$$

$$\delta'' \le 40\ 5\delta^4$$

for $m = 10$, $s = 3$, $l = 1$ (equality holds only when the corresponding distance ε or δ is equal to zero). Thus if $0 \le \varepsilon \le \sqrt[3]{1/6075} \approx 0.054\ 8$ and $0 \le \delta \le \sqrt[3]{1/405} \approx 0.1352$ the probabilities $P_n(l)$ and $Q_n(l)$ will converge as desired. Our values of $P_0(l) = p = 1 - \varepsilon = 0.95$ and $Q_0(l) = \delta = q = 0.1$ lie in this region of convergence.

When $s = 2$, a minimax search corresponding to $l = 0$ yields the fastest convergence for the P_n and Q_n. Moreover, the region of convergence for $l = 1$ lies inside that for $l = 0$, depicted in Figure 36. For $l \ge 2$ the probabilities $P_n(l)$ and $Q_n(l)$ do not in general converge to the desired limits. Nevertheless, for some values of p and q, employing the equations for a non-minimax search in finding the model scores at positions of high rank may turn out to be useful. For instance, set $m = 10$, $s = 2$, $P_{n,n} = p = 0.95$, $Q_{n,n} = 0.005$. The values p and q lie outside the region of convergence depicted in Figure 36. There is no convergence even when we apply a non-minimax search with $l = 1$.

However, if we use the equations for a non-minimax search with $l = 1$ to find the model scores for positions of rank $n - 1$ the probabilities $P_{n,n-1}$ and $Q_{n,n-1}$ will have the respective values 0.9937 and 0.0862. The point $(P_{n,n-1}, Q_{n,n-1})$ lies in our region of convergence for a minimax search, so that the model scores msc(A_k) for positions A of rank $k < n-1$ may be determined by Zermelo's formula, and as k decreases they tend to their desired values.

The case $p = 0.995$, $q = 0.2$ requires a somewhat more complex treatment. The point (p, q) of the PQ- plane lies outside of the region of convergence depicted in Figure 36. Suppose we are dealing with a Shannon model of depth n. If we apply Zermelo's formula to determine the model scores for positions of rank $n-1$, we find that $P_{n,n-1} = 0.9600$, $Q_{n,n-1} = 0.0050$. If we apply a non- minimax method with $l = 1$ to the positions of rank $n-2$ we find $P_{n,n-2} = 0.9928$, $Q_{n,n-2} = 0.0581$.

The standard Zermelo formula should be applied to determine the model scores for positions of lower rank.

In these last two examples an even better result can be obtained by drawing into the determination of the model score at a position A not only the model scores msc(B) at the positions B reached by the moves $\Psi = (A, B) \in \mathfrak{A}$ but also the value of the evaluation function $f(A)$. Let us look at such a rule:

(1) if two or more moves lead from A to positions B with msc(B) = $|1 - \text{col } A|$, then msc(A) = $|1 - \text{col } A|$;
(2) if there are no such moves, then msc(A) = col A;
(3) if one and only one move leads to a position B with msc(B) = $|1 - \text{col } A|$ then msc(A) = $f(A)$.

In a Shannon model with the condition that the model scores of all positions of higher rank are defined in the same way, the use of the above rules in computing the model scores for positions of rank k leads to the following recursion equations:

$$1 - P_{n,k} = Q_{n,k+1}^s P_{n,k+1}^{m-s} + \left(s Q_{n,k+1}^{s-1}(1 - Q_{n,k+1}) \right.$$
$$\left. \times P_{n,k+1}^{m-s} + (m-s)Q_{n,k+1}^s P_{n,k+1}^{m-s-1}(1 - P_{n,k+1}) \right)(1-p),$$

$$1 - Q_{n,k} = P_{n,k+1}^m + m P_{n,k+1}^{m-1}(1 - P_{n,k+1})(1-q).$$

Since normally $(1 - p)$ is small and $(1 - q)$ more or less near to 1 the probability $P_{n,k+1}$ is near to the value obtained by a minimax search, and $Q_{n,k+1}$ is near the value obtained by our earlier method with $l = 1$. If we apply the above rules to determine the model scores of all positions, the region of convergence of the P_n and Q_n to their desired values contains a portion lying outside the region depicted in Figure 36 for a minimax search. For example, with $m = 10$, $s = 2$ the point $p = 0.95$, $q = 0.15$ lies in the region of convergence.

We now consider probabilistic postulates relating to the ordering of moves in our game with respect to their expected quality. As before, we suppose that White and Black move alternately, that the true scores sc(A) and the evaluation function $f(A)$ take on the values 0 and 1, and that the model scores msc(A) of positions A in the tree of the model game $\mathfrak{A}_\mu$ are defined by the customary minimax search using the the values $f(F)$ of the evaluation function $f(F)$ of the terminal positions F in the model game $\mathfrak{A}_\mu$ (if F is also terminal in the original game $\mathfrak{A}$, then naturally $f(F) = \mathrm{sc}(F)$). We stipulate that the following probabilistic postulates are satisfied:

(1) The conditional probabilities $p(F)$ and $q(F)$ that $f(F) = |1 - \mathrm{col}\,F|$ for terminal positions F in the model game $\mathfrak{A}_\mu$, under the respective hypotheses that $\mathrm{sc}(F) = |1 - \mathrm{col}\,F|$ and that $\mathrm{sc}(F) = \mathrm{col}\,F$, are mutually independent.

(2) Also mutually independent for non- terminal positions $A \in \mathfrak{A}_\mu$ are the probabilities $\pi(A, \Omega)$ that if $\mathrm{sc}(A) = |1 - \mathrm{col}\,A|$ then the moves $\Psi = (A, B) \in \Omega \subset M(A)$, and only those moves, are winning moves.

$$\pi(A, \Omega) := P\!\left(\mathrm{sc}(B) = \left\{ \begin{array}{ll} |1 - \mathrm{col}\,A|, & \text{if } \Psi \in \Omega \\ \mathrm{col}\,A, & \text{if } \Psi \notin \Omega \end{array} \right\} \middle| \mathrm{sc}(A) = |1 - \mathrm{col}\,A| \right).$$

For non-terminal positions A in the model game $\mathfrak{A}_\mu$, provided the subtrees $\mathfrak{A}_\mu(A)$ do not intersect, these postulates imply the mutual independence of the conditional probabilities $P(A)$ and $Q(A)$ that the model scores will have the value $\mathrm{msc}(A) = |1 - \mathrm{col}\,A|$ under the respective hypotheses that $\mathrm{sc}(A) = |1 - \mathrm{col}\,A|$ and that $\mathrm{sc}(A) = \mathrm{col}\,A$. Then the probabilities in question are given by:

$$P(A) = 1 - \sum_{\Omega \in M(A)} \pi(A, \Omega) \prod_{\Psi = (A,B) \in \Omega} Q(B) \prod_{\Psi = (A,B) \in M(A) \setminus \Omega} P(B),$$

$$Q(A) = 1 - \prod_{\Psi = (A,B) \in M(A)} P(B).$$

We may use dynamic programming in constructing the models $\mathfrak{A}_\mu$. Suppose that we have constructed the conditionally optimal models $\mathfrak{A}_{p,q}(B_i)$ for the positions B_i reached by the moves $\Psi_i = (A, B_i) \in M(A)$, with the probabilities $P(B_i) = p_i$, $Q(B_i) = q_i$, and with the smallest set $V(p_i, q_i, B_i)$ of positions that satisfy these equations. We consider a model with the tree

$$\mathfrak{A}_\mu(A) = A \cup M(A) \cup \bigcup_{\Psi_i = (A,B) \in M(A)} \mathfrak{A}_{p_i, q_i}(B_i).$$

For this model

$$P(A) = 1 - \sum_{\Omega \in M(A)} \pi(A, \Omega) \prod_{\Psi_i = (A,B_i) \in \Omega} q_i \prod_{\Psi_i = (A,B_i) \in M(A) \setminus \Omega} p_i,$$

$$Q(A) = 1 - \prod_{\Psi_i = (A,B_i) \in M(A)} p_i,$$

$$V(\mathfrak{A}_\mu) = \sum_{\Psi_i = (A,B_i) \in M(A)} V(p_i, q_i, B_i) + 1.$$

Thus, to construct the conditionally optimal models $\mathfrak{A}_{p,q}(A)$ with minimum number of positions, under the proviso that $p(A) = p$, $q(A) = q$, it suffices to consider only the conditionally optimal models $\mathfrak{A}_{p_i, q_i}(B_i)$, $\Psi_i = (A, B_i) \in M(A)$.

Of course, there may be very many such models. However, if we are content with approximate optimality the number of competing models can be decreased. We may for instance consider only those models $\mathfrak{A}_{p_i, q_i}(B_i$ for which the number of positions $V(p_i, q_i, B_i)$ is minimal for $p_i \geq p$ or $q_i \leq q$. Or we may consider such models, not for all values of p and q but only for values in a more or less particularized net. Finally, disregarding the fact that the set of values of p and q for which conditionally extremal models $\mathfrak{A}_{p, q}(B_i)$ exist is discrete, we may construct by interpolation continuous functions $V(p, q, B_i)$ and apply known techniques for finding conditional extrema of functions of several variables.

Much computation is required for the determination of the values of $P(A)$ by the use of the above formula. To save computation we may limit the number of sets Ω of winning moves, by considering only those for which the corresponding probabilities $\pi(A, \Omega)$ are not too small. We may also impose a natural requirement that the values of these probabilities should be determined by a relatively small number of parameters. With these postulates the computation of the $P(A)$ is greatly simplified, and it becomes possible to make effective use of dynamic programming.

Research on methods for shortening the search by the use of various probabilistic assumptions is only just beginning, and we may expect interesting results from it. In particular, great interest attaches to a combination of the methods of non-minimax search, construction of models, and use of *a posteriori* information about results obtained earlier in the search. Also of great interest are statistical studies needed for the solution of problems about the applicability of various postulates concerning real games, about the usable evaluation functions and other parameters that define the character of a position in the tree of a model game, and also on the concrete numerical values of various probability characteristics.

Remarks on the Propriety of the Probabilistic Approach

The use of probabilistic postulates as a basis for choosing a model game and a search method may meet with principial objections, even though these postulates are supported by substantial statistical evidence in connection with such games as, for example, chess. In fact, every chess position has a true score established by a deterministic process. Also determined are the values of an arbitrary evaluation function applied in a model game. Therefore 0 and 1 are the only possible values for the probabilities of the occurrence of an elementary event, e.g. that a given position A has the true score $sc(A) = x$, or that the evaluation function $f(A) = y$, or that the value of some deterministically defined parameter $t(A) = t$. This means that the probabilities of compound events can have only the values 0 and 1, and are interrelated deterministically.

Analogous objections may be raised against many applications of probability theory, to which everyone has long since become accustomed. Clearly, the objections are not less convincing merely because of this historical acceptance. They were raised by the eighteenth century founders of probability theory. At that time, as a basis for the applications of the theory, it was argued that events governed by sufficiently complex laws behave in many relationships like random variables. A. N. Kolmogorov [20] has put forward an algorithmic approach to the concept of complexity, and has to some extent supported this argument.

We shall not give Kolmogorov's exact definition of complexity. We note only that the complexity of a finite subset Ω of a set $\mathfrak{M}$ of objects each having a finite description is defined as the length of the shortest program that will generate the subset Ω. There is a valid theorem to the effect that the values of sufficiently simple functions $\varphi(\Omega)$ defined on sufficiently complex subsets of the set $\mathfrak{M}$ will be near the correspondingly determined probabilities when Ω is taken as a set of random elements.

When we are studying the results of a minimax search in a Shannon model of depth n the finite subsets Ω are the subsets of positions in the tree of the game $\mathfrak{A}$ that satisfy the following conditions: their rank is n, their true score has a given value, and they belong to given subtrees of the Shannon model $\mathfrak{A}_\nu$. However, such subsets are not complex in the sense of Kolmogorov. To define them one need only search all positions in the game $\mathfrak{A}$, compute the true scores of all positions, and define the corresponding subsets Ω. A comparatively short program will accomplish this task.

On the other hand, in the course of developmental work on Kolmogorov's algorithmic approach to the concept of complexity, another definition of complexity has been formulated (cf. [23]). In this formulation the complexity of a program is taken as the sum of its length and the logarithm of the running time; the complexity of a subset Ω is the minimal complexity of programs that generate Ω. Thus the known algorithms for determining the true scores of positions turn out to be complex for the majority of the positions that have interested humans or game-playing programs since the invention of chess (but not for positions in chess problems requiring mate in a given number of moves). At the same time, the values of the evaluation functions and model scores for different search methods require relatively simple functions.

For the concept of complexity defined above, there are valid theorems on the values of simple functions of complex subsets, which can be formulated in the same way as the theorems based on the Kolmogorov definition. Apparently they are sufficient for the proof of the theorem on the relations between model scores and true scores, under the conditions that there exist no algorithms for defining true scores of positions more effective than those now known, and that the corresponding relationships between the values of the evaluation function and the true scores are valid.

If the first assumption holds, the second can be tested by standard statistical devices, good or bad, just as in the testing of any probabilistic hypothesis. The situation respecting the first assumption is notably worse. To validate it one must almost unavoidably inspect all algorithms displaying the desired effectiveness and prove that none will determine a true score, i.e. find a position whose score as computed by the given algorithm is incorrect (or, more exactly, one must show that for almost all tasks that may in principle be posed, the situation is exactly as described).

Thus the admissibility of the probabilistic approach to a determination of the objective quality of a program that uses a model game can be based only on an assumption about the non-existence of sufficiently simple algorithms that solve correctly the problem of computing true scores. Any method for defining complexity, if it is to be well supported by experiment, cannot make use of a highly complex exact algorithm, and the basic postulate for such a method may well turn out to be practically impossible to validate.

A similar situation has been met in discrete mathematics. A way out was found in the formulation and proof of theorems of a conditional character. A number of problems were proved to be universal: either there exists no algorithm for their solution having its running time dependent on a power of the dimension of the problem, i.e. the length of the description of a concrete instance, else such algorithms exist for all exactly posed discrete problems (a description is a word of finite length by which the object, about which the problem is posed, is uniquely established, so that if the running time is unlimited the problem can be unconditionally solved).

Conditional theorems on probabilistic models of games may be highly optimistic, e.g. theorems on the high quality of determinations of true scores based on the validity of some precisely formulated assertions about some set of objects having finite descriptions.

Nevertheless, without awaiting the formulation and proof of such theorems, we may use the probabilistic approach to the construction of new types of model games, search methods, algorithms for computing evaluation functions, and the selection of parameters defining the work of game-playing programs.

Appendix

A Sketch of Past Work on Algorithms for Games

Game programming, as one of the fundamental tasks for artificial intelligence, has drawn the attention of of many research workers, with diverse aims in regard to both the creation of game-playing algorithms and developmental work on them. The approaches employed, and the successes achieved, are correspondingly diverse.

At the dawn of the computer age, when the power of computers was not as evident as it is now, various games played by computers served more than once to demonstrate the wide scope of their powers. Examples can be found in the work of J. v. Neumann and O. Morgenstern [129] and N. Wiener [165].

C. E. Shannon devoted two papers [39,40] to the foundations of a chess algorithm; in these he sketched an approach so general that all current chess-playing algorithms appear as implementations of it—to be sure, with modifications.

At about the same time A. M. Turing [162], devising the same scheme in somewhat more detail, undertook an experiment using a manual simulation of a chess program. This simulation resembled a game played by a rather weak player. (We omit further consideration of the results of this and other games played by chess programs. Well-annotated texts of many such games may be found in a book by M. Newborn [131]).

In 1957 a program was developed at Los Alamos that played on a reduced board having 6×6 squares: it played three games. And, finally, in 1958 A. Bernstein and others [68,69] wrote a program that played normal chess. Although it played very poorly, it settled in the affirmative the question of whether machines could play chess.

In the sixties, many algorithms were written for various games, e.g. dominoes [30], noughts-and-crosses [31], bridge [59], and others. The general opinion was that the establishment of programs that would outplay the strongest players was only a matter of technique and a small amount of time. In a few years it became clear that the development of actually strong programs would require a substantial amount of work.

Precisely at this time there arose a legend, in our opinion not valid, but still widely accepted even now, that the development of strong algorithms for games required the participation of strong players. Among the adherents of this view are some who take a direct part in the programming—M. M. Botvinnik [7–9], H. Berliner [67], A. Steiner [32], M. Newborn—and others who take no part in programming but make up for it by their acknowledged authority in the game itself—M. Tal, R. Fischer, and others.

Disregarding the fact that in our group there are well-qualified chess players, we believe (and hope that this book is sufficiently eloquent testimony to our belief) that game programming is a matter of cybernetics and that the development of game-playing programs demands no less specific work on the algorithms than it requires knowledge of the specific game. The latter may even turn out to be detrimental, since it is not always well formalized.

A second legend says that the computer should think like and 'in the image of' the thought processes of a human, and that the thesis mentioned above is unconditionally correct. However, if our aim is to learn how to use computers for the solution of complex intellectual problems, we unjustifiably complicate the task if we first solve the problem itself, perhaps more difficult than the problem of developing computer methods.

All these legends, and the many debates arising in connection with game programming, stem from the fact that its methods have not been established nor has the topic been defined. Only one aim is clear—to create a program that will play a popular and complex game (primarily chess) better than the human player.

A notable group of theoreticians takes its departure point from experiments in the use of game programs to refine a model of human thought processes. Essentially, this work lies in the field of psychology—A. de Groot [88, 89], O. Tikhomirov [34], N. Charness [77], study the techniques of human chess players and attempt to reduce them to programs.

Botvinnik's work [7–9] sides ideologically with this trend. It is not the human thought processes in chess, but rather Botvinik's own thought processes (as he presents them) that form the basis for his construction of strong chess programs. This work is distinguished by the fact that it is accompanied by an actual program implementing the algorithm developed by Botvinnik's methods. The creative work is in its final stages, and one must wait until the program is put into play before appraising it.

A principal group of theoreticians in the field of game-playing programs constructs speculative concepts of the way humans think when engaged in playing games, and attempts to support these concepts by creating the corresponding algorithms. The majority of the investigators choose games as the intellectual activity of the computer and test the effectiveness of the general algorithm on game models.

The first such work was the (today classic) program 'General Problem Solver' of A. Newell, H. Simon, and J. Shaw [29] dating from 1958. Putting forward the hypothesis that thought is exhaustive, they wrote a program implementing the general exhaustive enumeration scheme: one interesting implementation of the scheme was a chess-playing program.

D. Michie [124] chose games as a model for working out the possibilities for machine representation of knowledge. This work has the insufficiency characterizing all work in this direction.

It is fashionable to choose a trivial game, so trivial that the advantages of the proposed concept remain unclear. (For the 25 years in which the problem of artificial intelligence has existed, the most popular model has up to now been the puzzle called the 'Tower of Hanoi'.) For Michie the model is the endgame with King and Rook against King. He is now completing his work with the more complicated endgame of King and Rook against King and Knight.

A large part of the work on game programming is devoted to methods of pattern recognition for various games. In essence these papers express a faith in the necessity of including such procedures in game-playing programs. There exists only one chess-playing program making use of pattern recognition. This is a program by A. Zobrist, F. Carlson, and K. Kalma [168–170]. It has taken part in the ACM Computer Chess Tournaments, but has invariably ended in next to last place.

Much work has been done on learning programs. Learning was applied, for instance, in a program by Kh. Brutyan et al. [12] to play simple endgames.

A. Samuel credits learning with the successes of his program for playing American checkers. [33,179]. This program is already a legend on, so to speak, a local scale. Learning in Samuel's program reduces to local variation in the weights of various factors in the evaluation function. The program carries out a complete search to a depth of nine plies with an exhaustive inspection of forced variations. In all probability, such a search depth is beyond the capabilities of even highly qualified players, and the evaluation function can vary within wide limits. For comparison, we note that the strongest chess programs use a search depth of six plies.

The work of H. Berliner [62] stands alone; he uses a so-called tactical analyzer—an algorithm that analyzes the tactical causes of failure of the plans developed by the program, and attempts to find a way around these failures. Berliner has written a program including this algorithm, but because of an ineffective representation of information, the program is a mediocre player.

J. Pitrat [137,138] developed the same idea in a program for finding combinations. He proposed a unique and very important simplification—the program knows what exactly it has to win; its task is 'find the combination winning the exchange'.

J. McCarthy [122] used chess as a model of commonsense.

R. Gadzhiev [13] used the principles of situational management to construct endgames in chess.

A group of papers has been devoted to search strategies. We note only the work of R. Banerji [49,50], K.Church and R. Church [79], and R. Atkin [45,46] (not to be confused with one of the authors of CHESS-4).

The remaining group seeks specific machine methods for the solution. Even Newell, Simon, and Shaw did not refrain from including in their programs various search methods, in particular the α, β-procedure, regardless of whether humans use such methods.

A characteristic feature of this direction is that the work is based on functioning programs. Unfortunately, however, not every functioning program forms a basis for serious research. A significant number of programs owe their existence to the enthusiasm of their creators, and in essence contribute nothing new to the body of research on game-playing algorithms. Nevertheless, the majority of playing programs, and especially the strong ones, are the fruit of serious scientific work.

The first example of such work, on which one would want to dwell in detail, is the Odnomastka program of A. Brudno and I. Landau [11]. Its creation was accompanied by theoretical developments. The correctness of all the methods employed for shortening the search was mathematically proved. In particular, the α, β-heuristic was first theoretically based in the course of this work. Also, their program possessed an ability—unique up till now—to set traps, i.e. to choose, from among equally valued moves, one that required of the opponent the only correct reply. Together with the high technical level of the programming, all this led to the creation of an absolutely invincible program. The only regrettable aspect of the work is that the chosen game is practically never played by humans, and the excellent quality of the program is not accorded the dignity due it.

An interesting method for realizing Shannon's scheme for chess, together with a number of effective heuristics, was applied in the chess program ITEF [2]. This publication contains the proof of an important theoretical result on the optimality of the α, β-procedure for an arbitrary game.

The majority of the American programs in the 60's [121] were devoted to the method of evaluation functions. In this method, not all moves from intermediate nodes in the search are made, but only those that are best according to some evaluation function. Usually they employ a function of the score that arises after a move to the position.The best program employing this principle is due to R. Greenblatt [104]. Moves are evaluated by an essentially different function, applied to the move itself and not to the position arising after the move.

Today the two strongest programs in the chess world, CHESS-4 [153] and KAISSA [15], are based on serious theoretical development of methods for

Table 8. North American Championship Tournament,
First International Championship for Chess Programs: 1974 List of Participants

Program	Authors	Computer	Program Development Locus	Country
1. KAISSA	Adelson-Velsky, Arlazarov, Donskoy	ICL-4/70	Institute of Control Problems Moscow	USSR
2. CHESS 4.0	Slate, Atkin	CDC-6600	Northwestern University, Illinois	USA
3. RIBBIT	Parry, Hansen, Crook	Honeywell 6000 ASEA	University of Waterloo	Canada
4. CHAOS	Ruben, Swarz, Winograd, Berman, Toikka	Univac 1110	Univac Corporation	USA
5. TECH II	Baisley	DEC PDP-10	MIT	USA
6. OSTRICH	Arnold, Newborn	Data General Nova 2	Columbia University	USA
7. FRANTZ	Wolf	Univac 494	Computer Center Graz	Austria
8. MASTER	Kent, Birmingham	IBM-370/195	Atlas Laboratory	England
9. BEAL	Beal	CDC-6400	Queen Mary College, London	England
10. FREEDOM	Barricelli	CDC Cyber-74	University of Oslo	Norway
11. TELL	Joss	HP-2100	Federal Technical University Zurich	Switzerland
12. A16CHS	Prinsen	GCS Alpha-16	International Data Systems	England
13. PAPA	Reiner, Almasi	CDC Cyber-73	Budapest University	Hungary

Table 9. North American Championship Tournament for Chess Programs: 1975

Program	Authors	Computer	Program Development Locus	Country
1. TREE FROG	Hansen, Kalnek, Crook	Honeywell 6080	University of Waterloo	Canada
2. CHESS 4.4	Slate, Atkin	CDC-6400	Northwestern University	USA
3. CHAOS	Ruben, Swartz, Winograd, Berman, Toikka	Amdahl 470	Amdahl	USA
4. OSTRICH	Newborn, Arnold	Data General Nova 2	McGill University	Canada
5. DUCHESS	Wright, Truscott	IBM 370/165	Duke University, North Carolina	USA
6. CHUTE 1.2	Valenti, Vraneshitch	IBM 370/165	University of Toronto	Canada
7. TYRO	Zobrist, Carlson	IBM 370/158	Univ. of South California	USA
8. SORTIE	Becker, Mammon, Anderson, Mann, Egan, Swingle	SIGMA-7	Bucknell University	USA
9. VITA	Marsland	IBM 360/67	University of Alberta	Canada
10. ETAOIN SHRDLU	Courtois	Nova-1200	University of Colorado	USA
11. BLACK KNIGHT	Sogg, Maltsen, Losov, Prouse	Univac 1110	UNIVAC, St. Paul, Minnesota	USA
12. IRON FISH	Buss, Mandstock	CDC Cyber-74	University of Minnesota	USA

Table 10. European Championship for Chess Programs: 1976

Program	Authors	Computer	Program Development Locus	Country
1. TELL	Joss	HP-2115	Federal Technical University Zurich	Switzerland
2. DEJA	Jan, Zagler	TR-440	Leibnitz University, Munich	FRG
3. SCHACH MB 5.6	Richter	TR-4, TR-440	Inst. of Informatics, Hamburg	FRG
4. ORWELL 3	Nitsche	UNIVAC 1106/2	University of Freiburg Computation Center	FRG
5. FISHER-SCHNEIDER	Fisher, Schneider	TR-4, TR-440	University of Stuttgart	FRG
6. SAMILL	Klein, Kruger	IBM 370/168	Bonn University Computation Center	FRG
7. PROSCHA	Hewitt, Appelrath, Franzen, Schulz, Schulz, Teschers, Fauriberger	IBM 370/158	Computing Center of the GKhK Dortmund	FRG
8. CHARLY	Keil	Siemens 4004/45	Gymnasium sv. Anny, Augsburg Computing Center	FRG

Table 11. Second World Championship Tournament for Chess Programs: 1977 List of Participants

Program	Authors	Computer	Program Development Locus	Country
1. CHESS 4.6	Slate, Atkin	CDC Cyber-176	Northwestern University, Illinois	USA
2. DUCHESS	Truscott, Wright, Jensen	IBM 370/165	Duke University, North Carolina	USA
3. KAISSA	Adelson-Velsky, Arlazarov, Donskoy	IBM 370/168	VNIISI, Moscow	USSR
4. BLITZ 5	Hyatt	Sigma 9	University of Southern Mississippi	USA
5. MASTER	Birmingham, Kent	IBM 370/168	Rutherford Laboratory, Harwell	England
6. TELL	Joss	DEC K 110	Federal Technical University Zurich	Switzerland
7. BELLE	Thompson, Condon	PDP-11	Bell Laboratories, New Jersey	USA
8. VITA	Marsland	Amdahl 470 V/6	Toronto University	Canada
9. OSTRICH	Newborn, Arnold	Supernova	McGill University	Canada
10. DARK HORSE	Ratsman	CDC-6600	Eriksen Telephone Company, Stockholm	Sweden
11. BCP	Beal	CDC-6400	Queen Mary College, London	England
12. ELSA	Zagler	TR-440	Technological University, Munich	FRG
13. CHAOS	Alexander, Macbride, Swarz, Toikka, Berman, Winograd	Amdahl 470 V/6	University of Michigan	USA
14. BLACK KNIGHT	Sogg, Prouse, Malzen, Leban, Adams	Univac 1110	Sperry-Univac, Minnesota	USA
15. CHUTE 1.2	Valenti, Vraneshitch	Amdahl 470 V/6	Toronto University	Canada
16. BS'66'76	Svets	IBM 370/168	Unaffiliated	Netherlands

shortening the search. Both programs painstakingly avoid prunings that might change the results of the search.

The theory of searching has also attracted other investigators. D. Knuth [115] and M. Newborn [133] have studied the theory of the α, β-procedure.

L. Harris [108,109] proposed some new schemes, unfortunately not supported by the development of programs. J. Birmingham and O. Kent [71], authors of the strong West European program 'Master', developed and implemented in it several interesting heuristics.

Here we should like to eliminate some misunderstandings, essentially of a terminological nature. It is currently customary to refer to a search without pruning as a full search. We, on the other hand, continually emphasize the fact that KAISSA employs a full search, meaning in our case that the results of the search agree at all positions. In other words, the program does not make erroneous prunings.

Control by search has its shortcomings. The basic one was remarked by H. Berliner [67]—the approximate character of the model employed. When the search parameters have been chosen we have defined only an approximation of chess. For a small search depth this approximation is very rough. With an increase in the depth of search the quality of the approximation improves, but apparently not as rapidly as one would like.

We should note yet another paper in chess programming. E. Fredkin and R. Greenblatt developed at MIT a specialized microprocessor implementing the technical side of a chess game. Thus people having ideas as to how to construct a chess program but not wishing to spend years in writing a program to prove the correctness of their ideas, may test them in a short time without having to spend time in tedious routine work. Experience has shown that without a trial in an actual program it is impossible to test the fruitfulness of one's ideas.

Today the number of functioning game programs is satisfactorily large. Tournaments for the games of Go and Noughts-and-crosses are held regularly, and computer chess tournaments have been held under the aegis of IFIP. Tournaments have been held for two World Championships, seven North American Championships, and a European Championship. Tables 8–11 list the official data on the entrants in the World Championship, the 1975 North American Championship, and the European Championship.

Material pertaining to game programming is published regularly in the journals Artificial Intelligence (Holland), Man–Machine Studies (USA) and in the Proceedings of Biennial Conferences on Artificial Intelligence.

Summary of Notations

References

1. Adelson-Velsky G. M., Arlazarov V. L. Metody usileniya shakhmatnykh programm. (Methods of strengthening chess programs.)–Problemy kibernetiki, 1974, **29**, 167–168.
2. Adelson-Velsky G. M., Arlazarov V. L., Bitman A. R., Uskov A. V. and others, O programmirovanii igry vychislitelnoi mashiny v shakhmaty. (On programming computers to play chess.)–UMN, 1970, **25**, vol. 2, (152),221–260.
3. Adelson-Velsky G. M., Landis E. M. Odin algoritm organizatsii informatsii. (An algorithm for organizing information.)–DAN SSSR, 1962, **146**, No. 2, 263–266.
4. Arlazarov V. L., Futer A. L. and others. Obrabotka bol'shykh massivov informatsii na primere analiza ladeinovo endshpielya. (Processing large volumes of data for the analysis of endgames with Castles.)–Programmirovanie, 1977, No. 4, 45–54.
5. Bellman, R. Dinamicheskoe programmirovanie. (Dynamic Programming)–M.: IL, 1960.
6. Bongard M. M. Problema uznavaniya. (The recognition problem.)–M.: Nauka, 1967.
7. Botvinnik M. M. Algoritm igry v shakhmaty.–M.: Nauka, 1967.
8. Botvinnik M. M. Blok-skhema algoritma igry v shakhmaty. (Block diagram of an algorithm for chess.)–M.: Sov. Radio, 1972.
9. Botvinnik M. M. O kiberneticheskoe tseli igry. (On the cybernetic goals of a game.) M.: Sov. Radio, 1975.
10. Brudno A. L. Grani i otsenki dlya sokrashcheniya perebora variantov. (Bounds and estimates for shortening the search of variations.) Problemy kibernetiki, 1963, **10**, 141–150.
11. Brudno A. L., Landau I. Ya. Odnomastka (Programmirovanie igrovoi zadachi.) (Odnomastka: programming a game.) Problemy kibernetiki, 1965, **13**, 141–160.
12. Brutyan Kh. K., Zaslavskii I. D., Mkrtchyan L. V. O nekotorykh metodakh matematicheskogo sinteza positsionnykh strategii v igrakh. (On some methods for mathematical synthesis of positional strategies in games.)–Problemy kibernetiki, 1967, **19**, 141–175.
13. Gadzhiev R. E. Eksperimental'noe issledovanie protsessa prinyatiya resheniya chelovekom i ego modelirovanie s pomoshch'yu metoda situatsionnogo up-

ravleniya (na primere shakhmatnogo endshpielya.) (An experimental study of
human decision-making process and a model of it using the method of
situational control (with endgames in chess as an example.)–M., 1975, 17–27.
(Trudy 4-i Mezhdunarodnoi konferentsii po isskustvennomu intellektu, **10**.
(Proceedings of the 4th International Conference on Artificial Intelligence, **10**.)

14. Gol'fand Ya. Yu., Futer A. L. Realizatsiya debyutnoi spravochnoi dlya
    shakhmatnoi programmy. (Implementation of a reference facility for openings
    in chess programs.) Problemy kibernetiki, 1974, **29**, 201–210.

15. Donskoi M. V. O programme, igrayushchei v shakhmaty. (On a chess-playing
    program.)–Problemy kibernetiki, 1974, **29**, 169–200.

16. Evgrafov M. A., Zadykhailo I. V. Nekotorye soobrazheniya o programmiro-
    vanii shakhmatnoi igry. (Some observations on algorithms for chess.)–Prob-
    lemy kibernetiki, 1965, **15**, 135–156.

17. Ershov A. P. O programmirovanii arifmeticheskikh operatorov. (On program-
    ming arithmetic operators.)–DAN SSSR, 1958, **118**, No. 3, 427–430.

18. Efimov E. I. Modelirovanie shakhmatnykh okonchanii na osnove avtomati-
    zatsii dokazatel'stva teorem. (Modelling chess endings using automatic theo-
    rem proving.)–Izv. AN SSSR. Ser. tech. cybern., 1977, No. 2, 47–60.

19. Knut D. Isskustvo programmirovaniya dlya EVM. (D. E. Knuth. The Art of
    Computer Programming, v. 2, Seminumerical Algorithms.) M.: Mir, 1977.

20. Kolmogorov A. N. K logicheskim osnovam teorii informatsii i teorii veroyat-
    nostei. (Toward the logical foundations of information theory and probability
    theory.)–Problemy peredachi informatsii, 1969, **5**, No. 3, 3–7.

21. Komissarchik E. A., Futer A. L. Ob analize ferzevogo endshpilya pri pomoshchi
    EVM. (Computer-assisted analysis of Queen endgames.)–Problemy kiber-
    netiki, 1974, **29**, 211–210.

22. Kronrod V. A. Krestiki-noliki na pole 5×5. (Noughts and crosses on a 5×5
    board.)–In: O nekotorykh voprosakh teoreticheskoi kibernetikii algoritmakh
    programmirovaniya. (On some problems in theoretical cybernetics, and on
    programming algorithms.) Novosibirsk, 1971, 185–210.

23. Levin L. A. Universal'nye zadachi perebora. (Universal search prob-
    lems.)–Problemy peredachi informatsii, 1973, **9**, No. 3, 115–116.

24. Maizlin I. E. Ob odnom sposobe poiska informatsii i ego primenenii dlya
    realizatsii na EVM algoritma nakhozhdeniya kriticheskogo puti. (On a method
    for information search and an application to a computer implementation of an
    algorithm for finding the critical path.)–DAN SSSR, 1964, **159**, No. 4, 761–763.

25. Minskii M., Peipert S. Perseptrony. (M. Minsky, S. Papert. Perceptrons, An
    Introduction to Computational Geometry. MIT Press, Cambridge 1969.)–M.:
    Mir, 1971.

26. Neiman J. fon. K teorii statisticheskhykh igr. (v. Neumann J. Toward a theory
    of statistical games.)–V kn.: Matrichnye igry. (In the book: Matrix games.)
    M.: Fizmatgiz, 1961, 173–204.

27. Nil'son N. J. Iskusstvennyi intellekt. Metody poiska reshenii. (Nilsson N. J.
    Artificial intelligence. Methods of search for solutions.)–M.: Mir, 1973.

28. N'yuell A., Saimon G. GPS-programm, modeliruyushchaya protsess chelo-
    veshcheskogo myshleniya.–V. kn.: Vychislitel'nye mashini i myshlenie. (Newell
    A., Simon H. GPS, a program that simulates the process of human thought. In
    the book: Computers and Thought.) M.: Mir, 1967, 283–301.

29. N'yuell A., Shou J., Saimon G. Programma dlya igry v shakhmaty i problema
    slozhnosti–V kn.: Vychislitel'nye mashiny i myshlenie. (Newell A., Shaw J.,
    Simon H. A program for chess and the problem of complexity.) Ibid., 33–70.

30. Pervin Yu. A. Ob algoritmizatsii i programmirovanii igry v domino. (On algorithms and programs for the game of dominoes.)–Problemy kibernetiki, 1960, **3**, 171–180.

31. Srapyan Sh. O., Ter-Mikaelyan T. M. Ob odnom metode otsenki situatsii v krestiki i noliki. (On a method for assessing a position in the game of noughts-and-crosses.)–Problemy kibernetiki, 1963, **9**, 171–176.

32. Steiner A. M. BRIBIP-programma, osushchestvlyushchaya torgovlyu v bridzhe. (The program BRIBIP, implementing the bidding process in bridge.)–(Trudy 4-i mezhdunarodnoi obedinennoi konferentsii poiskustvennomu intellektu, **3**) (Proceedings of the 4th International Conference on Artificial Intelligence.)

33. Semyuel A. Nekotorye issledovaniya vozmozhnosti obucheniya mashin na primere igry v shashki.–V kn: Vychislitel'nye mashiny i myshlenie. (Samuel A. Some studies in machine learning using the game of checkers.–In: Computers and Thought) M.: Mir, 1967, 71–111.

34. Tikhomirov O. K., Poznanzkaya E. D. Issledovanie visual'nogo poiska kak sredstvo analizirovaniya evristik. (Studies of visual search as a means for analyzing heuristics.)–Voprosy psikhologii, 1966, **2**, 39–53.

35. Zermelo E. O primeneniya teorii mnozhestv k teorii shakhmatnoi igry.–V kn: Matrichnye igry. (On the application of set theory to the theory of chess.–In: Matrix games.) M.: Fizmatgiz, 1961, 167–172. (See also English translation of German paper at 5th Int. Cong. of Math. 1912.—Firbush News, **6**, July 1976, 37–42.)

36. Chikul V. M. Metod tochnechnykh baz dlya sokrashcheniya perebora variantov. (The method of point bases for shortening the search of variations.)–Voprosy kibernetiki i vychislitel'noi matematiki, Tashkent, 1966, **5**, 35–45.

37. Chikul V. M. Ob evristicheskom programmirovanii intellektual'nykh igry. (On heuristically programmed intellectual games.)–Voprosy kibernetiki i vychislitel'noi matematiki, Tashkent, 1968, **16**, 54–72.

38. Chikul V. M. Universal'naya evristicheskaya igrovaya programma. (A universal heuristic game-playing program.)–V kn: Konf. po teorii avtomatov i iskusstv. myshl. Tashkent, 1968, 11–13.

39. Shennon K. E. Mashina dlya igry v shakhmaty. (Shannon C. E. A chess-playing machine.)–V kn: Shennon K. E. Raboty po kibernetiki i teorii informatsii. (In: Shannon C. E. Papers on cybernetics and information theory.) M.: Fizmatgiz, 1956, 180–191. (Scientific American, v. 182, Feb. 1950, 48–51).

40. ———. Igrayushchie mashiny. (Game-playing machines.) Ibid, 216–223.

41. Adelson-Velsky G., Arlazarov V., Donskoy M. On the structure of an important class of exhaustive problems and on ways of search reduction for them.–Advances in Computer Chess, 1977, **1**, 1–5.

42. ———. Some methods of chess play programming.–Artificial Intelligence, 1975, **6**, 361–376.

43. ———. More commentary on the Cichelli heuristics. SIGART, 1974, **45**, 12.

44. Advances in Computer Chess, Ed. Clarke M.–Edinburgh Univ. Press, 1977, v. 1.

45. Atkin R. H., Whitten I. H. A multi-dimensional approach to positional chess.–Int. J. Man-Machine Studies, 1975, **7**, 727–750.

46. Atkin R. H., Hartston W., Whitten I. H. Fred CHAMP, Positional chess analyst.–Int. J. Man-Machine Studies, 1976, **8**, 517–529.

47. Atkin R. H. Positional play in chess by computer.–Advances in Computer Chess, 1977, **1**, 60–73.

48. Atkin R. H. Multidimensional structure in the game of chess.–Int. J. Man-Machine Studies, 1972, **4**, 341–362.
49. Banerji R. B., Ernst G. W. Changes in representation which preserve strategies in games.–In: Proc. of IJCA12, 1971, 651–658.
50. Banerji R. B. Game playing programs: An approach and an overview.–In: Theoretical approaches to non-numerical problem solving/Ed. Banerji R. B., Mesarovic M. D.–In: Proc. of the IV Systems Symposium at Case Western Reserve Univ. New York: Springer-Verlag, 1970.
51. Baylor G. W., Simon H. A. A chess mating combinations program.–Proc. of AFIPS, SJCC, 1966, **28**, 431–447.
52. Bell A. C. Techniques for playing the endgame.–Computer Weekly, 1965, April 10.
53. Bell A. G. Computer chess experiments.–See [84].
54. Bell A. G. Kalah on Atlas.–Machine Intelligence, 1967, **3**, 181–194.
55. Bell A. G. Algorithm 50: How to program a computer to play legal chess.–Computer Journal, 1970, **13**, No. 2, 208–219.
56. Bell A. G. Games playing with computers.–Allen and Unwin, London, 1972.
57. Bellman R. Stratification and control of large systems with applications to chess and checkers.–Information Science, 1968, **1**.
58. Bellman R. On the application of dynamic programming to the determination of optimum play in chess and checkers.–Proc. Natl. Academy of Sciences, 1965, **53**, 244–247.
59. Berlekamp F. Program for double dummy bridge problems.–Journal of ACM, 1963, **10**, 357–364.
60. Berliner H. J. A new subfield of computer chess.–SIGART, 1975, **53**, 20–1.
61. ――――. Computer chess.–SIGART, 1975, **55**, 14–15.
62. ――――. A representation and some mechanisms for a problem solving chess program.–Advances in Computer Chess. Ed. Clarke M. Edinburgh Univ. Press, 1977, 7–29.
63. ――――. A comment on improvement of chess playing programs.–SIGART, 1974, **48**, 16.
64. ――――. Outstanding performance by CHESS 4.5 against human opposition.–SIGART, 1976, **60**, 12–13.
65. ――――. Man against machine 1974.–Firbush News, 1976, **6**, 63–70.
66. ――――. Chess playing programs. SIGART, 1969, **17**, 19–20.
67. ――――. Some necessary conditions for a master chess program.–In: Proc. 3rd IJCAI, Stanford, 1973, 77–85.
68. Bernstein A., Roberts M. de V. Computer vs Chess Player.–Scientific American, June 1958, **198**, 96–105.
69. Bernstein A., Roberts M. de V., Arbuckle T., Belsky M. A. A chess playing program for the IBM 704.–In: Proc. WJCC, 1958, 157–159.
70. Binet A., Psychologie des grands calculateurs et des joueurs d'échecs.–P.: Hachette, 1894. (See also Mnemonic Virtuosity: A Study of Chess Players. Genetic Psych. Monog., v. 74, 1966, 127–162. [Translation of 1893 paper in "Revue des Deux Mondes, v. 117"]).
71. Birmingham J. A., Kent P. Tree-searching and tree-pruning techniques.–See [44], **1**, 89–107.
72. Bond A. H. Psychology and computer chess.–See [84], 29–30.
73. Bond A. H. Descriptor index.–See [84], 95–112.
74. Boos G., Cooper D., Gillogly J., Levy D., Raymond H., Slate D., Smith R., Mittman B. Computer Chess programs: a panel discussion.–Proc. 1971 Annual ACM Conference 25, 97–102.
75. Bramer M. A. Computer chess: The knowledge approach.–Chess, 1976, **41**, 347–349.

76. Bratko I., Trancig P., Trancig S. Some new aspects of chess board reconstruction experiments.–In: 3rd European meeting on Cybernetics and Systems Research. Vienna 1976.
77. Charness N. Human chess skill.–See [78], 34–53.
78. Chess skill in man and machine. Ed. Frey P.–NY.: Springer-Verlag, 1977.
79. Church R. M., Church K. W. Plans, goals, and search strategies for the selection of a move in chess.–See [78], 131–156.
80. Cichelli R. J. Research progress report in computer chess.–SIGART, 1973, **41**, 32–36.
81. ———. Preliminary testing of the effectiveness of the Cichelli Depth–2 and refutation heuristics.–SIGART, 1973, **42**, 49–52.
82. Clarke M. R. B. A quantitative study of King and Pawn against King.–See [44], **1**, 108–116.
83. ———. Some ideas for a chess compiler.–In: Artificial and Human Thinking, Elithorn and Jones (eds.), Elsevier, 1973, 189–198.
84. Computer Chess. Ed. Bell. L.: Atlas Labs, 1973.
85. Cooper R., Elithorn A. The organization of search procedures.–In: Artificial and Human Thinking. Elsevier, 1973, 199–213.
86. Coriat I. H. The unconscious motives of interest in chess.–Psychoanalytic Review, 1941, **28**, 30–36.
87. De Groot A. D. Chess playing programs.–Proc. Koninkl. Nederlands Akad. Wetensch. Ser. A–67, Amsterdam, 1964, 385–398.
88. ———. Thought and Choice in Chess. Hague, Mouton, 1965.
89. ———. Perception and memory versus thought: some old ideas and recent findings.–In: Problem Solving. ed. Kleinmuntz, B. NY.: John Wiley, 1966.
90. Dutka J., King K., Newborn M. A review of the first United States computer chess championship.–SIGART, June 1971, **28**, 14–23.
91. Eisenstadt M., Kareev Y. Toward a model of human game playing.–In: 3rd IJCAI, 1973, 458–463.
92. Elithorn A., Telford A. Computer analysis of intellectual skills.–Int. J. Man-Machine Studies, 1969, **1**, 189–209.
94. Euwe M. Computers and chess.–In: The Encyclopedia of Chess. St. Martin's Press, 1970.
95. Faster than Thought. Ed. Bowden B. V. L.: Pitman, 1963.
96. Findler N. V. Computer experiments on the formation and optimization of heuristic rules.–In: Artificial and Human Thinking. Elsevier, 1973, 177–188.
97. Frey P. An introduction to computer chess.–See [78], 54–81.
98. Gillogly J. Reader commentary on the Cichelli heuristics.–SIGART, 1973, **43**, 27–28.
99. ———. The Technology chess program.–Artificial Intelligence, 1972, **3**, 145–164.
100. Good I. J. Dynamic probability, computer chess, and the measurement of knowledge.–Firbush News, 1976, **6**, 43–62.
101. ———. The mystery of GO.–New Scientist, 1965, **427**, 172–174.
102. ———. A five year plan for automatic chess.–Machine Intelligence, 1966, **2**, 89–118.
103. ———. Analysis of the machine chess game J. Scott (White), ICI–1900 vs. R. D. Greenblatt, PDP–10.–Machine Intelligence, 1969, **4**, 267–269.
104. Greenblatt R. D., Eastlake D. E., Crocker S. D. The Greenblatt chess program.–In: Proc. FJCC, 1967, **31**, 801–810.
105. Griffith A. K. Empirical exploration of the performance of the alpha-beta tree search heuristic.–In: IEEE Trans. on Computers, Jan. 1976, 6–10.
106. ———. A comparison and evaluation of three machine learning procedures as applied to the game of checkers.–Artificial Intelligence, 1974, **5**,137–148.

107. Harris L. R. Heuristic search under conditions of error, and plan oriented play.–Artificial Intelligence, 1974, **5**, No. 3, 217–239.

108. ———. The bandwidth heuristic search.–In: Proc. 3rd IJCAI, Stanford, 1973, 23–29.

109. ———. The heuristic search: An alternative to the alpha-beta minimax procedure.–See [78], 157–166.

110.. Hayes J. E., Levy D. N. L. The world computer chess championship.–Edinburgh Univ. Press, 1976.

111. Hearst E. Man and Machine: Chess achievements and chess thinking.–See [78], 167–200.

112. Hunt E. B. Artificial Intelligence.–N.Y.: Academic Press, 1976.

113. Kent P. A simple working model.–See [84], 15–27.

114. Kister J. et al. Experiments in chess.–Journ. ACM, 1957, **4**, No. 2, 174–177.

115. Knuth D. E., Moore R. An analysis of alpha-beta pruning.–Artificial Intelligence, 1975, **6**, 293–326.

116. Kozdrowicki E. W., Licwinko J. S., Cooper D. W. Algorithms for a minimal chess player: A Blitz Player.–Int. J. Man–Machine Studies, 1961, **3**, 141–165.

117. Kozdrowicki E. W., Cooper D. W. COKO III: The Cooper–Kozdrowicki chess program.–Int. J. Man–Machine Studies, 1974, **6**, 627–699.

118. Kozdrowicki E. W. A practical application of machine learning: use of learning in an interpreter for a tree searching language.–In: Proc. IEEE Systems Science and Cybernetics Conf., San Francisco, 1968, 250–257.

119. Levy D. N. L. Computer chess–a case study on the CDC 6600.–Machine Intelligence, 1971, **6**, 151–164.

120. Malik R. Observations.–See [84], 89–94.

121. Marsland T. A., Rushton P. G. A study of techniques for game-playing programs.–Int. J. Computer Science, 1973, **4**, No. 2, 26–30.

122. McCarthy J. Programs with common sense.–In: Semantic Information Processing. MIT Press, 1968, 403–418.

123. Michie D. ALI: A package for generating strategies from tables.–SIGART, 1976, **9**, 12–15.

124. ———. King and Rook against King: historical background and a problem on the infinite board.–See [44], 30–59.

125. ———. On Machine Intelligence.–In: Edinburgh Univ. Press, 1974, 31–49, 135–142, 186–192.

126. Mittman B. A brief history of computer chess tournaments 1970–1975.–See [78], 1–33.

127. ———. First world computer chess championship at IFIP Congress. Stockholm, August 1974.–Comm. ACM, 1974, **17**, 604.

128. Mittman B., Newborn M. Results of the fourth annual U. S. computer chess tournament.–SIGART, Oct. 1973,**42**, 36–48.

129. v. Neumann J., Morgenstern O. Theory of games and economic behavior.–Princeton Univ. Press, 1944.

130. Newborn M. A summary of the third United States computer chess championship.–SIGART, **36**, Oct. 1972, 9–26.

131. ———. Computer Chess.–.: Academic Press, 1975.

132. ———. Peasant: An endgame program for Kings and Pawns.–See [78], 119–130.

133. ———. The efficiency of the alpha-beta search on trees with branch-dependent terminal node scores–Artificial Intelligence, 1977, **8**, 137–153.

134. Newell A., Simon H. A. Human Problem Solving.–Prentice-Hall, 1972.

135. ———. Computer simulation of human thinking.–Science, 1961, **134**, 2011–2017.

136. Nilsson N. J. A new method for searching problems and game playing trees.–In: Proc. of IFIP Congress, 1968, 1556–1562.

137. Pitrat J. Realisation of a general game-playing program.–In: Proc. of IFIP Congress, 1968, 1570–1574.

138. ———. A general game-playing program.–In: Artificial Intelligence and Heuristic Programming. Edinburgh Univ. Press, 1971.

139. Samuel A. L. Some studies in machine learning using the game of checkers II–Recent progress.–IBM Journal, 1967, 11, 601–617.

140. ———. Programming computers to play games.–Advances in Computers, 1960, 1, 165–192.

141. Scott J. J. A chess playing program.–Machine Intelligence, 1969, 4, 255–266.

142. ———. Lancaster vs MACKHAC.–SIGART, 1969, 16, 9–11.

143. Scurrah M. J., Wagner D. A. Cognitive model of problem solving in chess.–Cognitive Psychology, 1971, 2, 454–478.

144. Shannon C. E. A chess playing machine.–Scientific American, 1950, 182, 48–51.

145. Silver R. The group of automorphisms of the game of 3-dimensional tic-tac-toe.–American Mathematical Monthly, 1967, 74, 247–254.

146. Simon H. A., Simon P. A. Trial and error search in solving difficult problems: Evidence from the game of chess.–Behavioral Sci., 1962, 7, No. 4, 425–429.

147. Simon H. A., Barenfield M. Information processing analysis of perceptual processes in problem solving.–Psych. Rev. 1969, 76, 473–483.

148. Simon H. A., Chase W. G. Skill in chess.–American Scientist, 1973, 61, No. 4.

149. Slagle J. R. Artificial Intelligence; The heuristic programming approach.–N.Y.: McGraw-Hill, 1971.

150. Slagle J. R., Dixon J. K. Experiments with some programs which search game trees.–J. ACM, 1969, 16, 189–207.

151. Slagle J. R., Bursky P. Experiments with a multi-purpose theorem-proving heuristic program.–J. ACM, 1968, 15, No. 1, 85–99.

152. Slagle S. R. Heuristic search programs.–See [159], 246–273.

153. Slate D. J., Atkin L. R. CHESS 4.5–The Northwestern University chess program.–See [78], 82–118.

154. Smith R. C. The Schach chess program.–SIGART, 1969, 15, 8–12.

155. Soule S., Marsland T. A. Canadian computer chess tournament.–SIGART, 1975, 54, 12–13.

156. Tan S. T. A knowledge based program to play chess end games.–See [84], 81–88.

157. ———. The winning program.–Firbush News, 1975, 5, 38–45

158. ———. Describing Pawn structures.–See [44], 1, 74–88.

159. Theoretical Approaches to Non-Numerical Problem-Solving/Ed. Banerji R. B., Mesarovic M. D.–In: Proc. of the IV Systems Symposium at Case Western Reserve University. N.Y.: Springer-Verlag, 1979.

160. Thorp E., Walden W. A partial analysis of GO.–Computer Journal, 1964–65, 7, 203–207.

161. Thorp E. A computer-assisted study of GO on M×N boards.–See [159], 303–343.

162. Turing A. M. Digital computers applied to games.–In: Faster Than Thought, 1953, 286–310.

163. Waterman D. Generalisation learning techniques for automating the learning of heuristics.–Artificial Intelligence, 1970, 1, 121–170.

164. Weizenbaum J. How to make a computer appear intelligent: Five-In-A-Row Offers No Guarantee.–Datamation, 1962, 24–26.

165. Wiener N. Cybernetics–N.Y.: Wiley, 1948.
166. Wolf G. Implementation of a dynamic tree-searching algorithm in a chess program.–In: Proc. ACM 73, Atlanta, 206–208.
167. Zielinski G. Heuristics for computer chess.–Applied Math.and Physics Dept., Warsaw Tech. Univ. 1975.
168. Zobrist A. L., Carlson F. R. An advice-taking chess computer.–Scientific American, June 1973, 92–105.
169. Zobrist A. L. A model of visual organization for the game of Go.–SJCC, 1969, 103–112.
170. Zobrist A. L., Carlson F. R. The USC chess program.–In: Proc. ACM 1973, Atlanta, 209–212.

Subject Index

RESCUE
AT
THE
REEF

RESCUE
AT
THE
REEF

THE MIRACULOUS TRUE STORY OF
A LITTLE BOY WITH BIG FAITH

JAMESON REEDER AND
MARY CATHERINE REEDER
WITH MATT MIKALATOS

New York   Nashville

Worthy
Hachette Book Group
1290 Avenue of the Americas, New York, NY 10104
worthypublishing.com
@WorthyPub

First edition: July 2025

Worthy is a division of Hachette Book Group, Inc. The Worthy name and logo are registered trademarks of Hachette Book Group, Inc.

The publisher is not responsible for websites (or their content) that are not owned by the publisher.

Worthy Books may be purchased in bulk for business, educational, or promotional use. For information, please contact your local bookseller or the Hachette Book Group Special Markets Department at special.markets@hbgusa.com.

Library of Congress Cataloging-in-Publication Data has been applied for.

ISBNs: 9781546007807 (hardcover), 9781546007821 (ebook)

Printed in the United States of America

LSC-C

Printing 1, 2025

To our firstborn, Jameson Jr.,

*When you were ten years old, we began writing this story
in a hospital room. Back then, you were a little boy, full
of bright dreams, bold ambitions, and an unshakable faith.
Today, after that tragic day at the reef, we're in awe of
your big faith in our great God. He has been your compass,
guiding you forward, helping you face your fears, even when
returning to the very reef where a shark took your leg.
Like David standing before Goliath, you face challenges
by focusing not on their size but on the strength of our Savior.*

*As your parents, we're humbled by your willingness to
share such a deeply personal story with the world and to
be an ambassador of hope and healing through it. Taking
what life has given you and allowing God to use it for good
is the mark of a true leader. Walking this journey with you,
sharing this story together, has been a profound privilege.*

*It is with great joy that we dedicate this book to you,
Jameson Glenn Reeder, Jr.*

With all our love,

Dad and Mom

Contents

Spiritual Reflections

Foreword

Some friendships are forged by time, some by trial, and some by a shared mission greater than ourselves. My friendship with Jameson Reeder has been shaped by all three. For more than thirty years, I have walked alongside him—not just as a friend, but also as a brother. I have had the privilege of knowing his family, seeing his faith tested, and watching him walk through the fire only to emerge stronger, more resolute, and more determined to see lives transformed by the power of Jesus Christ.

As a pastor, I have seen the weight of suffering up close. I have held hands with those in the midst of loss. I have prayed over parents who have buried their children, including my own beloved daughter Lilly. Pain and loss have a way of sharpening our vision, making us see what truly matters. And what matters—what has always mattered—is that we do not drift through life unaware of the rescue that has already been offered to us in Christ. Jameson and Mary understand this truth because they have lived it. Their hearts beat for the lost, for those caught in the currents of this world, unaware that a Savior stands ready to pull them from the depths.

Rescue at the Reef reminds me that miracles still happen. Jameson Jr., at just ten years old, faced one of life's most terrifying moments with an unshakable faith that belied his years.

That kind of courage doesn't just happen—it's the result of being anchored in something much greater than ourselves. It's about trusting God, even when life throws us into the depths of the unknown.

What strikes me most about this story is not just the miraculous way Jameson was rescued, but how his faith and hope became a lifeline for everyone around him. His unwavering belief in Jesus, even in the face of unimaginable pain, is a testament to the power of a childlike faith—faith that moves mountains, or in this case, calms the storm of a life-changing moment.

Through *Rescue at the Reef,* Jameson Sr. and Mary invite us into their journey as they lead their family through a time of immense trauma and loss to pursue hope and find restoration. Even more than that, they invite us to look beyond our own struggles and see how God is working in the midst of them. Whether you've faced a shark attack, a life-altering challenge, or a moment when the world seemed to collapse around you, this story reminds us all that God is faithful.

Jameson Jr.'s life and testimony are proof that even when the waves threaten to overwhelm us, we have a Savior who walks on water. He turns our pain into purpose, our trials into triumphs, and our scars into stories of His goodness.

As someone who has experienced the life-changing power of God's grace in the face of personal tragedy, I can confidently say that Jameson Jr.'s journey will inspire you, challenge you, and remind you of the hope we have in Jesus.

Dive into this incredible story and let it remind you that even when life takes an unexpected turn, God's plan is always good.

—Rob Pacienza, senior pastor of Coral Ridge
Presbyterian Church, Fort Lauderdale, Florida

Preface

It's natural that we as human beings make plans, whether it's what's for dinner tonight or who we're meeting for coffee next Tuesday at 3 p.m. Sometimes we speak our plans aloud, share them with others. Sometimes we hold them so deep and quiet in our hearts that we ourselves are surprised to discover what our own expectations are.

When those plans are disrupted, when something happens that changes everything, it can be a small annoyance, like there being more traffic than anticipated, or a life-changing phone call from a doctor with news about that test from last week. Our response in the moment reveals so much about us: our expectations, how prepared we are, our emotional state, our character, the extent of our network of friends.

And when that disruption is a life-changing crisis, it's all of those things to the extreme. A crisis like that changes everything and its ripples continue for decades. Not just what's for dinner or canceling plans with our friends, but the whole focus of our lives.

We never expected our day visiting a reef would so radically transform all of our lives. Not just us and our kids, especially Jameson Jr., but also the folks we met along the way who helped Jameson—the first responders, the doctors and nurses, the social workers, and the people who donated money to help—as well as the thousands of people who have responded to his story.

That's the other thing about these unexpected moments of change. Good things can come from them too. New friendships. New found wells of bravery in our lives. Good deeds done by strangers that create more good deeds that create more good deeds, as people gift each other with kindness, all of it spinning out of a single painful, difficult moment.

But in the moment, what we so often feel is only overwhelming fear and the certainty of our own helplessness. There's nothing we can do that is enough. There's no way we can turn back the clock to yesterday and make a different choice—swim at a different reef, skip the beach and go to the movies, stay home and play board games—and hope for a different outcome.

That's the terrible temptation, especially in situations like ours, where it's not anyone's fault that this thing happened. Yet you can't help but keep running it over and over in your mind, trying to figure out what you could have done differently, what you could have changed. Some part of you knows there's no answer to that question, and another part is still running the question through your mind once again.

Meanwhile, the entire matrix of our present and future plans is shifting, transforming into new shapes that we've never anticipated. It's stressful. It can make you angry or scared or confused. Or maybe, for some of us, there's a little glimmer of excitement at the possibilities that are opening up. Maybe we'll get there in time.

But in the moment of crisis—when your child is in the water and screaming—there's no time to think. Not like that. There's only time to dive in and act, to move more from instinct than from thought.

In this book, we're going to share honestly about our experience—and the experience of our family—in this unprecedented and unexpected moment of crisis. Not just the moments of being overwhelmed, or frightened, or angry, but also the moments of joy, surprise, and the miraculous.

And while this book is for people of all faiths and backgrounds, we thought it would be good to say up front that our story does intersect with our own faith in a powerful way... We believe that God is a Rescuer, and that God was involved in saving the life of our son in a dramatic and unmistakable way. We've included a few spiritual reflections in the back of the book for those who are interested.

We also take turns telling different parts of the story throughout the book, both so you can see our unique points of view, and also so you can understand firsthand the parts of the story where only one of us was present. When we switch who's speaking, we simply put one of our names at the beginning of that section so it's clear who's telling that bit of the story.

There are a lot of really personal things we share in this book about ourselves and our family. We do that because we believe our story might bring you hope when you are experiencing a dark moment, or peace in the midst of trouble. We all have moments of unexpected crisis in our lives... but there is a way through to better days. Let's push through together to find them!

—*Jameson and Mary*

Timeline

2022

- August 11: Arrival at the Sugar Shack in the Lower Florida Keys.
- August 13: Shark attack at Looe Key Reef. Airlift to Nicklaus Children's Hospital in Miami.
- August 15: Zack Spurlock finds the GoPro.
- August 17: Jameson says, "Dad, take me back to the reef so I can face my fear."
- August 29: We surprised Jameson for his eleventh birthday with a surprise party in the hospital.
- September 3: Jameson is discharged from the Nicklaus Children's Hospital (after four surgeries in three weeks).
- September 7: Interview with *Good Morning America*.
- October 2: We leave Miami and head back to the Florida Keys, where we will be for the next month.
- October 3: Jameson and family get on the same boat since the attack and get in the water for the first time at Marvin Key. (This is when we encountered sharks.)
- October 15: Little Smiles Stars Ball in West Palm Beach, Florida.

- October 22: Return to the reef, ten weeks after the shark attack, Looe Key Reef.
- November 8: Jameson received his first prosthetic leg, Miami, Florida.
- December 3: Surfing for the first time since the shark attack, Melbourne, Florida.
- December 18: Return home to Charlotte, North Carolina.

2023

- March 23: First time playing baseball since the shark attack.
- August 13: For the one-year anniversary of the shark attack, we returned to the reef to dive and enjoy the reef.
- November 5: *National Geographic* shoot in the Keys.
- December 5: Met Bethany Hamilton and family and Mike Coots in Hawaii.

2024

- July: Cooperstown, New York, for a weeklong baseball tournament.
- August 13: Fifth surgery (first surgery since the initial surgeries).
- September 16: We found out we were pregnant with our fifth child.

PART ONE

Survival

CHAPTER ONE

It Came Out of Nowhere

JAMESON SR.

It was a clear, quiet day with clear, calm water.

Mary and I have spent our whole lives by and in the water, and so have our kids. Swimming, surfing, snorkeling, skiing, you name it. We love the ocean, especially the turquoise blue waters of the Florida Keys. When I had a chance to do some work in Florida, we knew we'd pack up the kids and leave our home in North Carolina to go to the beach for two glorious weeks.

Mary and I both grew up in Fort Lauderdale, so we didn't want to pass up an opportunity to do what we love, and this time, we would let the kids experience some of the laid-back, coastal Florida life. I went to Orlando to work for a few days, and Mary and the kids stayed in Deerfield Beach with her family. Her mom and dad, Michael and Donna, and her sister and brother-in-law, Danielle and Anthony, all live in Deerfield Beach, so it was a great chance to connect with family while I was working. When my work was done, we headed down to the Florida Keys to settle into the beach house called the Sugar Shack that was owned by our friends, Mike and Tammy

Nolan. We'd also be borrowing their boat to reach a little bit of paradise.

Among the 1,700 islands of the Keys is Looe Key, an old stomping ground of mine. It's about five square miles of coral reef inside a national marine sanctuary, the home to over 50 species of coral and 150 species of fish: angelfish, yellowtail, barracuda, goliath groupers, turtles, even a moray eel or a ray if you're lucky. I had seen large fish, even my first shark there when I was a kid. When the waters don't have strong currents, they often are so still, they look like a natural mirror that reflects the sky.

August 13, 2022, began as an ordinary day. We ate breakfast and had our family time of daily devotionals and prayers. The kids are homeschooled, so we also watched some videos about reefs and the proper etiquette when you're in a national marine reserve. Then we gathered our things for a day on the water. We were headed nine miles off-shore to one of the top locations in North America for diving and snorkeling.

The surf had been a bit choppy the few days before, but when we pulled the boat up to a buoy in the sanctuary that morning, the sun was shining, and the water was exceptionally calm. Dozens of families had already moored their boats and were splashing and diving into the beauty of the ocean. I got the last available mooring ball on the inside of the reef and tied us off. I'm a certified lifeguard and experienced waterman, so after a safety talk (stay close to the boat, stay together, remember the distress signals and the okay signal) and putting the diver down flag off the back of the boat, we jumped in.

It is no small feat to get six people out the door, especially when four of them are your children. You always forget something or someone gets sick or you're running late. Mary and I navigate our family chaos together, raising ten-year-old

Jameson Jr., eight-year-old Noah, six-year-old Eliana, and three-year-old Nehemiah. All but the youngest swim like fish, and we keep Nehemiah in a life jacket so he can get in the water and have a little freedom too.

We started exploring the reef in water about eight feet deep. The kids were diving, splashing, just having the best time. We swam for a while before deciding we should all get out to eat and hydrate.

Mary had made sandwiches for us, but in the morning shuffle, she had forgotten to put meat on hers and mine. We laughed it off. Those things happen and it would take a whole lot more than that to ruin our day in this beautiful place.

After lunch, a sea turtle surfaced by the boat and caught Jameson's eye. He asked if he could take our GoPro and follow it around, and soon enough, I watched him dive to the sandy floor, gliding along, and watching this magnificent creature as it went about its day. Jameson is an excellent and strong swimmer, and he was in a wildlife refuge where novice divers train in the shallows, taking in vibrant colors and so many different fish.

Noah and Eliana decided to stay on the boat with Mary while I got into the water with Nehemiah, keeping an eye on Jameson. We kept our distance so he could enjoy the turtle, and as I watched them swimming over beautiful coral and meadows of seagrass, I couldn't help thinking how surreal it looked. There was something spiritual about it, a connection between them.

For a while Jameson Jr. was swimming down to the ocean floor, then popping up to show what he'd found to delight Nehemiah and me. The water was shallow enough that I could see down to the bottom. There's plenty to see, and Looe Key Reef is a safe, gentle place for novices to learn how to dive. Of course, it's amazing for snorkelers too.

As Nehemiah and I swam together, I looked down and saw a queen conch shell, a big one. I thought Nehemiah would like to see it, and he was safe in his life jacket. I dove down into the peaceful silence of the cool blue, grinning to myself and thinking of Nehemiah's face when I showed him the shell. I kicked for the surface.

Before my head broke water, I heard the screaming.

MARY

I heard Jameson Jr. screaming from where I was in the back of the boat.

I was cleaning up from lunch with Noah and Eliana and turned to see Jameson Jr. waving his arms. We've always taught the kids to wave their arms if they are in distress in the water, but I could see that Jameson was terrified. I immediately started screaming for Jameson Sr., who was underwater at that moment. Noah and Eliana both started crying, scared to hear Junior screaming like that.

When Jameson surfaced, he heard Junior and swam with Nehemiah to the boat. As soon as they were close enough, I grabbed Nehemiah and handed him to Noah, who took him and his sister to the front of the boat. Jameson Sr. boarded and started untying us from the mooring ball.

"Let me know when the flag is in!"

I started pulling on the string of the diver down flag as fast as I could while Jameson got the boat ready. The second it was in, I yelled, "Go!" and held on as we sped toward our son.

As we approached, I could see that Jameson Jr. was sitting up on his pool noodle—not hanging his arms over it, sitting on it—and he wasn't screaming anymore. In fact, he had gone completely silent. He looked like he was sitting in an invisible chair, or the hand of God was holding him out of the water. He was much higher in the water than his little yellow pool noodle should have been able to hold him, something I didn't think about until a few days later.

When I saw my son's face, he was white as a sheet, his eyes barely open, his hand reaching out toward me. He was silently mouthing, "Help me." He looked almost lifeless. I jumped off the back of the boat and swam toward my son. After he had let out that frightened scream for help, I couldn't get to him fast enough. My mind was sorting through some confusion because I couldn't see what was wrong. Nothing in the water. Maybe he'd been stung by a jellyfish? That was the extent of the danger I imagined in those few moments. But his scream? It was something more than terror; it was the sound of a child in mortal danger, the kind of scream that sends chills to the very center of your soul.

I couldn't tell what was wrong even when I wrapped my arm around him. He wasn't speaking or moving. I slid my arm under his arms and pulled him to the boat. Jameson Sr. met us at the back to lift Jameson out of the water.

As I handed him up to Jameson Sr., Junior's leg slid past my face. I saw some of his ragged skin and then just bone—way too much bone—and it seemed to keep going and going until his bloody foot rose past me. His lower leg was all bone. Just below the knee, all the way to the ankle. And then the foot.

A feeling of pure hopelessness slammed into me at that moment. I felt an overwhelming fear that we only had a few more minutes with him, as if he were already dead, that there was nothing we could do. I'd never felt fear like that before and I hope I never will again.

We knew right away it was a shark attack. Nothing else could do that kind of damage. Jameson had been mauled. The flesh was gone. When Jameson Sr. saw Junior's leg—the kneecap to the foot, nothing but bone—he shouted, "No! No!" It was a horrific, almost unreal sight. And that terrifying realization was still settling into my chest. *I'm going to lose my son!* We were nine miles from the coast and who knows how far from a hospital if we could even get him to the beach.

I became hyperfocused, my mind racing to find solutions while the whole world slowed around me. But what do you do when your child is bleeding to death, his leg half gone?

It was Junior who amazed us. He said, "Daddy, it's going to be okay. Jesus will save me."

Not Jesus might save me. Not, "Please, Jesus, save me." But, "Jesus will save me." Calming words. It wasn't a prayer of hope; it was a profession of his faith. And it jolted us out of our shock. We needed to get to work, and we did.

Jameson Sr. laid our son on the back of the boat, and I knew I was out of my depth. I immediately started to pray. "Jesus, please don't take my son." I made sure Jameson was awake and breathing and looking at me. Senior already had the boat in motion, headed toward the closest boats. "CPR," Jameson Sr. yelled out, and I went to give it but called back, "He's breathing!"

"Tourniquet," Senior shouted back.

I took the sleeve of a rash guard and tied it tightly around Jameson Jr.'s thigh, a makeshift tourniquet to stop the bleeding. Our other three kids were wailing and crying.

We knew we needed to get to shore as quickly as possible, needed to call 911, needed to let the hospital know we were coming, needed an ambulance to meet us at the shore.

The boat closest to us was much larger than ours. We would call the thirty-six-foot boat *The Invincible* because it was an Invincible brand boat. The vessel's actual name was *The Reel Screamer*, and it had arrived a few minutes before the shark attack. Todd and Lisa Grooms owned the boat. Their son Tyler was with them, along with their friends Kyle and Kristin. Kyle jumped over to us and immediately saw that my makeshift tourniquet wasn't enough. He grabbed some rope, and he and Jameson made a tighter one around Junior's leg.

Jameson Sr. shouted that we needed medical help. Tyler was already calling 911. He had heard us shouting before we reached them. "Shark attack! Help! Child! Doctor!"

JAMESON SR.

As we finished tying the rope on Jameson's leg, I looked to my right and saw a woman toppling onto our boat. I reached out to help her and asked who she was.

"I'm a nurse," she said, and immediately leaned down to check on Jameson.

Jennifer had been on another nearby boat—one of many miracles that day. She had immediately dived into the water to swim to us, without a thought about whether the shark might still be nearby. She brought a dry bag with medical supplies and began to work with quick, competent, professional grace to save our son, wrapping his leg in a towel and then putting ice packs on his chest and the back of his neck. Jennifer's best friend, Anneliese, was the captain of the other boat, and she'd just put a new medical-grade first aid kit, ice packs, and ACE bandages on the boat the night before, which were all in use. Jennifer used the bandages to make another makeshift tourniquet, one much better than the rash guard or the rope we had used earlier.

MARY

Emergency services were trying to figure out the best place to meet us. Jameson heard them talking about sending the Coast Guard out to us, but he knew that would take too long. We needed to start moving our son toward help. So Jameson asked Kyle if he knew which dock the ambulance was headed to. He did. Jameson immediately picked up our boy, gave him a kiss on the forehead, and handed him to Kyle in the big boat. "We can't wait," he said. "You have to run him in right away and save his life."

Before jumping over to *The Invincible* I reached over to put my hand on Jennifer's, then looked her in the eyes and said, "Please don't leave me." She was the closest thing to a medical professional we had, and I knew I didn't want her to send us off to shore alone. She said yes and came with us to help. I knew from the look on Jameson Sr.'s face exactly what he was thinking: *I might have just said good-bye to my boy for the last time.*

We didn't know exactly what had happened when we pulled Jameson from the water. We knew it was a shark by looking at the extent of the damage. We were told later that it was almost certainly a bull shark, eight to ten feet long, weighing between three hundred and five hundred pounds. We hadn't seen it. Not even in the nearly transparent water. Not even after all our lessons about ocean and boat safety. Not even while we watched over our children.

Later, Jameson said, "it came out of nowhere." A quarter-ton animal twice as long as our kid had snuck up on us.

But even if someone had told us there was a bull shark nearby, it probably wouldn't have kept us out of the water. We knew that shark attacks were extremely rare and don't even make the top ten for animals most likely to kill someone in the United States. You have a greater chance of being struck by lightning five different times than being attacked by a shark once. On top of that, there hadn't been a shark attack at Looe Key Reef in recorded history, that is, until Jameson's.

If we had seen the shark, we would have had a chance to react, even if only to know how to act. If we had seen it, we might have been able to better understand the danger and then the attack in some way. But we didn't see it coming, had no idea it was there. It was sudden and unexpected, so we couldn't protect Jameson or defend ourselves.

We don't know why it attacked Jameson, but we do know it wasn't anyone's fault, wasn't the result of a bad decision, wasn't caused by risky behavior. In fact, because Looe Key is a marine sanctuary, it's a protected reef. No fishing is allowed. No one is chumming the water. Jameson wasn't even wearing flippers when he was attacked!

We think back to that day and the days since, and we are surprised by so many things, not the least of which was the kindness of people who didn't know us at all but still went out of their way to do what they could. They spent time and money and resources to help us and our son. We were surprised in great ways by the people who worked hard for Jameson or who sent encouraging notes or gifts. We were surprised by the number of people who wanted to know the story, who wanted to know what was going on.

But more than anything, we were surprised by the seemingly endless stories of compassion and bravery that Jameson Jr. showed to us and the people around him and of his ability to uplift and inspire those around him.

MARY

The Invincible was flying toward shore at seventy miles per hour, racing toward the marina. An ambulance should be waiting there. I knew somewhere behind us that Jameson Sr. was getting the kids settled and doing his best to follow us in our smaller borrowed boat, desperate for news.

Jameson Jr. was lying on a large beanbag toward the rear of the boat. Everyone was trying to keep him comfortable. Someone brought a big towel to shield him from the sun as we hurried to shore. The boat slammed in and out of the water so hard, I'd have bruises on my knees for days from bending down beside Jameson. Jennifer was checking his vitals as the wind whipped around us, the towel snapping overhead, doing everything she could with the limited resources on the boat, hoping to keep him alive.

He didn't look good, but how could he? He had lost so much blood, more than I'd ever seen. This little boy who had been so full of energy just minutes before—leaping off the boat, diving into the depths, splashing around with his siblings, laughing as he brought shells and interesting rocks to his brother—was now fighting for his life. I was trying to keep him awake, to get him to keep his eyes open, praying that he wouldn't die. Sometimes, in situations like this, that's all a mother can do. Try to comfort your child and pray! pray! pray!

Suddenly, over the roar of the engine and the wind, I thought I heard something. I leaned down...Was that singing? Weak. Quiet. But certain. "Are you singing, Jameson?"

He was. He was singing a song from church, one of his favorites that he had just learned to play on the piano. "Here I Am to Worship." Somehow, he had the strength to sing. Some people might have responded with crying or yelling, asking why this was happening to them. And those would be understandable reactions, even healthy. But here was my little boy, bandages barely attached to his mutilated leg that was still bleeding, ice packed around that leg that he would surely lose if we couldn't get him to the hospital in time. And he was singing?!

I couldn't believe it.

I had been trying to get him to keep his eyes open, to look at me, and here he was, using what little strength he had to sing.

I didn't know what to do. So I leaned in close, held his hand, and sang along.

CHAPTER TWO

"I'll Hold Him Until You Get Here"

JAMESON SR.

As Mary and Jameson disappeared onto someone else's boat, I tried to focus on what I should do next. I wanted to take off after them, but our kids were crying and I was feeling overwhelmed and heartbroken. I was certain my son was dying. As I calmed the kids, I realized that Jennifer—who had dived into the water and swum over to help without any hesitation—hadn't had even a moment to tell her friends what was going on.

I pulled our boat close enough to the boat she had been on and explained that Jennifer was headed to Dolphin Marina and that my son was gravely injured. They could pick her up there.

Dolphin Marina is a private marina, but that's where the ambulance would be waiting. There was a lot of confusion and uncertainty from everyone. No one had all the details about what exactly was happening, and I tried to tell them as quickly and clearly as I could. In the midst of all that was happening, I wasn't 100 percent sure how to get to Dolphin Marina, so Anneliese offered to go ahead of us and show us the way.

Noah, Eliana, and Nehemiah were on the bow of the boat, watching their brother and mom get farther away. Their boat was bigger and more powerful than ours, so we would never be able to keep up. I felt a quick moment of thankfulness in the middle of the terror because Anneliese was sticking close and leading us to the marina. The children were crying, but as our boat picked up speed, Noah closed the door to the bow and huddled with his siblings. I wouldn't hear until later what the three of them were doing because, right then, I needed to focus on getting us to shore. I could tell they were doing their best, and I knew they were safe. I was grateful for that, at least.

At top speed, our boat would take at least twice as long to get to the marina as the boat Junior was on. That would mean thirty-five to forty minutes for us, pushing this boat to the limits. And there was nothing I could do to help my son.

Nothing but pray and hope. Sometimes when I am praying, I hear God speak. Not in an audible voice—I don't actually hear anything—but a word or a phrase might come to me, something internal that I recognize as God's voice responding to my prayers. I had one of those moments as I was pouring out my heart about Jameson.

"I'm going to hold him until you get here," God said.

I believed that God meant He was going to hold on to my son eternally. He was saying that my son would be cared for, watched over in heaven. That when I died someday, God would have been holding my son, waiting for me. A reunion was coming, and God was telling me it was okay. That I could trust Jameson was in good hands.

"You can let him go," God said.

I don't know how to explain this, but a sense of peace settled on me. Not that I wasn't still scared. Not that I wasn't still desperately pushing the boat as fast as I could toward shore. Not that I wasn't worried or angry or confused or running through hundreds of scenarios in my head. But there was something deeper too. A sureness. A certainty that if God said He'd take care of my son, what did I have to worry about? There was nothing I could do. But God loved me, loved my family, and actually had the power to do something about it all. So if I trusted God, and if God said He was holding my son, then I could let go. Let go of my worries, my doubts, my sense of helplessness in this moment. I could trust that God would do the best thing.

MARY

The whole way in to Dolphin Marina I just kept thinking, *I need to make sure that Jameson keeps his eyes open.* I was on my knees beside Jameson, and I kept saying to him, "Look at Mommy. Keep your eyes open. Keep breathing." At the same time I constantly repeated, like a mantra, "Jesus, please don't take my son."

We were going at least seventy-five miles per hour, the boat slamming hard through the waves while I was on my knees, the floor of the boat like sandpaper on my bare skin. Jennifer was across from me, also leaning over Jameson, taking his vitals, talking to him, talking to me, shouting over the roar of the engine and the wind.

There wasn't room in my mind for anything but Jameson's words and my repeated prayer: "Oh, Jesus, please don't take my son."

We got to Dolphin Marina in about eleven minutes, which felt like forever. Eleven minutes to cover nine miles of open water and come screaming into the marina, throwing waves through the no-wake zone. People shouted, "This is a no-wake zone!" and "Slow down!" Someone on our boat shouted back, "We have a shark bite!"

The ambulance was waiting, paramedics ready the moment we docked. Gurney out. I don't remember how exactly I got off the boat. I think the paramedics told me to get out to make room for them. Within seconds I was standing on the dock and the paramedics were in the boat, working on Jameson.

Something about hitting land took a weight off my chest. We were on land, not in the middle of the ocean anymore. I felt a sense of peace. A spiritual calm. I'd done everything I could do, and now the paramedics would do everything they could do. They were getting him onto the gurney, maybe putting an IV in.

A man in a beige uniform came up to me, pulling my attention away from Jameson. He was a Florida Fish and Wildlife official, the police of the ocean. He needed to ask me some questions. I remember so clearly his uniform, the fact that he wore a bulletproof vest and had a small black camera lens on his chest, recording our interaction.

"I'm so sorry," he said. I remember that. And then he asked, "What's your name? What happened out there?" The other questions I don't remember, but I wondered why we needed to do this at this moment.

Everyone was rushing between the boat and the ambulance, moving so fast yet also so slowly. I don't remember answering questions, but I know I went to one of the paramedics and asked, "I'm going to go with you guys, right?"

"Let me check," he said.

I realized I wasn't wearing shoes. I had walked across a gravel parking area to talk to the paramedic without noticing the sharp jabs in my feet, and now I was going back across it to reach *The Invincible* to get shoes.

"Does anyone have any shoes I can take?" I was wearing only my swimsuit and a towel.

Lisa handed me her sandals, and I hurried back to the ambulance. "You're going to ride with us up front," one of the paramedics said. They were getting Junior situated in the back.

I opened the passenger door and climbed in. Somehow I had my phone with me. When I had been jumping from one boat to the other, someone had yelled to me, "Grab your phone." No shoes, no clothes, but I had a way to get ahold of other people.

My family wasn't far away. I called my mom from the ambulance. She was at my sister's house, and when she answered her phone, I quickly told her as much as I knew, shouting over the blaring siren: Jameson had been hurt, a shark attack, getting into an ambulance, heading for a hospital somewhere. I didn't know which hospital. Could she bring some clothes and meet us?

"Just start driving south," I told her.

JAMESON SR.

The engine was wide open, the wind loud as I pushed toward shore as fast as our boat could go. I started making phone calls—shouting and crying over the engine noise and the wind—wanting to make sure my family knew what was going on. We were going to need help, and they needed to know what was happening. Mary and I both are really close with our families, so my parents were the first people I thought to call.

They didn't pick up. I would learn later that my parents were at my nephew Jack's birthday party celebrating with the family. My brother and sister-in-law, Justin and Jennifer, had a new farm, and the cell signal there was weak.

Of course I didn't know that then. I only knew I needed to get ahold of someone in the family. The first person I talked to was my brother Joshua, who lives in Hawaii. Shouting over the wind and the engine, I told him, "Jameson has been attacked by a shark. His leg is gone."

Joshua laughed. He thought I was joking. To be fair, we're always clowning around with one another. And our whole family—Joshua, me, and our younger brother Justin—we've spent our whole lives in the water. We knew shark attacks were rare— mostly unheard of—in the area where we were.

Then the severity of the news registered for Joshua: the urgency in my voice, the sound of the engine pushed to the max. Maybe he could tell I was crying as I shouted the details to him. This was real. Joshua immediately grabbed his wife, Jacki, and they prayed for me and Jameson and our family right then, over the phone.

"I can't get ahold of Mom and Dad and Justin," I told him. "You have to call them." I knew he'd take care of it, and I could focus on getting to Mary and Jameson. I was playing a literal game of telephone when none of us had all the information.

A few minutes later, as we got closer to shore, I saw a fire truck screaming across the South Pine Channel Bridge. An ambulance was close behind.

MARY

"Jameson, Mommy's here."

The sirens were so loud, I couldn't hear anything. I wasn't sure Jameson could hear me. I was shouting to him through a little window between the front compartment and the rear. We were following a fire truck over a bridge.

Ten minutes. We were in the ambulance for ten minutes. It felt like a year.

The ambulance tore into the parking lot of St. Peter the Fisherman Catholic Church on Big Pine Key. A helicopter was waiting in the open field behind the church, the back doors open. The paramedics jumped into action.

"I'm going with Jameson, right?" I asked.

"I'm not sure if you can go," one of the paramedics said. "Let me see."

JAMESON SR.

When we finally got to Dolphin Marina, Lieutenant Dodd Bulger was waiting for me. He was with the Florida Fish and Wildlife Conservation Commission and wanted to ask me some questions.

Shark attacks, again, are so rare in this area. They just don't happen. It made sense the first thing the officials would think was that we were doing something we shouldn't have. You can't chum the water near Looe Key Reef. You can't fish either. It's a marine preserve. If we had been doing something like that, it meant we would be in trouble—and also meant it might make sense why a shark had attacked a child in a place where no shark attack had ever been recorded before.

Lieutenant Bulger said, "Your son's good. Don't worry, he's good."

"He's good? What does that mean?" There's no way he was good. He'd just been mauled by a massive bull shark!

The lieutenant said Jameson was alive and breathing. He was stable and being rushed toward an airlift. A helicopter would take him the rest of the way. He also said how sorry he was, and I remember the gentle kindness in his voice.

Jennifer was still at the marina, and she and Anneliese turned their attention toward my other kids. Checking on them. Talking with them. Praying with them. Nehemiah didn't fully understand what had happened, so he had mostly recovered on the ride in from the reef. But the other two were old enough to know what was going on. They were in shock and broke into sobs every few minutes.

"Your son is remarkable, man," Lieutenant Bulger interrupted my thoughts. And then he described how Jameson had been so quiet, so calm, awake and peaceful, while the paramedics worked on him.

~

There were things to be done. Paperwork to be filled out. Lieutenant Bulger and his partner, John Hettel, needed to check the boat. They checked things over, looking I guess for an empty chum bucket or fishing gear, which of course we didn't have. It's a big deal if you use those things at Looe Key Reef. You aren't even allowed to drop anchor.

"It's not your fault," they said. "It's nothing you did." There was no evidence in our boat that we'd done anything to encourage a shark attack. I knew that already, but it was a relief to hear it from someone else's mouth.

"Hey," a firefighter, Jason Pearson, called out and pointed into the sky, drawing my eyes to a helicopter rising in the distance. "That's your son's airlift."

MARY

They were loading Jameson into the helicopter. Everything seemed to take forever. *Why are we still sitting here?* But they had to move him from the ambulance, and there were a lot of straps and safety harnesses to be fastened, equipment to be moved, vital signs to be checked. We had to fly to Miami because there was no hospital in the Florida Keys that could treat Jameson's severe wounds.

Someone came over and said they needed to ask me some questions. I started telling them, "I have to go with my son. I'm not afraid of flying. I promise you I won't pass out. I've skydived before. I can do this. Just don't separate me from my son."

"No, no," the paramedic said. "Not those kinds of questions. I need to know how much you weigh."

~

In emergency situations, there are so many unexpected but ordinary questions. What's your name? Address? Where are you staying? Sometimes maybe someone will ask you how much you weigh to make sure there's room for you on the helicopter.

It's interesting that both of us had a moment, even in the first hour after the attack, when we had a feeling of peace. A deep certainty that, well, not that everything was going to be okay, but a certainty that we had done everything we could and we needed to give the rest of it over to someone else: the paramedics, the doctors, God. It would be easy to be completely overwhelmed by the enormity of everything that was happening (and we had those moments). But we are so thankful that on this day, in the midst of everything else, we also had a few moments of peace.

JAMESON SR.

I didn't know if the boat could be driven back to the house where we were staying.

Where we'd pulled Jameson out of the water was on the opposite side of the reef from the other boats. I knew that to drive all the way around the reef would have taken precious minutes and the other boats were *right there*. So I'd decided, a split-second call, to drive our boat over the shallow reef. I would have never, ever done this if it wasn't a life-and-death situation.

Now at the marina, I knew the entire bottom of the boat might be damaged. I had tried to find a break in the reef where the boat was least likely to run aground.

The FWC guys—Dodd and John—looked at the boat with me. It was pristine. Somehow we had managed our way over the reef. Looked like maybe the skeg had lost some paint and gained a couple of scratches, but everything else was perfectly fine.

MARY

The helicopter lifted off at 5:09 p.m. It had been less than an hour since the attack. Despite how slow it felt at different moments, everyone had been moving with lightning speed, whether it was our new friends out at the reef, the paramedics and firefighters, or the crew of the helicopter.

They let me ride beside Jameson, sitting up near his head. We all had headphones on, including Jameson, but it wasn't so we could hear each other; it was just to minimize some of the helicopter noise.

I found myself trying to keep Jameson awake again. I desperately wanted to make sure he stayed conscious. It's amazing that he didn't pass out and that he kept responding, kept engaging with us.

Anytime it looked like he was drifting off, I'd tap him gently between the eyebrows, just reminding him that I was there and to keep his eyes open. "Look at Mommy. Keep your eyes open." He couldn't hear me, so I was doing that through gentle touch, and I would mouth "I love you" to him. He'd also gesture for me to lightly touch his arm... He loves it when I stroke my fingers on his forearm so that he gets that tingling sensation. But how was he able to ask about this right now?

The paramedics were doing paperwork, checking their laptops, monitoring the machines hooked up to Jameson. That seemed like a good sign to me.

At one point it looked as if Jameson's eyes were rolling back into his head, and I motioned for the paramedics. Later, Jameson told me he had just been so thirsty. Delirious, he was reaching for his IV bag, looking for something he could drink. Every few minutes he would reach down toward his leg and just say, "Ow!"

How much longer did we have to go?

Every time I asked, I got the same answer. Ten minutes. We were almost there. Ten more minutes to go. All communicated with hand signals and flashing fingers for minutes.

It was a thirty-minute flight. I'm sure they wanted to keep us from panicking, wanted to give us hope when they told us it was only ten minutes.

The door on my side slid open, and a paramedic ushered me out of the helicopter, both of us keeping our heads low. He took me to an elevator door, motioned for me to stay put, then ran back to the helicopter to help them get my son.

CHAPTER THREE

When the Shark Makes the Decision for You

JAMESON SR.

The kids and I were at Dolphin Marina and we needed to take the boat back to where we were staying, at our friends' house, the Sugar Shack. The boat was theirs too. Although we had checked the boat for damage, and it was good to go, I dreaded the thought of getting back into it and piloting it the half hour or so back to the house.

Lieutenant Bulger had already thought of that. Hettel, his partner, would pilot the boat and take the kids and me home, and Bulger would meet us at the Sugar Shack. This gave me my first opportunity to focus on the kids and catch my breath.

I was thankful for that and probably still in shock over the attack. I kept seeing the scene in my head of Jameson as I was pulling him out of the water. I kept thinking, what if there was a chunk of his flesh still left in the boat? The image of his shredded skin and muscle still attached to his body was seared into my mind. It didn't make a lot of sense, but I couldn't stop wondering if a piece of his leg had fallen off him and I needed to get it. Should we look in the back of the boat? Try to save it in some way?

There would have been no way to reattach it. A lot of time had passed, and although I didn't know it yet, there was no way to save his leg. But that didn't stop me from turning this thought over and over, trying to save my son.

As we made our way toward the Sugar Shack, I could see my own bloody handprint where I had been piloting the boat and the red-streaked remains where my son had been hauled into the back. I stared at those spots.

We moored at the dock. It's a very long dock. I carried Nehemiah, and the kids and John grabbed some of our things as we made our way to the house. Bulger was already there at the dock, and he jumped in to help as well.

Inside, I started the motions of helping the kids get into their familiar routines. They needed to use the bathroom. We were still in our swimsuits and covered in salt. I have a memory that I made sure everyone showered, but when I looked at some pictures later, we were all still in our suits. We must have rinsed off in our suits.

Feeling deeply unsettled, I gathered the kids and we prayed together. We sang some songs. None of us wanted to be far from each other. I got some tangerines and water for the kids, and we all sat at the dining room table. Bulger helped the kids peel the tangerines, and then he and Hettel went to clean the boat.

I was trying to decide whether we should pack up and head for the hospital. It was almost dinnertime. We'd been at the house for a couple of days prior to the attack. And packing everything into the car and driving for hours to get to Miami, with no clear idea of where we would stay or even if I could get into the hospital, was a lot.

And there was some paperwork to be done. The FWC needed a full report. Bulger gave me the form and said, "Whatever you can remember, write it down." In between taking care of the kids, fielding calls from the family, and waiting to hear more from Mary, I sat at the table and wrote down the events of the day.

No one had seen anything like this. Not the FWC guys. Not the locals I talked to. Some dive shops the next day were telling people, "They had to be chumming the water." It just didn't make sense that a shark would attack a kid. Not here. Elsewhere in Florida, sure, it happened. But Looe Key Reef? That was unheard of.

The kids were calming down but still wanted to be within reach. We put some program on the television, and they settled onto the couch that was directly in front of the table where I was working. I ordered pizza for dinner. Then I got another call from Mary. A new crisis had emerged that made me focus on trying to do the best thing for Jameson medically. This involved several phone calls and a hard decision that needed to be made quickly. And here I was, hours away, trying to help make the decision.

MARY

The paramedics unstrapped and unclipped Jameson, disconnecting him from machines and moving his gurney across the roof of the hospital. Then we rode the elevator straight to the trauma floor. When the door opened, waiting for us were people in gowns, gloves, and masks, all lined up, standing along both sides of the hallway. At first, I thought they were just getting out of our way. But they were lined up, ready for us.

We went down the people-lined hallway, and as we passed, they all fell in step behind us then followed us into the trauma room. It was like a well-oiled machine. No big conversation, no shouting. We rolled in and they got to work. No one was in a hurry. No one was slow. Everyone was on point, talking to each other calmly, professionally. Then the trauma nurse opened the wrap on Jameson's leg and immediately called the surgeon on duty.

It's strange, looking back, I wondered why someone wouldn't have already contacted a surgeon on call when they knew there had been a shark attack and the child was being life-flighted to your hospital. But I guess they need to see what they're working with first. Acute care surgeons are on call, but doctors don't call them until they know for sure what the situation is. It was clear that once they saw Jameson's leg, they knew what needed to be done.

I had seen Junior's leg when Senior had first lifted him out of the water. In that moment of horror I had seen bone—so much bone—rise past my face. I had wrapped it right away with my makeshift tourniquet, then someone made another, a better one. The paramedics also had wrapped his leg better when they took over, but I hadn't seen the leg since that moment.

Just below the knee, Jameson's tan skin had disappeared, replaced with a thin white layer of skin and a yellow layer of fat and then just bone, bone, bone, all the way down to his perfect and whole foot. A thick hunk of red muscle hung loosely beside the leg. I couldn't tell at first glance where it belonged.

I had been worried that my son might lose his life. Then I'd been worried that Jameson would lose his leg, but in this moment I realized he had already lost it.

They wanted to do an x-ray of his chest to make sure there was no water in his lungs, and as they prepared him for that, a hospital employee pulled me into the hallway to ask me about insurance, Jameson's age, and his blood type (I didn't know it at that time).

As the assessment finished, they started moving Jameson to an operating theater, with me walking alongside as they moved. There were things they needed to do to make sure Jameson would survive the night, the major one being that they needed to get him a blood transfusion and make sure he didn't lose any more. It was time to get it done. I walked alongside him, talking to him the whole way.

Then a nurse said, "This is where we say good-bye. You can't go any farther."

I hadn't expected that. I didn't know that was about to happen. I wasn't ready, wasn't prepared to say good-bye to him.

I focused on Junior. "They're going to take care of you," I said to my son. "You're going into surgery, and I'm going to be right here. Are you okay?"

Jameson simply said, "Yes."

His being okay was why I was okay. I remembered when he was two and had to be put to sleep for a minor surgery. They had given him "sleepy juice" so he just fell asleep. When he woke up, everything was fine. He was fine. Because of Jameson's peace, I believed he was fine. If he had been screaming and crying as they took him away, I would have been a hot mess.

As I was led through a maze of hallways to get to the waiting room, I realized I hadn't prayed with him as they wheeled him away from me and into surgery.

I just didn't think of it in the moment.

I was separated from Jameson Sr. and our other kids by a three-and-a-half-hour drive and wanted to let Jameson know what was happening to Junior now.

I called, and Jameson steeled himself for the worst, thinking I would say that Junior was gone. But instead I said, "Honey, you're not going to believe it. He's doing great. He just went back for surgery. "

In the shock of it all, I'd said our son was doing great—and I guess he was for a kid who had just been mauled by a shark.

~

I walked to a waiting room: a little room with chairs, tables, and a vending machine. My mom was there and I remember being so glad to have her with me. Sarah, a family friend, arrived soon after.

The hospital staff were looking for the paperwork. There was so much paperwork. They were having a hard time finding it.

A guy in a Fish and Wildlife uniform joined us. *What is this guy doing here?* I wondered. Maybe he, too, was waiting for someone.

Turns out he was waiting for me.

In those few moments on the dock, I hadn't had time to tell FWC what happened, and he had come with the report form for me to fill out. "Just whatever you remember," he said. So I sat in the hospital waiting room in my swimsuit and towel and filled out some paperwork while my son was, I hoped, having lifesaving surgery. My mom had brought clothes for me but there was just no time to change…Everything felt so urgent. The FWC interview. A police officer needed a statement of what had happened too.

Then the hospital paperwork showed up. Someone told me to

write "right leg amputation" and then sign the form. I had seen what was left of Jameson's leg. I signed it all.

Finally, when it was all done, I went to the restroom and changed. It was a relief to be in regular clothes again, even if I couldn't rinse off all the blood, salt, and sand.

Then there was nothing to do but wait.

JAMESON SR.

I was still debating when to drive to Miami. I knew we needed to go, but I couldn't leave the kids, of course. Mike and Tammy, our friends who own the Sugar Shack, had offered to buy the kids and me tickets to fly to Miami right away. They would pack up the house, get all our things ready and make sure we had a car. It was such a generous offer. I just wasn't sure if it was the right thing to do right then.

Meanwhile, Jameson was in surgery, and Mary was dealing with all of this alone. Well, not exactly alone. Her mom and Sarah were there, but *without me*. It definitely seemed like I should be there with her, but at this point it would be the middle of the night before I could get there. When I got to Miami, I'd be closer, of course, but in a hotel room with three kids while Mary was in the hospital waiting room.

My parents had gotten on a plane to Miami. They would be there by midnight. And Mary had more family, lots of family nearby. So she wasn't alone. I kept reminding myself of that: she wasn't alone. And God had told me, "I'll hold him until you get there." Jameson Jr. wasn't alone either. I wasn't alone. We had each other and our family and our friends and God. We all needed to remember that.

MARY

Catching up my mom and Sarah on all the details took energy and time. I was exhausted. So worn out, so taken off guard that I had forgotten to pray with Jameson as he went into surgery.

At some point a doctor came out and told me, "Tell your doctor friends to stop calling me. We're doing our best here, and I can't give them any information anyway. It violates HIPAA to share your information."

I had no idea what he was talking about. But one thing you'll learn quickly about Jameson Sr. is that he knows everyone. I assumed he had called some of his doctor friends and they were trying to get more information.

Eventually a nurse, Lizzie, approached me in the waiting room. "I'm getting ready to go off shift," she said. "But I was the one who was in the room with your son, helping him to go to sleep." He was still in surgery, but Lizzie wanted to see us before she left for the night.

"He's okay," she said. "Everything's going well."

We were so thankful for the update and told her so, and she said she wanted to tell us something that had happened while Jameson was being prepped for surgery.

"Jameson asked me if I knew Jesus, and I said yes. And he said, 'Well, I thought maybe you would want to pray with me.'" Her first language was Spanish, and with Jameson, she prayed out loud for the first time in English. "He went to sleep and was just the most calm, most sweet boy." She said she had never seen anything like it in her life.

I told her more about what had happened—the shark attack, the strangers who helped us get here faster than anyone thought was possible, and Jameson singing on the boat as we raced for the ambulance—and she decided to stay. She got my mom's phone number, and instead of going home, she went back into surgery and texted updates to us.

She was already off-duty. She didn't have to do that, but she had heard my mamma's heart and went above and beyond to make sure we had immediate news of everything going on. I am still so grateful to her for that kindness.

The really amazing thing was that even though I—in a moment of being overwhelmed and surprised that I was being separated from Jameson—hadn't prayed with Jameson, God sent someone to be with him.

Lizzie had prayed with Jameson before he went under.

JAMESON SR.

As soon as they had Jameson stable, an army of medical professionals came into the waiting room. Mary called me on speakerphone so I could hear what was going on. The first thing they told us was that he was out of danger. The lifesaving part of the surgery had been successful. The blood transfusion had gone well, and they had stopped the bleeding.

But Jameson was still on the table. They wanted our permission to amputate the leg.

My first thought was, *I've got to get down there.* A three-and-a-half-hour drive. Have to load up all our things. Get the kids in the car. We're not coming back here. We're going to have to be in Miami for a while.

And then I thought, *We can't decide this right now.* We need a second opinion, make sure we're getting the right advice here. I wanted to be in the room when we made the call, not patched in over a speakerphone, unable to see the doctors' faces.

I told them, "We'll make that decision tomorrow."

One of the doctors said, "No, no, no. You don't understand. We need to make a decision in the next five minutes. We did the lifesaving surgery, yes. But if we keep him lying on the table in this state, we don't know if he'll make it."

MARY

Jameson hadn't had as clear a view of Junior's leg as I had at this point. He hadn't written "right leg amputation" down on a piece of paper and signed it. He hadn't seen the unrecognizable leg laid out on the trauma table. The bone. The barely attached piece of muscle. The bloodless foot. At the same time, it was really important that he be part of this decision. I needed to know he was onboard with this outcome.

I didn't even think they would come in and ask. I was thinking, *Save his life. I know his leg is going to be gone. We can't save his leg. There's nothing to save.*

But the doctors presented this to us as a choice that needed to be made. Or maybe it seemed that way because they wanted our permission. It gave us this illusion that there was a choice that was different from what it actually was. It was like they were asking for our blessing before doing the surgery.

The doctor who told Jameson Sr. we had five minutes, left the room as soon as he said that. He was headed back to the surgical theater. A doctor we would come to know and love well, Aaron Berger, stayed with us to explain. "If we try to save his leg, we don't know if he'll make it. He's lost too much blood, among other things, and we could try to save his leg and lose him in the process."

JAMESON SR.

No. There's no way. How can I trust this person's opinion when they're being so matter of fact about it? Maybe they don't have the skill. Someone else might. Maybe they missed something. We have to get advice from someone else. That's what I was thinking.

Jameson's an athlete. He loves sports. Many people I had talked to that night, people who knew us and knew Jameson, had said, "Don't let them take that leg."

"We're asking," Dr. Berger said, "but this is what needs to happen." He was so patient and kind. "I have two girls of my own, and if this was my own daughter, this is the decision I would make."

I kept thinking, what am I going to tell Jameson when he's sixteen? This is going to change his life forever. How do I explain this to him when he wakes up and has no leg? How do we explain this to his brothers and sister? I needed to be there, be in the room when we made the decision. If we were going to do this, I wanted to be there when Jameson woke up, so I could help him to understand what had happened.

Our friend Sarah knew that we had a hard decision to make quickly and suggested we call two doctors we knew well and get their opinions. She called one on her phone and patched the second through on a three-way call.

I was so glad Sarah had suggested this. While the other

doctors were still standing by Mary, these were talking to us via speakerphone.

Dr. Daniel Chan is the chief of orthopedic surgery and sports medicine at Memorial Healthcare System, as well as the medical director of orthopedic trauma. A perfect person for this conversation. And we also called my friend Dr. Chris Chen, the CEO of Chen Medical, a company that has medical centers across the United States. These are two guys I knew and trusted. Once they were both on the phone, we had a sort of giant party line, with them talking to me and Mary and the doctors, everyone all at once.

The doctors in the waiting room started laying out all the particulars of the case—how much mass was gone, how many centimeters of bone or skin, the state of the arteries, a rapid-fire exchange of medical information—everyone speaking as quickly and professionally as possible, sharing details I didn't always understand while my son was lying in the operating room, waiting for a decision. The lower leg has two bones, the larger bone had been completely stripped of tissue. No skin, no muscle, nothing. The smaller bone was completely gone. All of the nerves, the muscles, and the vessels responsible for keeping his foot alive were gone.

After a few moments, Dr. Chan said, "Jameson, this is an amazing thing. The most unbelievable thing. You don't have to make this decision. The shark made it for you. You're off the hook."

An enormous wave of relief washed over me. It was unbelievably freeing for me. Doctors whom I knew and trusted were telling me this wasn't really a choice, just an opportunity to give permission to do what needed doing. That's what I told them: "If this is what we have to do, let's do it."

MARY

The second Jameson agreed to removing Junior's leg, all of the medical personnel left the room. There was work to be done. They would remove our son's leg.

JAMESON SR.

After the call, I was lost in the emotional drain and chaos. I had been walking back and forth between the bedroom and the dining room. The bedroom, because there were things I was needing to say and questions I wanted to ask that I didn't want the other kids to overhear. And then back to the table, because getting too far away from the kids was still making them (and me) nervous.

But once the decision was made, I collapsed physically and emotionally. I lay down on the bed and cried.

My biggest challenge was not being there. I always want to be there for my kids, for Mary. I was really wrestling with the fact that I wasn't at Jameson's side, that I wouldn't be there when he woke up to say, "Buddy, I'm so sorry, we had to amputate."

I felt helpless. I hadn't been able to keep him from the shark, and now he had lost his leg. I knew in my mind that it wasn't my fault, but that didn't change how I felt. I wanted to go back in time and change everything, I wanted to magically teleport and be by his bedside in the hospital. There were things I couldn't change, and I wrestled with those things, trying to bring my emotions—what I felt was true—in line with reality and what I knew was true. I did everything I could; that's what was true.

But I kept turning it all over in my mind, trying to find something I could have done better, some way I could have changed it all.

Eventually—not in that moment—I knew I would have to come to a moment of acceptance. That if I wanted to do my best in the weeks and months to come, I would need to surrender to reality so I could best respond to the situation we were in.

That night I let all three kids pile into the bed with me. We could take comfort in being together, and I wanted to make sure they slept well.

I wasn't there with Jameson in the hospital.

I wasn't *there*, but I was *here*.

So I would do my best here, right where I was.

CHAPTER FOUR

Maybe Tomorrow

MARY

Jameson Jr. had gone back for surgery and I was waiting. It was coming up on 11:30 or midnight when someone came out to tell us that it went well and they had a room for Jameson and for us. They were taking him to the Pediatric Intensive Care Unit (PICU), and they took me, my mom, and Sarah to a little room where we could wait until he was out of recovery. Now that we knew he had made it through surgery, I called Jameson to update him.

JAMESON SR.

Around midnight our friends Marc and Holly came, and they'd shown up with pizza. I was exhausted, the kids were asleep, and it was so great to have friends who would drop everything and drive to be with us. You know they are real friends when they have your back in the middle of the night. They weren't the only ones either. I knew our friend Sarah was with Mary and her mom. And my parents were on the way.

Marc, Holly, and I talked, and we decided they'd come back in the morning and help the kids and me pack. We had to get to Miami. Now that we knew Jameson was safely through surgery, I wanted to get there and be with him and Mary.

MARY

Jameson had been wheeled into his room, but we couldn't go in yet. They wanted to get him completely settled and make sure he could sleep for a little while before we went in to see him.

I felt relief when I finally walked into his room. He had been intubated for surgery, so there was a tube coming out of his mouth, although he was breathing on his own. An IV was in his arm. Monitors and wires were going everywhere. But he was alive. I was thankful he was alive.

I started kissing him, holding his hand, touching his face, his hair. He was still out, deeply asleep, but he was alive, and I was more grateful than I could express.

It was a surreal moment. It had been less than twenty-four hours, less even than twelve hours. Twelve hours ago we had been swimming. We had been laughing about sandwiches. We had been diving in the water and enjoying the sun and sea together. Had this really happened? My son was attacked by a shark? The whole day so far—the frantic race to shore, the ambulance, the helicopter, the surgeries—seemed so strange. I felt like I was underwater, and the world above was hard to make out... with the sound and light distorted by the water. I couldn't focus, couldn't see things around me unless I really concentrated.

But now I was in the PICU, and my son was alive. It hadn't been so long ago that I wasn't sure that's how the day would end. I was thankful. I couldn't stop thinking about how thankful I was that he was here, that he was breathing.

I didn't want to take my eyes off him. I wanted to be able to see his face. There was a sort of bench on one side of the room where we could lie down, and I positioned myself so I could keep an eye on him. I was sad that Jameson Sr. and the other kids weren't here with us, but I could at least keep Junior where I could see him. Mom was exhausted, and she lay down on the other bench.

Jameson's parents, Bob and Celia, had flown in from North Carolina by this time. They took up positions near Junior, sitting near him by the bed. I knew that, in this moment, Jameson was as safe as he could be. I was here. My mom was here. His grandparents were here. The doctors were nearby if we needed them.

I closed my eyes and did my best to sleep.

JAMESON SR.

In the morning, there was so much to do.

We were three-and-a-half hours from Miami, and we needed to pack up the whole house. If you haven't moved into a vacation home for several days with your four young kids before, it's a big process to leave that place again. There are clothes in weird places. The stuffed animals and toys left all over after the kids had played with them. Games and luggage and food you brought along and water toys, all of which has to be collected and packed and then somehow put back into the car, leaving space for all the people who need to fit in there too.

I had a little bit of that feeling of the fog of war. Not really knowing what was going on. I don't know how to describe it exactly, but it was kind of like being at the front row of a concert. You're focused on the band. You don't really know what's going on behind you. You're feeling the pounding of the bass. You're listening to the music. You're shouting with your friends. Until you get far enough back from it, you're in a kind of twilight zone. Getting a handle on what exactly is happening, trying to stay focused on things like packing the car is hard, because thinking about Jameson is so central.

Friends were calling, wanting to know if Jameson had woken up yet. Really sweet calls. Our friends were crying with us on the phone about what was happening. And in the meantime, trying to be upbeat with the kids as everything was thrown into turmoil, we were leaving the house early, leaving vacation. There was some confusion among the kids about Mom not being here, about Junior not being with us too. They were a little disoriented, and I guess I was too.

Oh, and the media started calling. My brother Joshua had posted a quick update on social media, and now it wasn't just our family trying to get the whole story. It was strangers too. People wanting to know firsthand the details about the boy who had survived the gruesome shark attack. Even though shark attacks have been on the rise, they are still very uncommon. They usually happen at the shoreline to trap their prey in shallow water where there is no reef or rock to hide under. The media knew there had never been a documented shark bite, let alone shark attack, around Looe Key Reef.

Marc and Holly had returned to help us with the children and pack. The kids love them, so their presence also kept the kids positive and kept me moving. Holly called Mary and went from room to room to make sure we were only packing up our things, not things that belonged at the house.

~

Not being able to be there in the hospital was hard. And I couldn't help but replay things from yesterday over and over in my mind. For instance, when the attack happened, I was with Nehemiah. He couldn't swim, so I had to take him back to the boat before I could go to Jameson. I felt some regret about that, like maybe I could have gotten to Jameson sooner if I had decided to swim straight to him.

But at the same time, when I played it back, I felt so certain that God was leading every snap decision. Every moment felt like we had been guided into doing the best thing. Taking the boat over to Jameson had been a better call, once we knew what had happened. I felt regret, but at the same time I couldn't deny that if there was going to be a shark attack, things couldn't have gone better. A faster boat nearby. The emergency response. The ambulance driver had even called the helicopter ahead of time and told them to get in the air. And there was a nurse on the boat nearest to us. Unbelievable. I kept going over it in my mind and kept coming back to the fact that if we had to go through this, we experienced miracle after miracle after miracle.

MARY

Jameson never really woke up in the night. He kept stirring, reaching for the intubation tube, but he didn't wake up. He was obviously uncomfortable and wanted that thing out. We all kept talking to him. Touching him, trying to comfort him.

I knew Senior and the kids would be on their way as soon as possible, and I expected Senior to call at any moment. When he did call in the morning to check in, I took the phone and put it up close to Junior. Senior said, "Jameson, Daddy's here," and Jameson nodded. We all got so excited. It was the first conscious movement Jameson had made, the first real response we'd gotten from him since he'd been in surgery.

He had heard his father's voice and responded.

JAMESON SR.

One thing we've always told our kids is that even though we're responsible for them, we also believe they need to have freedom to go out and explore, to be in the world. God is there to watch over you even when we're not, and we try to teach them to recognize that the Lord is going out before them, that they can't go anywhere that God isn't already there. Even with our best intentions and all our abilities, we know God is our protector and caretaker even when we feel weak.

We've talked about Hebrews 2:1, which tells us that we have to pay close attention to what God has told us so that we don't drift away. We must continue to look to God by the truth of Scripture, or the current will cause us to drift with the ways of the world. This provided an opportunity to teach our children that the cultural current will continue to pull them, but they won't know how strong the current is unless they look back at their father. We can tell if we're drifting by looking at our Father and gauging the distance.

So even when we're out on the beach, we use that metaphor.

One of the things we teach the kids is by telling them, "I don't want to be calling out to you all the time, telling you that you're going too far. It's your job to watch for your father and come back." God's not a nagging parent, and we don't want to be either. If you're caught in the surf, you need to recognize it's not your dad who's moved. It's your job to recognize that and come back. (And of course, just as God does with us, we're keeping a close eye on our children to make sure they are safe!)

We talk to our kids a lot about how freedom in Christ means going with Christ, finding how God is moving, and paying attention when they're exploring their freedom. They need to be able to see us if we're waving them in from playing outside. They need to be able to hear our voice if we call them to come in from playing in the surf.

It's not quite the same thing, but I was so relieved and excited that when Jameson heard my voice in the hospital, when he knew I was there, he responded. We were all so excited for him to be waking up, because we could finally communicate with him.

MARY

When Jameson woke up, he really wanted that tube out. He was breathing on his own, and the medical staff agreed he could have it out, but it might be hours before they got to it.

Jameson couldn't talk, so he used hand signals. He let me know that he wanted a pen and paper, and we managed to get some. He started writing. Slowly and carefully. He was having a hard time, but he managed to slowly write, "It c—"

I knew what he was trying to say. He had said it on the boat. So I said it out loud for him: "It came out of nowhere, bud?"

He nodded yes. He wanted us to know that he had done everything we had trained him to do. We teach the kids to look around, to be aware of their surroundings, to pay attention. The ocean can be a dangerous place and you need to stay aware.

"I know, bud. I know," I told him. It wasn't his fault. It wasn't Dad's fault or my fault either, though emotionally it would take us a while to get there. But this is how it is when these unexpected harms come on us. We find ourselves trying to figure them out. What could we have done differently? How could we have dealt with this better? What other decisions could we have made?

JAMESON SR.

Packing up and keeping the kids fed, closing up the house, and everything else that needed doing before we could leave took most of the day. Marc and Holly were coming with us. Marc would drive our van so I didn't have to, and Holly would drive their car.

I said, "Let's take a minute to thank God in the midst of all of this." So we prayed together, Marc and Holly and the kids and I. Later, Marc would point out this was a great way to help the kids face their fears about these kinds of situations, to refocus on the good things that God was doing in the midst of it, that Jameson was alive, that he was waking up, that we were headed to see him now.

Before we left, we sang a song together, the same song Junior had been singing on the fast boat, "Here I Am to Worship." It was suddenly a much more personal song.

MARY

It took a while, but eventually the doctors came in to remove Jameson's tube. Actually, a whole lot of people came in at once. We were at Nicklaus Children's Hospital, Miami, a teaching hospital, which means a lot of times we'd have not just the doctors and nurses and techs but also interns and students. That's not a complaint. They did an amazing job; it's an extraordinary place. We never felt the teaching got in the way of the care, and maybe because they're showing students the right way things should be done, we never felt as if things were ever done poorly or without clear intention.

When Jameson started to talk a little, I breathed a sigh of relief. His throat was sore, and he was exhausted, of course. But despite all sorts of medications and painkillers, he could speak.

He had been out of surgery only about ten hours and didn't know he had lost his leg. He was awake and aware, but he was on a lot of medications, drifting in and out of sleep.

Jameson came out of his drug stupor long enough to say, "Mommy, I can't feel my toes." In that moment, I didn't know what to say. I knew Senior wanted to be here when we told him, but he was still hours away. I felt an extreme sadness. He knew he'd been hurt, so he must have been thinking that there had been a lot of damage, and maybe he just couldn't feel them yet. In fact, that's what he said next in his sweet voice: "Maybe I'll feel them tomorrow."

My heart was so heavy that I couldn't stop the negative questions: Why wasn't I right there next to him in the water? How could I have changed things so my little boy wouldn't be in a hospital room and telling me he couldn't feel his toes? I was his mother. I was supposed to protect him.

Junior had fallen back asleep, and I was caught up in that whiplash between the past and the future. What could we have done differently? But also, how are we going to deal with what's to come? Very soon, Jameson needed to be told that he had lost his leg so he could start to come to terms with everything.

~

I wish Jameson could feel his leg again, but that will never be. We can't help ourselves, though, wishing we could take the hard things that happen, the life-changing ones, and somehow reset our lives back to the moment before those things took place. We think maybe tomorrow we can feel things the way we used to. Maybe tomorrow we can set aside all of our fears and pain. Maybe tomorrow we can be repaired, be made whole.

But we can't undo a shark attack any more than someone can undo any natural disaster. We can't regrow body parts. We can't turn back time. We can't change the future by worrying, and we can't change the past by wishing.

JAMESON SR.

It was afternoon and everyone was in the car. I'd been able to speak to Jameson after his tube was out. His chipper voice had said, "Hey, Dad!" He sounded so much like his normal self that I was really surprised.

I should have suspected that would be true, but I was so wrapped up worrying for him that I forgot how he has this contagious joy you can't help but catch when you're with him.

Even thinking back to the first time I met him—the day he was born—I remember cutting the umbilical cord and the doctors wiping him down and then handing him to Mary. She was crying and in pain and so full of joy, and I started singing a hymn. I sang all my kids into the world, and with Jameson I sang the hymn "Great Is Thy Faithfulness."

I had been crying, and the nurses were crying, and I had held Mary's hand. As I bent down and reached over to touch my son for the first time, baby Jameson reached up and grabbed my finger. It was like an electric shock went through my body, unlike anything I've ever experienced. And he looked at me, really looked, which they'll tell you a baby can't do a minute after birth, but I can only tell you what happened. He looked right at me. There has been something in and on this child's life from the beginning. Undeniable. Unmistakable.

His whole life, Jameson has been tuned in to people around him. He can tell when someone is feeling down and needs some attention. When he was little, we'd go to the beach and, under our watchful eye, he'd walk around and visit people near us and talk to them. Sometimes people would be looking around like, "Who does this little boy belong to?" We'd wave to let them know we were there—to explain that he just loves saying hello to everyone.

Which is only to say, Jameson has always been a guy who brings a lot of joy to people. They feel seen when he's around. So I shouldn't have been surprised to find him cheerful and ready with a bright greeting when I walked into the hospital room. He was just delighted to see me, and it made me feel great.

Because Junior was so chipper, my dad brought up the possibility he might still be in shock. But weeks later he said, "We thought he was just in shock before, but if so he's *still* in shock!"

Jameson embraced the whole crisis with more acceptance, goodwill, and a better attitude than most people would, more than most adults too. As my mom said, "He made it easier for the rest of us." He just seemed to roll with everything that came his way.

While Marc drove the car, I made phone calls and talked to the kids, and at one point, I took a nap.

When we drove over the same bridge the ambulance had crossed with Jameson, it took me back to yesterday's events. I could barely believe it had only been yesterday. I started thinking about what was coming next. That chipper voice on the phone was a good clue. He was going to want to get back into the water. Just the thought of it made my stomach clench. I was going to need to face my own fear before Jameson could face his. There was a very real possibility that Jameson was going to want to "get back on the horse" and return to the reef.

I said to Marc, "I gotta figure out how I'm going to prepare to come back here if Jameson wants to." The thought of Jameson being too terrified to get back into the water filled me with a deep sadness. He just loves the water so much, but the idea of him not wanting to get on a boat or not wanting to get in the water filled me with fear.

"That's incredible," Marc said. "How are you even thinking about that right now?"

It was part of being a Reeder, projecting ourselves out into the future and all of its possibilities. Every time we go to the Keys, we start asking questions, thinking about opportunities, telling stories. What's out there in the water? What's under the water? What kind of exciting adventures are we going to have? What big fish are we going to see? What if there are pirates? (We are always talking about pirates. The kids love pirate stories.) We love to think about the future, to turn it over together and come up with ideas and thoughts about it.

As we crossed the South Pine Bridge, I started pointing things out to Marc, narrating yesterday's events to him. I showed him where our boat was as Jameson's ambulance was shooting across this very bridge. I was struck again by the improbability of how well things had gone. Not the shark attack itself, of course. But what if we had gotten stuck on that reef? We had to cross it to get to the other boat.

I thought, what if Nehemiah had already been on the boat? I would have swum directly to Jameson, and who knows what the shark might have done? I knew Mary was processing all of these what-ifs related to how we might have prevented the terrible part of the day. I was processing what-ifs related to how it could have been much, much worse. If one of the younger kids had been the one who had been attacked, for instance, would they have survived?

We stopped only once, to gas up and get something to eat. I was still fielding phone calls, and I talked to Mary a couple of times along the way.

We drove straight to the hospital, arriving there very late in the afternoon. All six of us—Marc and Holly, the kids and I—went straight to Jameson's room. I thought back to the moment I had handed him across to the other boat, hoping they could save his life as I wondered if I'd ever see my son alive again.

We walked into the room, but Jameson was not in his bed. Mary explained they'd wheeled his bed out to do a test. I was disappointed but happy to see Mary, and the kids were thrilled to have their grandparents there.

MARY

Even though he had come through surgery successfully, there were tests and checkups after the massive blood loss and shock to the body because of the loss of his leg. And it was clear that our couple of weeks in Florida would be stretched into many more weeks (actually months, as it turned out) of living away from home.

I thought of Jameson saying maybe he would feel his toes tomorrow, but I knew that wasn't true. A whole bunch of other maybes, however, were crowding in on us, other possibilities, other things that would need doing. One of which, I knew, would be finding a way to tell Jameson his leg was gone. The doctors had told us there is an emotional recovery that comes with the news and the physical recovery. How you tell someone about this big change in their lives can help move them toward recovery faster, and this hospital had people who could help us with this. People who, unlike us, had had to tell a child they'd lost a limb before.

JAMESON SR.

When they finally wheeled Jameson back in, the first thing I noticed was how swollen he was. I tried not to be distracted by his puffy face. The medications had made him almost unrecognizable.

The second thing I noticed was the big grin on his face. He was so happy to see us. His brothers and sister and me. Marc and Holly too. And, of course, his grandparents and Mary.

My parents took the other kids out of the room for a while to give them some space to run and breathe. And a little later, I got a few minutes alone with Jameson. No one else. No nurses. None of our family. Not even Mary.

"I'll never let anything like this happen to you again," I told him. "I'm so proud of you. You're so brave. You fought a Goliath and won."

We already suspected the shark had been a bull shark and later would know for sure when we recovered the GoPro, the compact video camera that we used underwater and that Jameson had been holding during the attack. It wasn't an exaggeration to say he had fought a Goliath and won. The shark was bigger, longer, heavier, stronger than Jameson.

The bull shark is one of the most ferocious sharks in the ocean, and one of the most dangerous to people. They tend to hunt in shallow, warm waters (the same places we like) and can survive in salt and fresh water. Coming across a bull shark is more intense than a lion or a great white, and pound for pound, the bull shark's bite force is higher than any other shark (1,330 pounds of force!).

Joy to be with my son was welling up inside of me, and a deep pride in my heart that he had done this, that he had encountered one of these magnificent creatures and come out alive. As I was telling him this, I started to cry and told him how sorry I was.

"It's okay, Dad," he said. "It's not your fault. There wasn't anything you could have done."

I had an overwhelming sense that this couldn't be real. That it was a dream, a terrible dream. Sitting across from my son, swollen from the drugs, he tried to comfort me and tell me what I had been trying to tell myself: it wasn't my fault. We did the best we could do, and it all turned out the best it could have turned out.

~

Junior's words that day only increased my pride. What an amazing person, that in this moment of unexpected harm, he was looking at me, at the people around him, and thinking about how he could comfort us.

How thankful I was for those words: It's not your fault. There wasn't anything that you could have done.

I was reminded again of what God had said to me on the reef. "I'll hold him until you get here." I was reminded of what we tell our kids all the time, that when they're out there without us, God is still there with them.

Whatever came tomorrow, I knew those things would still be true.

God is always with us and would take care of us, no matter what was still to come.

CHAPTER FIVE

The Gift of Giving

JAMESON SR.

Eventually, Jameson would be moved out of the PICU, and we'd get a little more of a regular schedule. There were a lot of recovery activities to fill the day. If you've been in a hospital before, and especially an intensive care unit, you know you barely get any sleep. Doctors and nurses come in all day and all night to check vitals and make sure all is well. On top of that, you have all the visitors, and only a certain number of people can be in the room, so you're switching people out so everyone can get a visit.

MARY

The hospital has a great program called the Child Life Program, which is a whole department dedicated to keeping the kids comfortable emotionally and physically, as well as taking care of the parents. We loved to see Karla Filosa come to Jameson's room. Karla works in the PICU, and with every kid who comes in, she makes sure they are comfortable and taken care of. She's not checking their vitals. She is giving them things to do, talking with them, and making sure they are okay.

Karla says her skill set is providing developmental education for parents to help them be a part of what's going on with their child in the hospital. She and her team provide psychosocial support to the patient and family, including siblings. They clear up misconceptions, diagnoses, and procedures to allow patients and families to cope and adjust to hospitalization easier.

What that translated to for us was someone who made our lives better, who was so focused on our needs and on Jameson and on our other kids too. She made sure to have tons of fun things for each of them based on their interests. She not only provided things for Jameson, she also asked me what the other three kids liked as well.

When Karla came in, you could tell, even through her medical mask, she was a happy, upbeat person. She had smiling eyes and was so positive, loving, and energetic, a great fit for our ten-year-old boy. (We were in the hospital during the COVID pandemic, so everyone was wearing masks when they came in. Our family didn't have to wear them in our hospital room, but anytime we went somewhere else in the hospital, we did.)

Before the attack, Jameson had been very active. He loves surfing and being in the water and baseball, but now he was bedbound, hooked up to machines, on all these medications. He was not feeling great, of course, but he was also bored. Not surprisingly, Karla instantly became one of his favorites. She brought goodies for Jameson. She asked what sorts of things he liked to do, what kinds of toys he played with, his favorite color, all sorts of questions like that. Then, after she knew Jameson a little better, she showed up with all these incredible gifts: coloring books, games, books, *Star Wars* toys, stuffed animals, even an e-book reader.

In those first hours in the hospital, Karla had pulled me into the hallway to talk about how Jameson didn't know about his leg yet. "I think we should do it all together," she said. "And I want him to hear it from the doctors as well." That sounded good to us.

When one of the doctors started to tell Jameson about his leg, the doctor got all teary-eyed and Karla stopped him.

"He can't be told like that," Karla told me. "I want Jameson to hear it from the doctor, but he can't be crying." She wanted the doctor to sound confident, and Jameson to hear it with as many of us around him as possible for support.

So she planned a full-on meeting with everyone.

We all gathered around Jameson's bed. I was by his head on one side, and Senior's dad, who the kids called Sabba, the Hebrew word for "grandfather," was on the other. Senior's mom who they call Cece, and my mom, who the kids called Grandma, were there too. The doctors and Karla were at the foot of the bed. Senior and the kids weren't there yet, but we couldn't wait... We didn't want Jameson to reach down and find out for himself what had happened. So we needed to tell him soon after he woke up.

Karla took the lead in this talk with Jameson. Just telling him we were all there for him, and we wanted to talk about his treatment and what was going on.

At some point, Jameson just asked her, "Why can't I feel my toes?"

"Do you remember why you're here?" Karla asked him.

"Yes," he said. "I was attacked by a shark."

And Karla did this wonderful thing, where she didn't say it straight out, she just asked him with such kindness and gentleness, "Why do you think you can't feel your toes?"

He didn't say anything. He was processing the question, trying to find his way from it to an impossible and horrible answer. But I couldn't take it. I said, "Buddy, do you remember what happened to Bethany Hamilton?"

He looked at me, his eyes so tired and worried, but also— like his eyes always are—kind. I was holding his hand, standing close to his head. There was a long silence. It felt like time was standing still. He asked, "So that's what happened to me?"

Jameson closed his eyes and grimaced. He didn't cry. He just seemed to be taking it in.

Across from me, tears rolled down Sabba's cheeks. Jameson had opened his eyes and saw his grandfather crying. He grabbed his hand and said, "It's okay, Sabba. It's going to be okay."

I was so struck by Jameson's sweet attention toward his grandfather. Once again, he was trying to comfort us during a really hard moment for him. That is his character. Sometimes I have to hold him back from trying to help people in inappropriate moments or ways.

Then the doctor took over and started to explain it all: why they had to amputate, the process, and some of the things that were still to come. We talked for the first time about prosthetics and how good they've become, how Jameson wouldn't have his own leg, but he'd be able to walk again, to run, to play with other kids. In fact, they had been able to save enough of his leg below the knee so he would be well situated for a good prosthesis. We told him, "You're going to get a really cool leg!"

It's important to know that we had recently watched the movie *Soul Surfer*. It's a true story about a young woman, Bethany Hamilton, who's a competitive surfer. While surfing one day, she lost an arm to a tiger shark. Eventually, through a series of hard choices and difficult decisions, she started to surf again. This movie is a family favorite, and we've watched it a number of times. Most recently, Junior had watched with Grandma just days before the attack.

About two million Americans live with limb loss, and about 185,000 people a year have to have an amputation. While children are often more adaptable to the change emotionally, they can struggle more with the change in their body image, especially in relationship to other kids. It makes sense really, because the physicality of going from a kid with all your limbs to missing a limb means changing how you interact with other kids and what games are easy to play and how good you are at playing them. So many things have to be learned afresh.

Some of the things that can help with the adjustment are things like speaking honestly about the change, talking about feelings, getting support from the people around you, and re-engaging with your purpose in life. For instance, Bethany Hamilton's story includes her coming to grips with the loss of her arm as she realizes she can use her story to help other people.

We were trying to capture some of those things. Showing Jameson that he was in a community where he could share his feelings. Making sure to answer his questions and be clear with him. Karla didn't want the doctor crying when he told Jameson about the amputation; she wanted him to talk about it calmly, without coloring an emotional response for Jameson.

JAMESON SR.

Once the kids and I arrived, Karla turned some attention toward them too, which was really kind. She brought toys for our kids and explained—in a way appropriate for each of their ages— what had happened and what kinds of things Jameson would be going through in the months to come. She had a model of a person with detachable parts so she could explain about his leg to them, and she talked with them about prosthetics and also how the kids could be good helpers to Jameson.

Jameson was so happy to see his siblings being cared for too. He's a high-energy guy, and it was a big deal to have some toys and things to do when he couldn't get out of bed, and he was glad the other kids were getting gifts too. He recognizes the needs people have, and he wants to fill that gap between what they have and what they're missing. In his best and brightest moments, he recognizes needs and finds a way to meet them. Sometimes it's just because he can be so bright and vibrant, so kind and comforting...even when he is in a hospital bed. Sometimes it's just that he notices a need and points it out to the people around us. "How can we help this person? How can we bring them something they need?"

MARY

Sometimes I have to tell Jameson no when he wants to help someone because it's not the right time. For example, it could be that a kid wanted something at the store, and his mother has told him no, so he's crying in the shopping cart. But Jameson wants to know, "Mom, can I go and ask that lady if I can give her son my bracelet?" He loves to give people things.

We were at the dentist a while back, and there was a girl next to him who was crying because she was getting an adjustment to her braces. While her mom was setting up the next appointment, Jameson went over and asked her for permission to give her daughter his bracelet. She said yes, and he handed it to her and said, "I hope you have a good day."

JAMESON SR.

Jameson did that over and over during his recovery. There was a cancer patient, Wayne Kerr, whose leg had been removed, and he was really struggling to accept it. After being with Jameson, though, he said, "If he can get through this, so can I." Random people we don't know have been inspired by Jameson's story, and meeting Jameson is another thing entirely. He so badly wants people to be well that I think people believe they will be. He does anything he can to help encourage.

He's still an ordinary boy. He sometimes argues with his siblings or gets cranky about schoolwork. He's a real person who just loves people so much. He always wants to give, and as we were learning, even though he was in the middle of a life-altering time, he didn't change.

MARY

Despite Jameson's difficulties, he turned toward the people around him instead of dwelling on his own problems. It would have been easy to focus on his own pain and suffering, his own losses, but instead he started seeing all these kids around him in the hospital and thinking, well, is there a way I could help all these other kids?

JAMESON SR.

Not to downplay at all how bad Jameson's wounds were—he almost died. In fact, he should have died, the doctors said. It was a miracle that he lived. But there were kids in the hospital who had cancer or blood diseases or a whole bunch of other things and were terminally ill. One day, Jameson would leave the hospital, but not all of them would. And there were those who were there for accidents, and they had stories that made you feel for them. There was a child there whose dad had rented a boat and lost control of the vessel when his kid was in the water. The boy was in the PICU because his back had been cut up by the propeller. The hospital was full of stories like that: kids who were sick or had been hurt in really awful ways. Jameson recognized this almost immediately and became really focused on what nice things he might be able to do to brighten the days of the other kids.

MARY

A question Karla asked Jameson was, "If there was any one thing you could get, what would it be?" Jameson's birthday was coming up, and he'd be spending it in the hospital, but that's not why she asked. She asked all the kids if they had an unrequited wish.

"A Disney cruise," Jameson said.

But this wasn't a Make-A-Wish Foundation kind of question; it was more about his time in the hospital. "Not that big," Karla said. "Think more of something like a laptop for school. A gift that can be given to you here in the hospital."

Jameson asked if he could have some time to think about it, and before he had a chance to tell her what he wanted, Jameson was moved out of the PICU and onto the surgical recovery floor. We didn't realize this meant Karla wouldn't visit anymore; she worked in the PICU exclusively. She had wanted to connect with our other kids, but their schedules never really worked. She had missed them.

Of course, within a couple of days, Jameson Jr. knew every nurse on the floor, and Jameson Sr. had already started making friends with some of the other families. We were still doing a lot of things every day. There were tests and therapy and appointments, doctors coming in to check on things, nurses checking on healing, and the wounds on Jameson's leg still took a lot of daily attention (and would continue to long after we left the hospital).

It took a few days for Jameson to answer Karla's question. But once he figured it out, I went to the PICU to see her and let her know what he wanted to do.

"Jameson wants to take some of the gifts he's received and deliver them door-to-door to share with the other patients." He wanted to do a sort of reverse trick-or-treat and drop off toys and gifts for the other kids in the hospital. He so badly wanted to share some of the joy of all the kind gifts he had received.

Karla was taken aback by that. She hadn't gotten a request

like that before. She told me quietly, off to the side, where Jameson couldn't hear, "I'm going to have to ask the administrators. I honestly don't know if we'll be allowed to do this."

I was thrown for a loop. It was such a sweet request, but as Karla pointed out, we were in the middle of a pandemic, and a lot of the kids had fragile health at best. I went back and told Jameson, "I don't think you're going to be able to do it, bud."

If you could have only seen his face. He was so disappointed. He had been exhausted, in pain, and sleeping a lot, but he had still been smiling all the time. Now he looked so sad for the first time... and not because of his leg, but because he wanted to spread some joy to the other kids in the hospital and was worried he wouldn't be able to. I told him to wait and see what they said. Maybe something good would come out of it. "I just want to go talk to people," he said. "I want to encourage them."

We didn't have to wait long for an answer. Karla came back the next day and said, "We can't let you go door-to-door and hand out gifts, but we think we've come up with a solution."

Every Wednesday the hospital hosts a farmer's market that the kids who were well enough could attend. It's conducted on the hospital grounds, right outside the main entrance, and local vendors bring in a variety of wonderful food: pizza, Mediterranean dishes, juices, tacos, cookies, fresh vegetables, nuts! The idea of the farmer's market is to give the patients and their families and the staff some options to the hospital cafeteria. The hospital suggested that while they couldn't let Jameson go from room to room, he could make something—some sort of craft—and give it away at the farmer's market if he wanted to.

And boy, did he want to!

Suddenly we had a mission. Every day there was a craft and activity time in the hospital lobby for the patients. Jameson wanted to go every day, mostly because he loved meeting the other patients, but also because he enjoys crafts. One day they had everything to make bead bracelets, and Jameson, his siblings, and all the cousins got to work making bracelets together. Dozens of them. Jameson was intent on making a lot of them and making them perfectly.

For a while he was able to focus on these little treats for the other patients. On the day of the farmer's market, he was very happy. His siblings came, and his cousins too, and we wheeled Jameson downstairs in his wheelchair. He happily handed out his bracelets to everyone. It was a really wonderful day.

That seems strange to say. A really wonderful day in the hospital with our son who'd lost his leg.

JAMESON SR.

Jameson had four surgeries in three weeks. Each involved surgery prep beforehand, recovering from surgery afterward, getting the meds and trying to get the right balance on the meds, and all of this while you're still getting used to the fact your family had experienced a shark attack that almost took the life of our son. We were getting more than two hundred texts a day from friends and family and the media. We wanted to respond to everyone and tried to set aside time in the evenings to text back, but we also wanted to spend time with Jameson and his guests and the people coming to see him. Mary wrote down the names of the people who brought or sent gifts or cards so we could send thank-you cards, but we still haven't managed to get through that list. (So if you're reading this, thank you! And we're sorry we didn't get a thank-you card to you yet!)

MARY

I was trying to be present through it all. Poor Jameson was in pain and on drugs and so, so bored. He wanted to play a game at night or watch a movie together, but starting early in the morning, the doctors would be coming in and we'd have a full day of work ahead of us.

I was very grateful my mom was there. I don't know what we would have done without our parents and family. My mom was always taking care of things in the background. She'd make my bed (the room we were in had been a maternity trauma room once upon a time, so it was a bit larger, and the hospital staff had brought in an actual bed for me). There were cards covering the walls, and Mom helped organize all the games and gifts and toys and food.

I was consumed with making sure everyone was thanked and okay, and that was a lot. And on top of that, so many family members wanted to be there, but we also needed some space.

JAMESON SR.

When it's your child who was bitten by a shark, you will see that it brings a lot of attention. Everyone wants to know the details. It's an unbelievable thing and rare to have a victim of a shark attack in a Miami hospital. And it was an attack, not just a shark bite. Jameson Jr. was violently attacked by an apex predator that was massive and fierce. The bull shark clamped down on his lower leg, violently shook it and tore away all of the muscle and flesh, and then broke and removed one main bone in the lower leg. Because so many reporters were trying to sneak in, the front desk staff called us the Plum family instead of the Reeder family. I'm not sure where that particular name came from, but the hospital told us this was the most media coverage they'd ever had. People had heard about Jameson and wanted to know more.

MARY

There were some weird moments. Like the one doctor who came in—he wasn't even one of Jameson's doctors—and asked, "Are we sure it was a shark attack? Could it have been a barracuda?" Which was a little ridiculous. Barracudas *can* attack people. But you're looking at stitches with that fish, not missing limbs. His leg was *gone*. He still had some teeth marks on him.

JAMESON SR.

Everyone was curious. Everyone wanted to know Jameson's story and how he was doing. But there was another conversation going on too. People were just amazed by Jameson's incredible attitude. In fact, that first day I was there, the paramedic who pulled Jameson off the helicopter asked if he could bring someone in to meet Jameson.

One of the hospital security guards, Mark Amador, wanted to meet Jameson, partly because he had heard so much about him, and partly because he wanted us to have a friendly face if we had a security issue. If a reporter or someone made it through without permission, for instance.

When people came to see Jameson for the first time after the attack, they expected to see a little boy who would be sad or depressed by the tragic situation he was in. Instead, every time you walked into his room, he'd be smiling, making jokes, making sure everyone else around him was happy. He didn't have an "I'm okay" attitude, but an "I'm going to be okay" attitude! No one expected that from a ten-year-old. I was amazed by this, and I would sometimes think to myself, *It's like he was born with one leg.* Not because he was up and around, but because he didn't seem as bothered by it as I would expect. The security guard, Mark, started dropping by to check on Jameson Jr. all the time. It was really sweet. Sometimes he would stay past his shift so he could say good-bye to us before he left. It wasn't about shark bites or trauma either. He just loved getting time with Jameson, catching up with him, talking to him. At one point, Mark asked me, "Why is your family so different?"

We were standing in the lobby, and right there I told him about how our family is centered around God, that we are always looking for what Jesus might be up to, even in hard situations like the one we were in now. He said he wanted to be that way too, and he prayed with me right there in the lobby, asking Jesus to come into his life.

CHAPTER SIX

A Whispered Wish

MARY

Life in the hospital became normal life in some ways. Jameson had physical therapy every day, trying to make sure that he was retaining use of his leg, bending his knee, and not just lying in bed all day.

Physical Therapy had him out of his bed on day one, and they came every day. He was still swollen from the steroids and painkillers, and they were there getting him into a wheelchair, encouraging him to use a walker. To Jameson's credit, he was committed. He didn't complain or fight it. He just asked, "What do you need me to do?" and he would do it.

In fact, he was always asking them, "Can I walk now?" and they had to tell him to slow down, to take it easy. "You have to take baby steps first."

Still, he was in a lot of pain. The hardest thing was trying to straighten his leg. His wound didn't have any skin on it until the very last surgery, at the end of our time in the hospital. They had placed a wound vac on the end of his leg where there was no skin. This amazing technology is a vacuum-like wrap, lined with silver, that uses constant suction to gently remove fluid from the wound over time. This replaces needing to have dressing changes each day. However, under the wound vac is an open wound, so you can imagine the kind of pain involved. Even though it was covered, any kind of pressure at all sent unbelievable pain through him. We had a rule that no one could tap his wheel-chair or the bed, because it just hurt him so badly.

The therapists didn't even want him to have a towel under his knee, which was really difficult because that meant his wound was in constant contact with the bed. That was the point, actually. They wanted his leg straightened as much as possible, but the hardest and most painful thing was straightening it. They were worried the muscles in his leg might atrophy if he didn't keep using it, especially because he couldn't put any weight on it.

They also were working on his upper-body strength, because whether it was a walker or crutches or something else, they knew it was going to be important that he could use his arms to do the work he needed to do.

It was difficult. Painful. Hard. I said the other day that Jameson liked some of the PT, that he thought some of it was fun, and he immediately corrected me. "Mom, none of it was fun. I didn't like the PT, but I loved the physical therapists, Amanda and Yesha. Everything else was hard."

Also—and this is kind of funny—I have a workout DVD at home called *Rockin' Body with Shaun T.* You know this kind of video—it's a dance-exercise DVD that promises you a "beach body on demand" and you can "party away the pounds." Anyway, a couple of months before we went to the Keys, I had pulled it out to try it. I had bought it for myself many years ago, and the kids wanted to try it too. They *loved* it. They wanted to dance and party with Shaun T, so I let them play it whenever they wanted. We're a homeschool family, so I figured, hey, it's like PE. It's actually a hard workout, though, so by the time Jameson was trying to do upper-body exercises with the therapists, he had already built up some significant strength.

That wasn't the only kind of therapy. This hospital was on top of making sure the kids in the hospital were processing the emotions that came with the changes that were happening in their bodies, whether they were in there for an accident, a regular treatment, or a terminal illness. There were a lot of feelings for kids to work through, which meant the hospital staff were always bringing in new ways to process those feelings. The hospital provided music therapy, activities, art therapy, even dog therapy. Or actually, they called the therapy with the dog "calming therapy." The dog was as big as Jameson! And he would come and just lie there with him.

On top of all that, we were making videos to keep people updated, sending texts to people, hosting people who were coming to see Jameson, keeping track of all the gifts, and sending notes that we received them. In the outpouring of love from so many people, there wasn't a lot of sitting around doing nothing.

JAMESON SR.

Over three weeks, Jameson had four surgeries. The doctors wanted to keep as much bone length as possible, but part of the issue was the skin. The shark had taken too much of it. Jameson's leg was wrapped, but underneath there was just muscle and bone.

The first surgery was to save Jameson's life and amputate his leg, and they told us from the beginning that the last surgery would be a skin graft. We didn't know how many surgeries would be needed between those two. Each time the doctors went in, they removed any dead or infected skin, trying to figure out how much more bone they'd have to take. We thought the third surgery might be the last one, but there was more skin dying so they had to cut his bone farther back.

MARY

After the third surgery, they thought they could finally do the skin graft. Dr. Berger came in to describe how it worked. He explained to Jameson, "We're going to take a very thin layer of skin off your right thigh...thinner than the hair on your head. We're going to take that skin and run it through a machine called a mesher."

The skin gets turned into latticework. Think of it like a net. They take a thin layer of skin and create a big net out of it, then place the net over the muscle where there isn't any skin. Because the human body is so amazing, the skin grows to fill in the empty spaces in the net. It's an incredible procedure, but the new skin needs a lot of medical attention. They continued with the wound vac to aid the healing process.

JAMESON SR.

There was still life outside the hospital as well. Both Mary and I had jobs to return to. Her boss gave her some time off to focus on the things that were happening in our family and was very supportive in many different ways, but there were still emails that we had to answer, phone calls to make. Not to mention that we have three other children who needed our attention, wanted to hear from us, have bedtime stories, needed some time to play together, needed to be fed, and so on. Both of our parents were a huge help with all those things, but there are some things that just keep going regardless of your situation. Bills are still due at the same time each month. The electric company isn't going to call to make sure you're doing okay emotionally if you miss a payment. But at the same time, maybe you don't feel like paying the bills or checking in at work when you get to the end of a day in the hospital, while watching your son struggle to move across a room. We were so thankful for the many ways our friends and family were supporting us during this time. Our friend Jimmy Miller even surprised us with a crowd funding campaign that took a lot of pressure off the financial burden we were suddenly facing.

Though we were basically living in the hospital, we liked the different activities that gave us a positive break. Sometimes they were created by the hospital, and sometimes they came from friends who volunteered to set something up or from people in the community who wanted to help. A good friend and retired member of the local SWAT team, Frank Briganti, showed up and made Jameson an honorary member of the team.

A longtime friend and local pastor of VOUS Church, Rich Wilkerson Jr., came and visited in the hospital. He brought his two sons, and they were all so encouraging and kind. The very next Sunday he preached a sermon titled "Stop the Bleeding, Start the Blessing" and told Jameson's story. And when we got out of the hospital, he provided us with a car for the entire time we were in the area.

One of my best friends, Dr. Rob Pacienza, came and spent time with us too. He had unexpectedly lost his young daughter, Lily, a few years before. He rallied to minister to us during the whole time of Jameson's recovery. He's the senior pastor at Coral Ridge Presbyterian Church, and our ministry, Urban Village, has partnered with his church.

Folks we didn't even know sent gifts or encouraging messages. Bethany Hamilton, the famous Soul Surfer who lost her arm, sent a video to Jameson, and so did the world's greatest surfer, Kelly Slater.

We also got a visit from a guy named Big John, a huge Power Team dude who came to the hospital and spent time with the kids.

Power Team is a group of Christian athletes who travel internationally to give inspiring talks while doing feats of strength. Big John had reached out and asked if he could volunteer to give the kids a presentation. We met in the lobby on our floor where Jameson and his siblings and cousins got to meet him and be part of his presentation.

When I say he's a huge dude, you need to know I mean he's a *huge* dude. His biceps are as big around as Jameson's chest. He looks like he could bench-press a car.

So for two hours Big John talked and demonstrated to our small group. He broke a baseball bat in half with his bare hands. He tore a phone book in half. Finally, he bent an iron horseshoe into the shape of a heart.

Afterward he stuck around and let Jameson get on his back while he did push-ups. He would lift a few kids at a time using one arm. He did other impressive feats too.

As we started talking more afterward and he got to know Jameson's story more, he began opening up about his own challenges. It turned out he was going to need a surgery of his own in the near future, and he was pretty concerned about it. He said, through tears, that he was so inspired by the way Jameson was dealing with his many surgeries. "I hope to find the courage, strength, and ability to face it with faith like you all have," he said.

It was just a beautiful moment, where he had come to encourage Jameson (which he did!) but left feeling encouraged as well. (I love that!)

MARY

Jameson was still in the hospital when his eleventh birthday came around. I don't know if there's a worse way to spend a birthday than in the hospital recovering from an amputation. So we were trying to think of some ways that we could make the best of our situation. I started thinking, we could just blast "Come to Jameson's Birthday" on social media. I was imagining the whole lobby full of our friends and family, and then wheeling Jameson down to be surprised. This was the best plan I could come up with, and I was processing the best way to do it.

I talked with Gabby, our social worker, about it, and she put the brakes on pretty fast. There were hospital rules to consider. There was a pandemic going on. "Let me talk to some people and see what we can do," she said, and I immediately regretted even bringing it up.

They eventually came back to us and said we could have fifteen people in the lobby. But that wouldn't even include all our family! So we asked if there was any way we could have a few more than that. They told us we could use the courtyard area outside, which would accommodate roughly thirty people. That was better, but this is Florida. It was muggy and hot, and the idea of getting thirty people in the courtyard seemed like it would probably be more miserable than fun.

The best solution we could come up with was to have it in the early evening. So we made plans. We'd have to select our friends and family to invite instead of an open invite, but that was okay. The hospital was actually great once we had a plan in place. The maintenance crew pressure-washed the whole area. They also moved some tables around and made it look really nice.

The night of the party, which we scheduled on the day before his birthday to keep it a surprise, everyone started showing up and sneaking outside. Jameson was in his room, of course, so there wasn't much chance he would see everyone.

Upstairs, we asked Jameson if he wanted to go on a walk. At this point, a walk just meant a ride around the hospital in his wheelchair. But again, he was so bored and he loves people so much that a ride around the hospital was almost always an exciting opportunity. Of course, he wanted to go.

We wheeled him to the elevator and went down to the main floor. We began pushing him toward the glass doors that went to the outside area where all the guests had gathered. Jameson told us later that he looked out the window and saw my dad out there and asked himself, *What is Grandpa doing out there?*

The sliding doors whooshed open, and we pushed Jameson out to be greeted by fifty (yes, I know they told us we could only invite thirty) of his closest friends and family, his PT and other staff, including Dr. Berger and Dr. Payares and her whole family. They were all lined up waiting for him. It reminded me of when we first arrived and witnessed all the nurses and doctors standing at attention, ready to do their work. But this time it was a whole lot of our loved ones, and they were all lined up to sing and party.

We all sang "Happy Birthday" and then the party started. We had ordered piles of pizza, there were stacks of cupcakes, and my best friend, Claudia, and her son Lenny made a *Star Wars* cake that looked like BB-8. All of the decorations were made and coordinated by another friend, Myriam. The whole thing was a huge community endeavor. It was two hours of fun, laughter, and connecting, and for a little while we didn't have to think much about sharks or legs or hospitals. We were with people we love and who love us, and we celebrated Jameson being born. It was a really special time.

JAMESON SR.

Junior's birthday was a happy day. We had our community around us to celebrate our son's life, and I got to see Junior enjoying it. That makes me think about a different moment with Junior, on the fourth day after the attack. I had just left the hospital, because I was going to see a nearby condo our friends, the Porter family, had offered us as a place where Mary and I could take a break from the hospital, sleep, or take a shower. At that point, we had no idea how long we were going to be in Miami or how long Jameson would be in the hospital.

Then my phone started to ring. I looked down and it was my dad.

Of course, my first thought was that something was wrong. I had just left, so why would he be calling me so soon? I hoped everything was okay, but I answered the phone right away.

It was Jameson, and he was whispering when he said hello.

"Hey, bud, everything okay? What's going on?"

There was a slight pause, and then the whisper again. "I don't want Mom to hear what I'm going to ask you. I'm hiding under the sheets."

I had no idea what he might possibly want to keep from Mary, so I asked him again, "Okay. What's up?"

And he asked, "Will you take me back to the reef?"

I was sleep deprived and exhausted. I hadn't showered. This question came out of nowhere just like that shark had. I thought back to Marc and me driving over the bridges that connect the Keys to the mainland. I had told him, "I'm thinking about how to face my own fear, and at some point I'll need to help the kids face their fear about getting back into this beautiful water that we love so much."

That scenic drive to the Keys always brought us so much joy and fun. The point of the Keys isn't to be a landlubber; you want to get in and under the water. I knew we'd want to come back here again, and I knew it would be up to me to figure out how to have those conversations with the kids. But I had been thinking *someday* I'd need to be ready for that conversation, and to be honest, I thought I would be the one to bring it up.

So, a return to the Keys was on my radar, but having that conversation this soon caught me off guard. And I didn't expect Jameson to be the one who would ask me about going back, especially now, only four days after the attack. Plus, we had a lot of other things to get through first, not the least of which were several more surgeries. But I couldn't help admiring Jameson's spirit and his incredible willpower.

I told him, "One day, son, we'll go back."

He could sense my trying to straddle the fence in my response.

"Not one day. I'm saying as soon as they let me out of the hospital. Will you take me back?"

I was speechless. How in the world was this his question already? The only thing I could figure was that he was trying to get me on the hook so that when we asked Mary, he could say, "Well, Dad said yes." He knew I loved the Keys and that I loved the comradery and connection our family found in being in the water together. I guess he had some innate sense that if he wanted to keep getting in the water, he'd have to face his fear.

I finally managed to say, "Jameson, are you sure you want to do this? Why do you want to do this? What's going on?" I was still surprised that this question was coming from him. I can imagine plenty of adults who would just be done with swimming in the ocean if they had lost a limb to a shark just days ago.

Jameson said, "Dad, I don't want the shark, this hospital, and my leg to be my worst nightmare. I want to face my fear and move forward."

He wasn't thinking about the surgeries that had already happened or the ones that were to come. Whenever the doctors came in and told him, "We need another surgery," Jameson would just say, "Okay, what's this one going to be?" Businesslike. I don't think he cried a single time about it. Yes, he was in pain. Yes, he was hurting. But what was on his mind was facing his own fear. And his answer was that he needed to get back into the water. He needed to go back to the same reef where he was attacked.

I still hadn't answered him.

But I already knew in my heart I would have to say yes. This was my boy, telling me what he needed to do to not be afraid, so we were going to do this. I didn't know how. I didn't even know how I was going to talk to Mary about it so it didn't sound insane.

It crossed my mind that some people were already going to think I was a terrible dad because I somehow let my kid be attacked by a shark. But to take him back to the scene of the accident? People would think I was a monster. I didn't know how to talk to Mary about this, but looking back, I realize as I was processing the idea in those few seconds on the phone that I wasn't sure how I felt about it. Was this normal? Was this wise? Was this right?

I think my response would have been different if I had jumped on a plane that first night after the attack. I would have flown to Miami in a panic, jumped deep into the hospital drama together with Mary. But instead, I had been given this gift where I had been able to go a little slower, gather the kids, drive the three hours with friends. If I hadn't already been crossing that bridge with Marc, looking out at the ocean, and recognizing that *someday I'll have to help my kids deal with their fears about this*, Jameson's request would have taken me completely off guard. Again, I was so thankful that it seemed like everything that happened, every moment since the shark attack, it felt like things had been orchestrated to be as good as they possibly could be from that moment forward.

Later, we talked with Bethany Hamilton, the young surfer who had survived a shark attack in Hawaii. There was something of a template here. Her recovery was quite a bit faster than Jameson's—my understanding is that her wound was a lot cleaner than Jameson's—but she wanted to get straight back in the water, just like Jameson. She was back in the water twenty-five days after her attack. So what Jameson was asking for wasn't unprecedented. That made me feel much more secure about the whole idea, once we learned that.

Regardless, at that moment I had to answer Jameson's question. Already I could sense my mind shifting, a different plan coming into place than the one I had been relying on so far. I'd been debating how soon we could get him back to North Carolina. His leg was so tender, and it was such a long drive. Now I realized that getting Jameson healed so he could get in the water and defeat his fears might be more important than getting him home quickly.

So I told him, "Yes. We're going to find a way. I'm going to put a plan together and we'll figure it out. We'll get you back in the water and back to the reef." From that moment on, we saw the overcoming spirit rise up in Jameson, and that set into motion a new way forward.

PART TWO

Courage and Hope

CHAPTER SEVEN

Finding a New Normal

Four months after the shark attack

JAMESON SR.

"Somebody left their leg over here!"

We were in Orlando. It was Christmastime, and we had gone to meet friends at Disney's Wilderness Lodge. Brian and Tami had their three kids, so our kids just loved it. They were roaming around together, while the adults were sitting around the Christmas tree in the lobby. And, well, some people over there found a leg.

MARY

I jumped up right away. "Where is it? That's our leg!"

Jameson had taken it off at some point because it was bothering him, and just left it behind. And that leg can be really surprising when you find it in different places. I felt bad for that family coming across it, and of course, Jameson shouldn't just be

85

throwing it around like a sweater or something. Nobody wants to come across an unexpected limb.

The foot freaks me out sometimes. I'll see a foot by itself in a place where a foot shouldn't be. We have a whole drawer of prosthetic pieces at our house now. It's like Halloween every day. Feet lying around. Legs. We're always cleaning disconnected feet, cleaning the pieces that go together.

If the foot gets wet or gets in the sand, you have to disassemble the whole thing. It's not easy to do. You have to take this huge metal shoehorn to pop the footshell out of its socket, then clean it and dry it. And of course the whole time you're cleaning it, Jameson is hopping around without a leg.

We didn't get the leg until Thanksgiving, so he'd only been wearing it for a few weeks. At first it's all about getting used to it. He started by just wearing it for ten minutes a day, and then every day he'd wear it a little longer.

JAMESON SR.

Once we got beyond the point where we were questioning the past and what we could have done differently and wishing for some other outcome, once we came to a place of acceptance, we had a really different job in front of us. We had to get back to our normal lives, but we didn't know exactly what normal was anymore. We couldn't go back to our old lives. Too much had changed. Once you have prosthetic legs lying around the house, you can't help but be reminded of what's happened and how much will need to change going forward.

Some of it was little things. Like, for instance, Jameson now needs a shower chair, which means he's going to have to use our bathroom instead of the kids' bathroom. That's normal now, but it's not what was normal before. We're discovering all of these new normals, and that takes time and attention and some getting used to.

MARY

Once we got home, we realized Jameson couldn't sleep on the top bunk anymore. Or at least not for a while anyway. It was too complicated, getting up and down the ladder.

JAMESON SR.

The dynamic was different and challenging. Mary was out three to four days a week doing appointments with Jameson. The other kids weren't seeing Mary as much as they were used to. She's always been the main teacher for homeschool, which we had largely put on hold while we were in Florida. But once we got back to North Carolina, it was time to start again, and things were just different. They had to be. There wasn't anything we could do about it. There were things that were necessary for Jameson's health even when they disrupted what had been the normal state of things for us before we left.

MARY

And while it felt like we'd been in Florida forever, the changes felt like they were instantaneous. The shark attack happened on

August 13. We were out of the hospital three weeks later, but we had to stay nearby for several physical therapy appointments, three days a week to start. Prosthetics can't be used until the leg heals, so we were waiting on that too. We have family in Florida, so it was good for us to be there. We stayed at Marc and Holly's place for a month, a month in the Sugar Shack in the Lower Keys, and all of November at our friends Jim and Darla's home in Plantation, Florida. Then Orlando for December, and all of this so we could get Jameson the best care possible in the situation.

But once we got back to North Carolina, we started counseling, physical therapy, chiropractor appointments, and laser therapy. Setting up each of those took lots of phone calls. I'd verify insurance and ask if they dealt with trauma and children and families. We also wanted a place that was biblically based. And counseling kids is a whole different thing compared to counseling adults, and counseling families is a different specialty too. So finding someone who could do all those things and who took our insurance required a lot of phone calls.

My dear friend Katie told her chiropractor, Dr. Josh Buck, our story and asked him to pray for us. He not only agreed to pray, but Josh said he'd take care of Jameson and all of his chiropractic needs when we got back to town. He also does laser therapy, which helps speed healing, and deals with scar tissue and pain. Jameson was hopping around on one leg all the time and it was messing things up in the way he held himself, his posture, the alignment of his bones. We were very thankful to be introduced to Dr. Josh. He took one look at Jameson and heard his story and said, "I'm going to take care of your son." Free adjustments for life. We've never paid a penny.

JAMESON SR.

We'd been away from home for *five* months, after we had planned to be away just a few weeks. We came home with more than we left with—toys and gifts, so much stuff that many generous people gave us. My parents were packing suitcases to bring things home for us. They needed five large suitcases. We were extremely blessed and deeply thankful. But then we had to reopen the house. Get our giant packet of mail and sort through it. And all of this was happening the week before Christmas. Now we're thinking, we need to deal with all of this, but also we need to put it aside because we need to get ready for Christmas.

MARY

Santa was coming in a week! But the kids had just been overwhelmed with a wave of generosity from all of the people who were sending us really kind and thoughtful gifts after the attack. Meanwhile, I was gearing up to get back into the routine of working too. I had started back to work in November while we were still in Florida. My boss had given me some assignments, and since I worked remotely, I could work in the evenings or whenever was convenient, with everything else that was going on. Now, we had come home to a house in need of a lot of attention. It had been summer when we left, and we'd returned home in winter. I had to dig up all the appropriate seasonal clothes, and of course, just because we'd been in crisis didn't mean the kids had stopped growing, so a lot of last year's clothes didn't even fit. We'd left at the end of July. Now it was a week before Christmas, and a lot had changed.

The kids were adjusting, and Jameson was figuring out new things. We had appointments that took up most of our time at least three times a week (sometimes more). And we still needed normal things like meals and laundry and dishes and home-schooling to be done. I felt so overwhelmed, I was tipping into depression. I just couldn't see how we were going to get through it all and come out to somewhere normal on the other side.

We're settling into this new reality, but it feels like it shouldn't be anything new because we're back in our home. New questions continue to pop up. Where do we keep all these prosthetic pieces? What can we find for Jameson to do that he's always done before? He can't sleep on the top bunk, but can he still do his chores? Can he still do the dishes when it's his turn?

JAMESON SR.

The answer to that last one was yes, but Jameson didn't really want to. We'd tell him, "Please go put your leg on and do the dishes, do your chores." But he'd tell us, "I just need a break." His leg was hurting or the fit on the prosthetic was bothering him. And sometimes that was true, and sometimes we were asking whether he was really in pain or just trying to get out of doing the dishes. We had to remember sometimes that, despite the amazing, incredibly mature, exceptionally others-focused kid Jameson is, he was still growing up. Still having bad days. Still squabbling occasionally with his siblings.

And he was still figuring things out too. He'd have moments

when he wanted to prove he could do things on his own, and at other times there were things he just couldn't do. Now he was asking for help for things he had never needed help with before. If he was using his walker or his crutches, he couldn't carry a glass of water back to his room. He'd ask us to do those things.

MARY

There were new chores too. The gel liner on his prosthetic leg has to be taken out and washed every night and then hung up to dry. But how is he going to do all that when he has to take off his leg to get to the gel liner? Then he has to get crutches to go wash it, and it's just a lot. It's not that there isn't a solution. There was a pretty easy one. We bought more gel liners! That way he could wash one while still wearing his leg. But it's amazing how long it takes you to figure out a simple solution when there are a hundred little things like this, and you're just trying to get through the day. We were constantly figuring things out.

And we've been talking so much about the trauma and recovery for Jameson, because he had the worst of it by far, but we had to keep remembering that this happened to all of the people around him. Not just Senior and me, but our other kids too. Jameson needed our attention. Not taking him to physical therapy was not an option. Not helping him figure out ways to deal with things like washing his liners or doing chores was not an option. But we couldn't just ignore the other kids, of course. They needed our attention just as much as Jameson did because they were each responding to the trauma and the aftermath in their own way.

JAMESON SR.

We both work hard to find ways to be present in each individual kid's life. I coach all their baseball teams, so we get a lot of quality time together that way. Mary takes the boys out one on one, and I take Eliana out on daughter dates. We knew this was just a season of life and it wouldn't always be this way. That didn't change how overwhelming it all was, but it helped us to be more thoughtful, more aware of making sure all four kids were getting attention.

MARY

Intentional. That was the word we kept using. Always trying to make sure that, as much as possible, the other kids weren't feeling left out. So we did counseling, not just with Jameson, but at least a few times with the whole family. We all had gone through some trauma too.

JAMESON SR.

Nehemiah brags about his brother. He's so little that he doesn't remember much about what happened. Even though he was in the water with Jameson and me at that moment, he didn't really see anything…though he does remember the screaming. In fact, he says he "felt pain" when he heard those screams, and that's about the best description any of us have for that moment.

He also remembers the blood all over the boat, and being

scared. Still, he brags about it to people. He just thinks it's so cool. "My brother got bit by a bull shark."

Noah has just been a huge champion for Jameson. He never wanted to leave his side at the hospital. Jameson is his big bro, and Noah made it clear that nothing's ever coming between the two of them. In every picture we have of Jameson with his siblings and cousins in the hospital, Noah is the one who is always by his brother's side. He's so proud of Jameson and always wants to be near him.

MARY

Eliana, though, doesn't even want to hear about it. She's creative, very loving, and very joyful, but she doesn't want to hear us talk about the day of the attack. Whenever we tell the story to someone, Eliana goes into another room or covers her ears or puts on headphones. She can talk about the hospital and the recovery time and anything since then. But the day of the attack itself, she just doesn't want to be around when that's being discussed with anyone. As the only girl, she has definitely learned how to hold her ground with her three brothers too.

JAMESON SR.

Nehemiah, our youngest, will steal the heart of anyone he meets with his outgoing, sweet, and hilarious personality. He liked dinosaurs and cars before the shark attack and now, interestingly, he *loves* sharks. He talks about them all the time. He knows their scientific names. He always wants to learn more about them, watch documentaries, read kids books about them. For

Christmas the one thing he wanted was monster truck toys with shark identities. His love of sharks doesn't seem to be completely connected to the reality of sharks... For instance, his favorite is the tiger shark, we think because he saw a monster truck shark toy that was painted orange and black, even though actual tiger sharks aren't orange. Jameson is not too keen on the idea of his little brother loving sharks so much, but he doesn't complain.

MARY

They're all just such amazing kids. But even so, this was an extraordinary situation. We were doing our best, but it was wearing on all of us. The new normal is so much better than things could have been, but it's still *new* and it's not necessarily *better*. There's a wheelchair in the house now and a walker and all the extra accessories, like Jameson's "sticks," as we call his crutches. They're everywhere. And when we want to go somewhere—especially before he was proficient with his prosthetic leg—we have to pack those up and fit them in the car. It was exhausting and new and discouraging at times.

And then there were the bandages. For the first three weeks after we left the hospital, a nurse came by every day to change them. But after that, it was just me. As I did it, every day I had this gory reminder of what my son went through and is going through: removing the bandage, cleaning the wound, putting a new bandage on.

Even though his leg is fully healed now, we will still need to go through it again. Because Jameson's not done growing, his bones will keep growing too. There are several more surgeries in our future because the bone will continue to grow out at the end of his leg. That means additional rounds of changing bandages, of keeping the area clean, of checking for infection. All of those things.

Our new normal isn't only about adjustment. It's partly that, but it's also about embracing some of the worst things we—and especially Jameson—are facing: surgeries, bandages, fitting and getting used to new prosthetics, and the pain that comes as a result of those things.

In the last week of winter break, Katie came to the house for four full days. My mind was in a hundred different places emotionally, mentally, spiritually, struggling to figure out our new normal. Katie left her family—she has four kids—and came to our house to help me with our four kids, who share two bedrooms between them. Katie helped me go through my kids' things, figuring out what was part of our old world and what was part of the new. It really helped me to stay upright and to feel more in control, more intentional. So new things don't have to be something bad or something less than what we had before.

Katie showed me in those four days what it meant to really be there for a friend. It was a huge sacrifice for her. She made us and our needs her priority for those four days. It really helped me feel more settled and was a good reminder that some things from the old normal would stay the same: our good friends, our family, our community, our faith.

JAMESON SR.

Our first family therapy appointment was really interesting. I've gone to a counselor in the past and I've done pastoral counseling, so I was looking out for certain things. For Mary, these sessions were helpful and intriguing since they connected with her career as a Board-Certified Assistant Behavior Analyst, but I wasn't sold at first that we all needed to do counseling. After the positive results, though, I was on board.

MARY

I had to make a lot of phone calls to find a counselor that I thought would be a good fit for us. Not every counselor took our insurance, and I wanted someone who was good with children and with trauma issues. The counselor we chose has been so helpful for us. At first the whole family went, and Senior, Junior, and I still go. In fact, the other three kids are always asking why they don't get to go. Our counselor did a lot of "play therapy" with them because they were so young, and they'd like to go play with her some more.

JAMESON SR.

For Jameson, life wasn't just entering a new normal; it also had new challenges. He was having terrible nightmares. We all were, but for Jameson they were really bad. Also, there was so much attention on him. Television and radio interviews and invitations to speak to groups of people. Some people wanted to know every detail of what had happened at the reef. It's not like Jameson wasn't already reminded all the time of what had happened to him and what had changed, but even strangers would come up to him in a grocery store and ask, "Why is your leg gone?" It's strange but true. That happens. People you don't know just feel like they can ask anything. So there are certain coping things to learn as you ask yourself, How do I deal with this? Do I just ignore them? Tell them the whole story? Make up a story or tell a joke?

Our friend Wayne Kerr helped with the constant questions. Wayne is also a BTK (a below-the-knee amputee). He has a bunch of funny T-shirts about being an amputee, and he gifted a shirt to Jameson, who absolutely loves it. The T-shirt says, "Before You Ask, It Was a Shark." This shows his (and Wayne's) sense of humor, and his courageous way of not shying away from what happened. I don't know that it makes people less curious, though, because now they want to hear the story.

MARY

There were calls to make to our insurance company, new bills to pay, and three-days-a-week doctors' appointments. We were having nightmares, and our kids were feeling like they weren't getting enough time with us. I needed to catch up at work. Jameson was in pain and worrying about his next surgery. But we kept coming back to one really important phrase, the one thing that kept us focused on being thankful, something that helped us to take every day as it came: *We still had our son.*

Not every parent can say that. Many have lost their child. Families in the same hospital that treated Jameson were losing their kids to accidents and diseases. We were spared that. It gave us even more compassion for those who lost their children, because we had come so close to losing Jameson.

Even on the worst days, I could say, "I still have my son."

CHAPTER EIGHT

The Worst and Best Day

MARY

The screams haunted me for days. For weeks. To be honest, even as I write this, they still do. For a while, I couldn't go to bed at night. I'd lie down and hear Jameson's screams, replaying them all in my head and having that horrible feeling again of not being able to get to him.

JAMESON SR.

These traumatic events are so painfully difficult to get past. We imagine ourselves in those situations and how we'd shake it off and go about our lives, but when we experience them in reality, it's not that simple. It's like having a wound, an actual wound, and thinking, *I'll just ignore it*. You can't do that. It'll still hurt; you'll still need attention. You need time and healing. Our bodies, our minds, our spirits are wrestling with all these things for a long time afterward, and we have to find ways to deal with our fear and trauma. For us as parents, we were trying to figure out how to deal with the fear our other kids couldn't shake too.

MARY

Jameson started having nightmares at the hospital, and that didn't change once we got into a house. He'd scream so loudly, it felt like it was coming all the way from Looe Key Reef. Like it was coming straight out of the past. I'd wake up panicked, my heart already in my stomach, and I'd jump up and run through the house to find him.

I've become much more sensitive to the kids getting hurt and definitely to screaming. Whenever I hear it, I think the worst thing imaginable is happening. How could I not, after what happened in the water that day?

Sometimes, I slept in his room. Or he'd sleep on the floor in our room. He just wanted to be near us. And of course, the other kids felt the same way, so they'd rotate into the room too. I'd be sleeping in Junior's room some nights, and all the rest of the kids would pile in with Senior. None of us were sleeping well; everyone was having nightmares. They would wake us multiple nights a week.

JAMESON SR.

I definitely had nightmares and terrible daydreams. My mind would wander off, back to the day of the attack, and I'd be trapped into thinking about it until something jolted me back to reality. It was so visceral, very real. One of the things I hated about this new trauma was the ocean had meant *healing* to me my whole life. The salt, the water, the sun, the sounds of the ocean. It was soothing. Happy. But now it was starting to represent a different *h* word to me: hurt.

I didn't feel the shark teeth in my leg. That was Jameson. That was my son. I kept asking myself if there was some way I could have traded places with him, so he wouldn't have to go through any of this. I have imagined that so many times and imagined how Jameson must have felt and was feeling.

The other thing that happens in moments like these is you start looking at other stories from your life, trying to make sense of it all. Trying to find anything that might connect or resonate with what's happening to you in the moment.

I remember years ago when Mary asked me to give up spearfishing. We already had Jameson Jr., and Mary was pregnant with Noah. I would spearfish with my friend Grant occasionally and sometimes it would attract sharks. Mary was always concerned I was unnecessarily putting myself in harm's way.

Another story that brought concern to Mary around the same time was a day we went to the beach with Jameson Jr. and I went pretty far out from the beach, carrying two-year-old Jameson on my shoulders. It was a beautiful, cool day, and the ocean was flat. Then the weirdest thing happened. Something swallowed my entire right foot.

A fish? A fish! It just put its entire mouth over my foot. I think it might have been a goliath grouper. It was very weird, very eccentric behavior. Not something you would expect. Plus, I couldn't see the fish because the water was too deep. Whatever it was, it swallowed my foot, up to my ankle. I was so startled, I jumped. And I dropped Jameson into the water. I quickly grabbed him and managed to shake the fish off my foot.

I was bleeding, though, which is not a great thing to be doing in the ocean when you're far from shore. Grabbing hold of Jameson, I started swimming back to the beach. Because Mary saw all this play out from the shore, she said, "I'm not raising these kids alone. You can't go out that far holding our son on your shoulders, and you need to stop spearfishing with Grant!" So I did quit spearfishing, even though the episode with the goliath grouper was a weird one-off event. Mary was right. I needed to be more careful and to think about our family.

It was this weird moment when I was "attacked" by a fish, and Jameson was there, and there are these strange similarities to the other story. But what could I learn from it? It didn't help me figure out how to stop the nightmares, but I couldn't stop thinking about it.

MARY

The thing about fear is you can't always tell when it's reasonable. Yes, there are sharks. But also, there were no recorded shark attacks at Looe Key Reef. None in recorded history until Jameson! It's something to be careful about and aware of, sure, but scared? I don't think so.

But now, sometimes in the night, I would wake up filled with fear. Just spilling over with sadness. And I would have to check on Jameson. I was so worried I was going to lose my son somehow. I'd lie down with him or just check on him, touch him, hold him, be close to him, make sure he was still here.

JAMESON SR.

Jameson has told us his nightmares are sometimes about shark attacks. Sharks attacking him, sharks attacking the other kids. But sometimes they're just weird and bad. Just nightmares, where he wakes up sweating, hot, screaming and scared, but he can't remember what exactly the nightmare was about. And he—just being a kid, though sometimes soaked in sweat—rolls over and goes back to sleep.

MARY

Not me, though. That screaming keeps me up. The screams haunt me. In fact, we have the GoPro that Jameson was holding when he was attacked. Some friends went to Looe Key and dove around the reef until they found it. We watched the footage, which is bad: a cloud of blood, a shark's fin, a shark's tooth floating to the ocean floor. But I've never listened to the audio. I don't want to hear it. I don't ever want to hear it again and yet I hear it every day.

JAMESON SR.

What's amazing to me is that just a comforting touch from his parents can calm Jameson down. When he starts screaming in the night and Mary shows up beside him, rubs his back, and helps him relax, he just slides right back to sleep. He feels secure. He feels safe, knowing Mary and I are there with him.

That never really goes away, I guess. I remember in the hospital when Jameson was still in the PICU, and we were crowded all around his bed for one of those meetings with the doctors. I couldn't get to the head of the bed, and I was feeling terrible, just upset and frightened and trying to figure out how to best help Jameson. And then my mom would put her arm around me. Just that little moment, that loving human touch from someone I know and trust really brought me back. I still think about that. The proximity of these loving relationships isn't just good, it's essential.

MARY

So much about our recovery is just about *time*. With time, our nightmares have become fewer and less often. With time, healing takes place, not just in Jameson's body, but in our minds and emotions too. When we first got out of the hospital, we all got sick. The whole family developed this deep chest cough. Someone told us, "Trauma comes out in the lungs," and I guess that was true for us. I hadn't eaten much in the hospital. I just couldn't let myself be off duty long enough for a meal. I mostly drank juices for calories. I didn't have any appetite, and I just felt sick to my stomach. That continued for a few weeks after we left the hospital. We just needed a little space, a little time to move beyond some of those things.

JAMESON SR.

It's interesting that we originally went down to the Keys for this trip because of a dream we had as a family, because of a vision we had been given through our ministry, Urban Village. It was called the RISE campaign, a Rich Investment in a Soul's Eternity. Our idea was to go to the major cities in the Western World and help equip local churches to build missional discipleship movements. We went to Florida to start the work. It's something our family had talked about, because it's going to require a lot of travel and focus over the next ten years. We went to Florida with this vision, this dream—supposedly for three weeks—and came back with nightmares. It's so stressful when you have something you're planning and then it goes in such an unexpected direction. There was a feeling that we had been derailed from the direction we knew we wanted to go. It took some time to recognize all of this would be permanent in some ways. We'll always be burdened with some bad dreams about this event. We'll always be working through some of these emotions, not just now, not just a year from now, but to some degree for years to come.

MARY

There was some fear, too, that we had lost this place that was so special to us. Jameson and I both grew up visiting the Florida Keys, and the thought of losing it forever because of this event really weighed on us both. Not as much as caring for the kids, of course, but we were mourning the possibility that we had lost this place, that Junior might never be able to bring himself to go down there again, or we as a family might find that returning to that area brought more bad memories than good experiences. We have so many friends and relationships down there. That potential loss felt like it was thrown on top of all the rest.

JAMESON SR.

The Keys are also a big part of our story, Mary and me. I used to run these youth group trips to the Keys, and those trips played a big part in our relationship. I was the next-generation pastor at the church Mary's family attended. Mary's brother, Dana, was involved in the youth program, and I was his pastor. Mary started coming too (she had a boyfriend at the time). In fact, I led her boyfriend to faith in Jesus! And while it was not something either Mary or I thought at the time, Mary's mom thought we would be great together.

MARY

From the first time she heard about him, my mom thought Jameson would be great as a husband. She thought he was so wonderful and would say, "He's a pastor!" She was always trying to get me to think about Jameson as romantic material, and to be honest, her love for Jameson sort of pushed me the other way.

JAMESON SR.

Back then, one of my mentors, Eric Jersted, helped me run these trips to the Keys. They were adventure trips called Survivor, just like the popular reality television series. We'd take youth groups down there and divide them into two teams, the Pirates and the Buccaneers. They competed, slept on a boat on the open water, and would catch their own food. Every evening, we would gather for worship and Bible teaching. It was a memorable time for all, students and chaperones. And Mary came along as a volunteer chaperone on several of these trips.

MARY

One night, Jameson was leading in the front of the room. I don't remember if he was singing or preaching or what it was, but I was praying in the back. I was no longer dating anyone at this point, and Jameson and I had become good friends. But there was no romantic spark there. Nothing. Nada.

I was in the back of the room in deep prayer. I wanted to get more serious about my relationship with the Lord. I wanted to stop dating and find God's will for my life. I wanted to move on to the next phase. I told God, "I want to find my husband and move forward in my life," and God said so clearly, "Jameson." I looked up at the front of the room and just said, "No, no, no! It's not Jameson." I wasn't attracted to him. But something about that conversation with God changed things. Over the next two days, I started to feel a spark. In fact, we were at the beach, and I saw him take his shirt off, and I was like, "Whoa, wait a minute! I've seen him take his shirt off before, but it was never like that!"

JAMESON SR.

My younger brother, Joshua, was engaged and getting married soon. And my grandfather was asking me, "What's wrong with you? Your younger brother is getting married, and you're not even dating anyone? You're not even bringing anyone to the wedding?" And I thought to myself, *Who could I bring to the wedding so that we could have a good time and it wouldn't be a weird date where she's wondering, "Am I going to get a call-back?" Who would I enjoy having around my family?* And then I thought, *Mary would be that girl.* So, I called her and asked her if she would go to the rehearsal dinner for my brother's wedding. Just as friends.

MARY

When Jameson asked me about being his plus one to his brother's wedding rehearsal, I told him no. I already had plans. Which was 100 percent true, I had plans already. There was no reason to cancel my plans to go to this event where I wouldn't know anyone—not even the bride and groom. I wasn't even invited to the wedding itself. It seemed a little strange.

JAMESON SR.

I was fine with Mary saying no to the invitation. It was my brother's wedding. I was going to be busy anyway. I just thought Mary would be fun to have around. Meanwhile, we were doing a baptism for a friend. This was before the wedding. My dad was there, and when Mary walked in, he grabbed me and asked, "Who is *that?*" I told him it was Mary, and he asked, "What are you waiting for?" Mary's mom and my dad both seemed to think we should be together.

MARY

The more I thought about it, the more I realized it could be fun to go to the rehearsal. I liked spending time with Jameson, so I canceled my plans and told him I'd go. But I barely saw him because he was the MC of the evening. I ended up sitting with his family most of the night. I remember, though, watching the slideshow of his brother Joshua and his fiancée, Jacki, and thinking, *I'd love to have that someday.* Not for a second was I thinking they would be my in-laws a year later! I would have never guessed I would marry Jameson.

JAMESON SR.

I had thought of Mary as a good friend. If I had taken the time to really reflect in the moment, I would have realized I was starting to have feelings for her, but I was afraid of wrecking our awesome friendship. We could laugh and joke around and just have fun together. A lot of our relationship was also connected through our time together serving in the church. I think part of me was scared of losing this deep, fun friendship with this great person by saying I had feelings for her.

Those feelings increased at the wedding rehearsal, though. I was pretty busy because it was my brother's wedding, but afterward when everyone was cleaning up, she and I sat by the fireplace for a little while and talked. A new album from Radiohead—one of my favorite bands—had just come out, and at one point I thought it would be funny to quote some of the lyrics. I can't remember why I thought that would be funny in that moment, but I quoted them anyway: "'I don't want to be your friend / I just want to be your lover.'"

Then I waited for the hilarious laugh, this big response to my joke, but Mary didn't laugh. She didn't know Radiohead, didn't know the song, didn't know it was a quote, and we were just looking at each other. Just then someone walked up and started talking to us, so I didn't get a chance to explain, and Mary hadn't really responded and it was a bit of an awkward moment.

MARY

After that, we went for a walk on the dock. It was a cool night and I hadn't brought a jacket, so Jameson gave me his and then put his arm around me while we were talking and looking at the moon on the water. And I remember thinking, "If he tries to kiss me, this is over." I wasn't "there" yet... but I was getting there.

Then we got interrupted again when his brothers and the groomsmen and the whole wedding party showed up and said, "Let's go downtown and find some live music somewhere. Where should we go?"

I got roped into that... Jameson drove the men, and I drove the women, so once again we were split up, and I was driving all these people I had just met who were best friends. But we all had a fun time dancing once we got downtown.

JAMESON SR.

All of this was a month before Mary was praying in the back of this meeting in the Keys. The Keys aren't just where Jameson was hurt, they're also where God told Mary she would be marrying me. They're a place where we've both had so many wonderful, beautiful experiences. And so have our kids, including Jameson! There was this fear—small compared to some other things—but nevertheless a fear that this might be something we would lose.

MARY

The things God reveals to us, we know will come to be. And sometimes they come fast. We had our first date in April. Jameson said up front that he wasn't interested in hanging around and dating. If we were going to date, he wanted to be serious. I wanted that too. In fact, I knew I'd have to change some things to get ready for this season of life. I used to stunt ride tandem on motorcycles, doing shows and events and competitions. This is basically doing acrobatics on a motorcycle while someone else drives. So the other person would do a wheelie while I flipped around on the bike, for instance. A hobby like that requires a lot of time—we practiced most every night—and it's not the safest hobby either. God was growing something new in my life: a desire for stability and a family. It was changing not just the way I saw the world around me, but also what I wanted. I felt like I couldn't have this dangerous, time-consuming hobby and fully embrace the new season we were starting together. I called and told my friends I wouldn't be doing that anymore as soon as I knew Jameson and I wanted to be serious.

Eight-year-old Jameson Jr. enjoying the crystal clear waters of the Lower Florida Keys *(All photos from the authors' personal collection unless otherwise noted)*

"Argh, Matey!" A night of pirate fun in Key West, Florida: Eliana (5), Noah (7), Nehemiah (2), Jameson (9)

Family vacation in the Florida Keys one year before the shark attack

Our children enjoying the crystal blue waters at Looe Key Reef moments before the shark attack

Still shot from Jameson Jr.'s GoPro footage of the shark during the attack!

Dinner at Roostica, one of our favorites in Key West, with our dear friends Mike and Tammy Nolan, owners of the Sugar Shack and the boat we took to Looe Key Reef; Jameson Jr., nine years old

Mark the Shark with a nine-foot bull shark similar to the one that attacked Jameson *(Mark the Shark)*

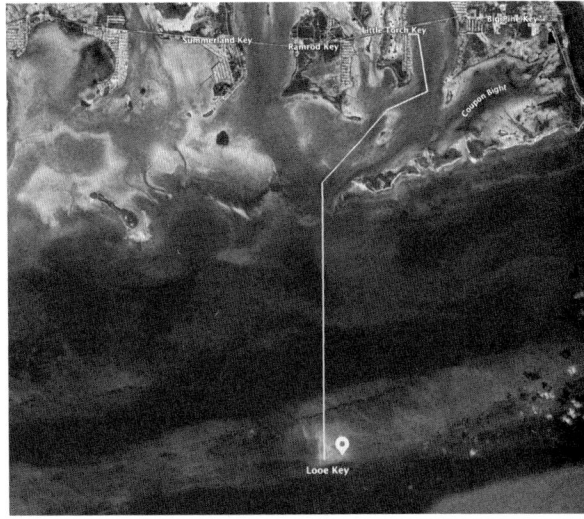

The nine "miracle miles" from Looe Key Reef to Dolphin Marina; *The Invincible* crossed it in ten minutes when taking Jameson to the awaiting ambulance

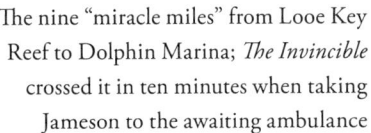

The Grooms family (Todd, Lisa, Tyler, and friend Lauren), owners of *The Invincible*, the boat that carried Jameson Jr. to shore in record time

Jennifer Cadwell (CNA), who swam to our boat to give aid, and Anneliese Dietrick, captain of the boat Jennifer swam from

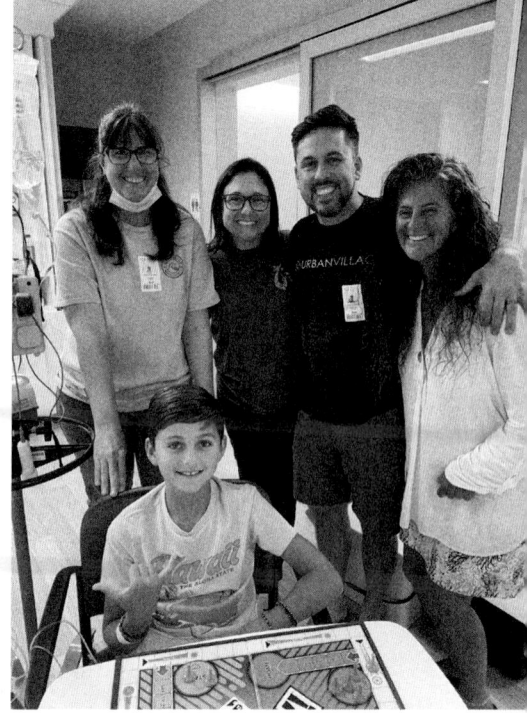

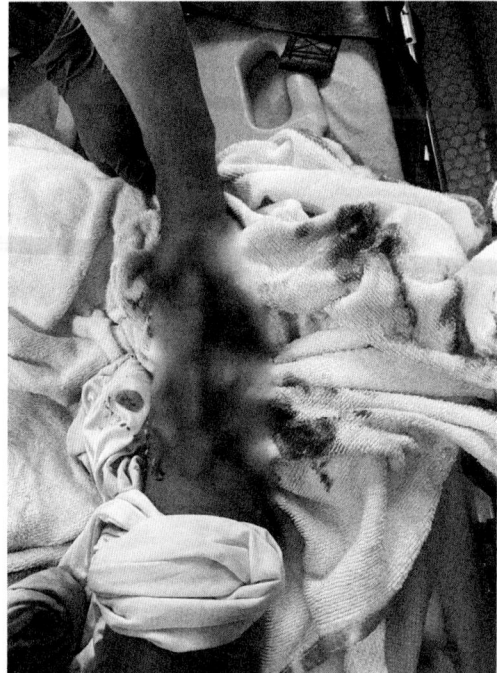

What was left of Jameson's leg after being mauled by a nine-foot bull shark

On the helicopter moments before takeoff for a thirty-minute flight to Nicklaus Children's Hospital for his life-saving surgery

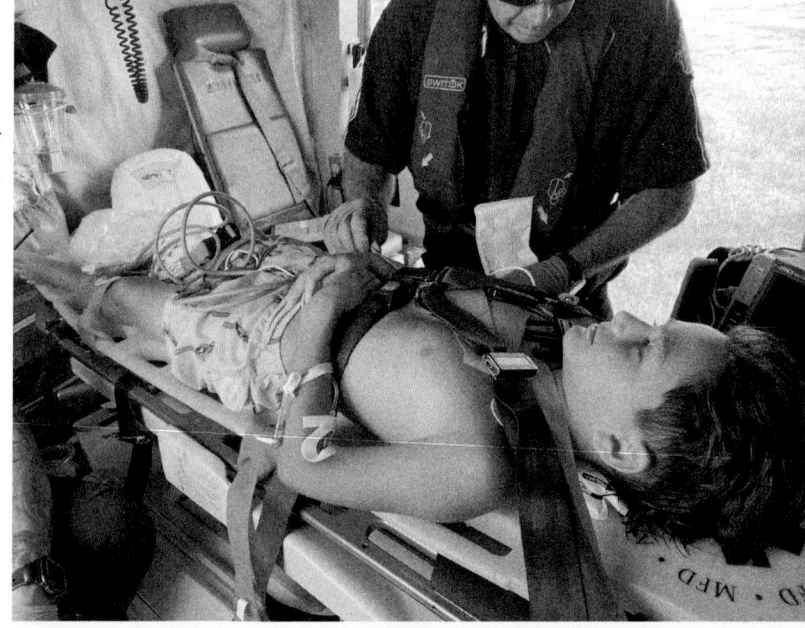

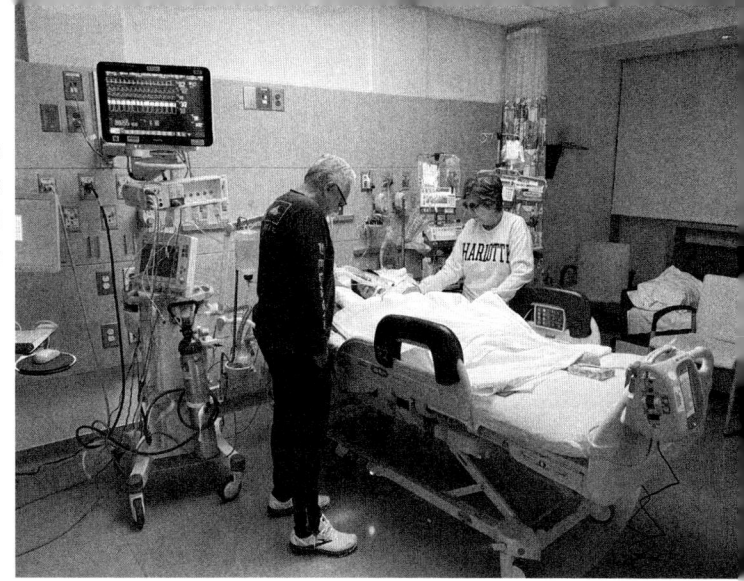

Right after surgery in the PICU with Jameson Jr.'s grandparents Bob and Celia Reeder

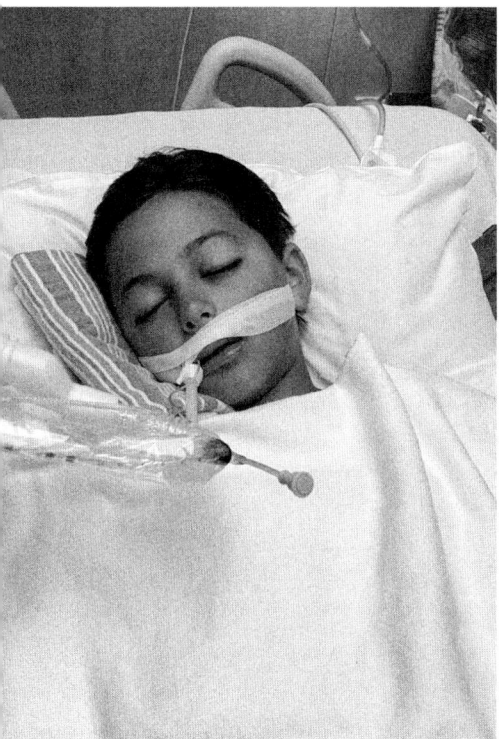

Moments after life-saving surgery, thankful our beautiful son is still alive

Our family, reunited after being separated by the shark attack

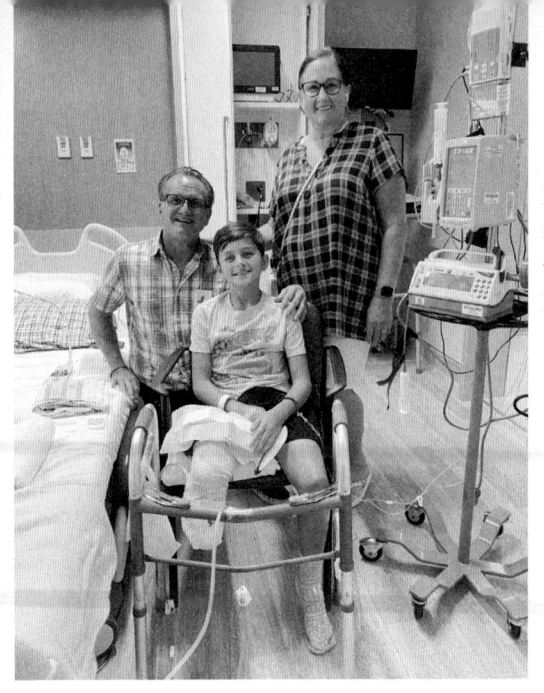

Moved out of the PICU to the surgical recovery floor; Jameson Jr. with his grandparents Michael and Donna Marchione

Jameson Jr.'s eleventh birthday celebration in the hospital with his surgeons, Dr. Berger and Dr. Payares

Back in the Florida Keys, Jameson Jr. holds the jaws of a nine-foot bull shark as he prepares to face his fears and return to the reef

"Brothers in arms": Noah and Jameson Jr. returning to the reef

"Return to the reef": back in the water, without a prosthetic, exactly ten weeks after the shark attack

Reeder family at the Little Smiles Stars Ball in West Palm Beach, Florida; Jameson Jr. represented Nicklaus Children's Hospital

Jameson Jr. playing baseball with his travel team, Storm, at Dreams Park in Cooperstown, New York, July 2024

Jameson Jr. in Kalihiwai Beach in Kaua'i with Bethany Hamilton Dirks, who inspired him to be an overcomer long before his own shark attack

Jameson surfing at Hanalei Bay in Kaua'i with Mike Coots in the background cheering him on; Mike lost his leg to a tiger shark in 1997

JAMESON SR.

Our first date was in April, and we were immediately serious. We had spent so much time growing as friends that once we were dating, it was with a pretty clear intent to not be dating long. It had taken a while to come to grips with our feelings, but now they were strong and moving fast. Honestly, I knew on that first real date. This wasn't just someone I had feelings for; this was someone I wanted to spend the rest of my life with.

In November, we traveled to New York City for Thanksgiving. We had a big day in the city, which started very early with the Thanksgiving Day Parade—Mary had always wanted to go to that parade—and a full, fun day in the city, followed by a great meal with Mary's family friend Luigi and his family in Brooklyn.

After the big Thanksgiving feast, Mary and I took the Number 5 train from Brooklyn back into the city. I had an engagement ring with me, and Mary had no idea. That rock was burning a hole through my pocket the whole trip. All I could think about was I couldn't let it get stolen. I was so nervous. I'd been to Times Square with my brothers years ago, and Justin and I had seen Joshua get his wallet stolen. I didn't want a repeat of that.

I said, "Let's get off the train and take a walk on the Brooklyn Heights Promenade. It has a beautiful view overlooking Lower Manhattan and the Financial District." I told her our friend Jon Tyson, who we were going to see the next morning, had told me it was a must-see view.

Mary is always game for an adventure, so we did. There was nobody there. The space was completely empty and silent, which was ruining my plan. I finally saw two ladies and asked them to take a video for us. Mary had no intention of being in a video and told me so, but I told her we needed some videos to send back home. She finally agreed to stay on camera with me.

That's when I got down on one knee. She didn't see it coming, and it was incredible to see her response. She started crying and kissing me and yelling too. She was so full of joy, and she loves to turn the volume up. I'm pretty sure the whole city heard us that night.

We got back on the train and headed for Central Park, where we took a horse-drawn carriage ride through the south end of the park. We went from Cherry Hill to Strawberry Fields. All the iconic spots. We relished being with each other while taking in the city, our first Thanksgiving together as a couple.

It's still the greatest decision I've ever made in my life, apart from following Jesus.

MARY

It was a great, great day. Such a wonderful day and an amazing memory. There are unexpected things in life that are wonderful, like Jameson and me getting married. We both thought it was ridiculous to even suggest we should date each other, and look at all the wonderful things that have come from the unexpected push from God that brought us together.

But then, there are these unexpected things like the shark attack, something that changes your life in a moment, and you have no say in it, no vote, no chance to say yes or no. It just happens and you're suddenly in a new phase of your life.

So even in the midst of the nightmares we were having in the aftermath of the attack, we were asking ourselves, could it be there's something good too? A dream that counterbalances the nightmare? A vision for something better on the other side of all of this? Some unexpected good that could come out of this really difficult moment?

JAMESON SR.

We were seeing already that Jameson's story was inspiring. It was changing people in some really cool ways. And we realized that we needed to be good caretakers of his story and to make sure that Jameson could talk about it in the way he wanted when the time came.

MARY

A few weeks before the attack, Jameson Junior was sharing Jesus with a new neighbor, but the neighbor asked him not to talk about it anymore. Jameson was a little upset by that. And I told him, "Buddy, be careful how you share Jesus with people. First, you need to be an example and build a relationship with that person; that opens the door to share Jesus with them. We don't hide it, but it has to be natural too." But in the hospital, we had a conversation that was completely different because there he could tell anyone about Jesus the first time he met them. He had an excuse to talk about God anytime he wanted because he had a miracle story about how God had shown up to protect him.

So do I. So does Senior. When I looked out in the water and heard that horrible screaming, he was riding so high in the water, as if a hand were holding him up, as if he were sitting in an invisible chair that was keeping him out of danger. Senior had a moment when God told him, "I'll hold him until you get here." The medical professionals all said it was a miracle. How did this kid lose this much blood and make it to us? How was there a nurse in the water next to you?

And Junior told us something amazing too. He said that he had really heard from God that everything was going to be okay. He wasn't trying to comfort us; that's what he had been told. He was just relaying the message. And later he told us that as the boat shot toward shore, he had seen a "person on fire" standing on the boat. "I knew it was Jesus," he told us. "I wish I could draw it."

He thought about it for a minute and then said, "It was the worst and best day of my life."

CHAPTER NINE

New Leg, New Challenges, New Goals

JAMESON SR.

When I was Jameson's age, I played football in the street with my brother Joshua. We did this all the time. On one particular day, our neighbor, an adult named Tom, was out throwing the ball with us, and his wife called him and said it was time for them to get going. I can't remember where they were headed, but he said good-bye to us and trotted over to his car.

I said to Joshua, "Let's grab our bikes and race them out of the neighborhood!" We ran for our bikes. Joshua grabbed his, but my bike, which was this cool chromed-out trick bike called a Dyno VFR, one of those really short bikes, was pinned against the wall by the rest of the family bicycles. Meanwhile, our neighbor was escaping, and I knew Tom would get away if I took the time to dig my ride out. Joshua was already pedaling out to the road, shouting at me to hurry up.

My mom's bike was right there, though. I grabbed it and ran, jumping on and pedaling hard to catch up with Joshua. My mom's bike was a big pink beach cruiser. It stood a good foot taller than my bike, so I was up high on it as we started racing after our neighbor.

Our parents had given us a boundary that we weren't allowed to cross in the community. We could wander free inside a territory around our house that was fair game, but we couldn't go past a stop sign that marked one edge of our boundary.

Well, as Tom pulled past that stop sign ahead of us, I knew I could catch up if we kept going. Joshua and I were both laughing as we came barreling up to the stop sign, and my brother immediately hit the brakes to stop at the sign. But I was not going to let Tom get away that easily. So I shot past the stop sign and kept going.

If I had just stopped like Joshua in that moment, that would have been the end of the story, and maybe we wouldn't even remember it. But I didn't. I chased Tom all the way to the edge of the community. As he turned out of the neighborhood, he honked at me and I waved back, still laughing, and then headed back toward home.

There was a median in the middle of the road, with palm trees and bushes all planted down the length of this one section of road. I sped back the way I had come. I burst across the median and through bushes but didn't see a pickup truck that was coming until just before it clobbered me.

I barely remember it. I saw headlights and then I woke up in the back of an ambulance.

The doctor said that being on my mom's beach cruiser probably saved my life. I was higher than usual. If I'd been on my own bike, the front of that truck would have hit me square and run me over, but instead it hit me a lot lower and sent me flying. But most of the impact wasn't directly on me. I needed stitches and I had a concussion, but I didn't break any bones.

The doctor could barely believe it. He told my parents, "Listen, that truck should have killed him. He should have been run over. Being on that higher bike saved his life. But I'm telling you, this kid has an assignment in life. Someone's watching over him."

I thought about that a lot during those hospital days with my son. So many things could have gone differently and we could have been making funeral arrangements instead of dealing with surgeries and physical therapy. I had no doubt that God had been watching over Jameson. And it seemed so clear that this was because God has a plan for his life and for his story.

MARY

We were so relieved the day Jameson finally was released from the hospital. We weren't going directly home, but at least we wouldn't be sleeping in a hospital room anymore, and we wouldn't be on such a rigorous schedule with doctors and health care providers talking with us day and night.

Jameson Sr.'s dad, Bob (Sabba), went to get the van while I filled some prescriptions for Junior. Junior had been transferred to a wheelchair and rolled down to the waiting area. Bob pulled up in the van and walked around to Junior. He slid the door open, and then he looked at Jameson and said, "I'm not sure how we're going to do this." We hadn't taken him anywhere in a vehicle at this point, of course, and wrangling him from the chair and into the van then folding up the wheelchair was a little daunting.

This is the sort of moment you dread as a parent, the moment when your kids are suddenly faced with new challenges and realize their abilities have changed pretty radically. Jameson was a self-sufficient kid, and the thought of him not being able to get into the van by himself was overwhelming.

Bob didn't know what to do. Try to pick Jameson up on his own? Use a hand to yank him upright and topple him into the van? It wasn't clear. Jameson, of course, had never done this either.

Jameson looked at the van, looked at his grandfather, and with a determined look on his face, he said, "It's okay, Sabba. I've got this."

He stood up on one leg, carefully holding on to the wheelchair at first and then the side of the van. He spun around and sat on the floor of the van, grabbed the seat, and pulled himself onto it.

A huge grin spread across his face, and he looked at his grandfather, joy in his eyes, and gave him two great big thumbs-ups. Then he said—as if he were in a movie or had just hit a home run—"I'm back!"

JAMESON SR.

There were so many new things to figure out. We didn't know how to do even simple things like get Jameson into the van. Or how to fold up the wheelchair and the walker and fit them in the van with the crutches. We were learning a lot of new skills.

We were so thankful for our family and friends, including my dad being there to help us get Jameson home. My parents had watched our other three kids the whole time Jameson was in the hospital. At the time of the attack, our friends' home in Miami was vacant. They offered it to us to stay in for as long as we needed, which ended up being two months. This allowed my family to stay together: my parents, our three children, my brothers and sisters-in-law, and all their kids.

MARY

When we brought Jameson to the house, he was so excited to see his siblings and cousins out front with their homemade "welcome home" signs. They made a big deal out of Jameson being out of the hospital, which was so sweet and encouraging.

JAMESON SR.

Jameson really wanted to play basketball with the cousins even though he was in his wheelchair, but with everything going on and all the new things we were doing, we kept putting that off. So on the last night they were there, before the cousins left, we all went outside—Jameson was still in his wheelchair—and we pushed Jameson around and played a big family game of basketball.

This was Jameson's first real outdoor experience in a wheelchair since the shark attack. It was such a powerful and life-giving moment, which almost seems strange to say. After all that time stuck in a bed, he was so happy to be wheeled around on that court, laughing and shouting with his siblings and his cousins. We could have dreaded this being a moment of trauma or sadness, but instead the family helped make Jameson's first sports activity in a wheelchair something that felt like fun and freedom.

It was the last night the cousins were there, so of course, it was really bittersweet. We'd had them with us for this experience that became an amazing time but they were headed home the next morning.

MARY

We didn't even think about it before we left the hospital, but we didn't have a handicapped parking permit. Trying to get a wheelchair out in a narrow parking spot is really hard! As Jameson got stronger, he could sort of hop around the vehicle when we got to where we were going, but the wheelchair was so difficult to navigate.

To get a permit, there's a lot of paperwork to fill out and red tape to cut through. Jameson was missing a leg—a pretty clear-cut argument for needing a handicapped pass—but we had to fill out forms to prove we needed it. It was more challenging and complicated than you would think.

On our first try, we apparently filled out the wrong form. The next time, when we got the right forms and had all the signatures we needed, the doctor's signature was on the wrong line. Which meant they couldn't issue a pass.

I finally told a woman in the office, "Look, my son just lost his leg in a shark attack. Can you please help us out here?" She took pity on us and I walked out with everything we needed to get our vehicles registered correctly.

JAMESON SR.

Some of the new challenges we were facing weren't even directly related to Jameson's leg. One was dealing with the media. We were getting a lot of media attention, and people were calling or trying to get meetings with us to talk about what had happened.

Before we left, the hospital had registered us under a fake name, and their security team watched for reporters trying to sneak in, but now we were on our own.

We had no idea how much media attention we were getting until the president of Nicklaus Children's Hospital came to visit us. He said they'd never had this much media attention for one patient in the history of the hospital.

This was a lot to handle on top of everything else. To be clear, the interviews we did were amazing, and all the people were great. But there were too many requests to fit in and still be able to prioritize Jameson's health, our family, and everything else that needed thought and contemplation. My brother Joshua took care of keeping people informed on social media, and he really went above and beyond and made "Brave Like Jameson" T-shirts and set up a new website, bravelikejameson .com, where people could get updates, make donations, and also funnel some of our media requests into a more manageable system.

When we left the hospital, we were quickly overwhelmed by the media, but thankfully, other people stepped in to help. They would field calls and make decisions that often had to be made fast. We got to do an interview with *Good Morning America*— and they did an amazing job sharing our story with the world. We are thankful for that opportunity and possibily sharing more in the future. One thing we've learned is that God has given us a story, but we have to be clear about why and how we are telling the story. We want to share it with our values of compassion, kindness, and community—and of being aware of the needs of people around us. We want to make sure the people who tell the story with us know that.

MARY

One of the challenges that emerged in daily life was there was no set schedule. Jameson had follow-up appointments and fittings for his prosthetics, which kept us on the move. And with everything in flux, we honestly weren't moving forward that much on our kids' homeschooling. People were so generous in lending us places to live and a car to drive and raising money to help with hospital bills. However, we were still living out of our suitcases with the few clothes we had brought for what was supposed to be—at least for the kids and me—a vacation.

In these emergency moments of unexpected harm, it's not surprising that we felt off-balance even when we were out of the hospital. Changing Jameson's bandages was really challenging too. It was painful for him and not for the faint of heart.

But every day, even on the days when it didn't seem like it or we couldn't see it clearly, Jameson was healing just a tiny bit. And so were we. Minute by minute, day by day, we were moving away from the traumatic experiences at the reef and toward building an ordinary life again. We started to turn our eyes slowly away from the past, but the present was pretty much all we could deal with. As Jameson and the rest of us started to heal, we could start to think about the future again.

Everyday life gets tangled up too. We were calling around to prosthetists, for instance, trying to find the best one for Jameson so he could achieve one of his new goals: learning to walk again.

JAMESON SR.

We started PT in South Florida before we were aware of the timeline for when a prosthetic leg would be available or even could happen. Jameson was learning how to walk again, first with a walker and then eventually with crutches. As that was all progressing, Jameson was also talking about how we were going to go back to the reef, and I was trying to make plans for that.

Because of what was happening, we were missing the baseball season back home. All of our kids are on teams, so, depending on the season, I usually coached a couple of their teams. We normally played in a recreational league run by an organization called Stewards of the Game, which focuses not just on creating a great baseball experience but also on teaching leadership skills and character building. But the whole family had lost track of baseball season. Then the league director, Casey Fitzsimons, called to hear how things were going. He and many others were concerned about Jameson. I told him Jameson was healing but I needed to be replaced as a coach.

Even though baseball had been a big part of our lives, we really weren't thinking about sports. We were just trying to make it through each day in the hospital, not planning for the future. The only time I had ever thought about baseball was in that moment when I had the phone call about the possible amputation, and the thought flashed across my mind. *Jameson might never get to play baseball again.* That was an emotional whiplash moment, because just hours before, I had been worried that my son might not live. Now I was adjusting to a much lesser loss, but it was still a loss. My son, the athlete, was losing this part of his life. He's so quick, so fast, so agile, and then, all of a sudden, that's all gone. All of it was hitting me at once, and I didn't have any context for limb loss at the time. I had never thought much about amputees.

Jameson hadn't mentioned sports either. He wasn't pushing for anything; he was just focused on what needed to happen each day. We didn't have a family discussion to figure out a timeline if Jameson decided to play baseball next season. No, we were talking to different prosthetists, trying to understand the process and find the right fit. We really liked Adam Finnieston, but he was in Miami. Other doctors were telling us, "After Christmas or sometime in the new year, we'll start to talk about fitting Jameson with a prosthetic." But Adam said, "I'll have him in a prosthetic by October or November." Because he knew Dr. Aaron Berger, one of Jameson's doctors, he already knew about us and said, "With your family's spirit and your son's spirit, honestly, we can definitely get this done before Christmas."

Adam was the only prosthetist who had visited us in the hospital instead of talking via video or telephone. He had grown up in the Keys, and many of his patients are younger. Most prosthetists primarily worked on adults, but Adam specialized in youth and in athletes, very active people. We had a lot of things in common: he loved diving, fishing, and baseball. He also lived and breathed prosthetics. Adam's dad and grandfather were also prosthetists, and his wife had a prosthetic leg.

The next baseball season started in February. If we waited until January, Jameson would definitely miss it. But if we could start trying a prosthetic in November or December, well, there was a chance at least that Jameson could play.

It was late September or early October, so our minds were

starting to reset and think about the future. We were still getting communications from back home about our baseball league, and Jameson began to mention it. He was starting to think that maybe he could play again in the new year.

Mary and I weren't so sure. In fact, we weren't just asking when Jameson could play again; we were still wondering whether going back to baseball was going to be in the cards at all. There was still a long road of recovery ahead. But Adam—who had been a professional baseball player—told us to expect that Jameson would play ball again. He would surf again. So our conversation started turning away from if he could play to "Is this something Jameson really wants to do?"

No one could commit to any timelines yet, and we were still talking to our insurance about what they could do concerning prosthetics—what was covered, what was out of pocket, all that sort of thing. We didn't want to flood Jameson's mind with questions that might not matter or promises that couldn't be kept.

Jameson got his first prosthetic in the first week of November, which he began to wear to get used to the feeling of it and to strengthen his leg. He would wear it for short periods of time when he was using the crutches. Juvenile prosthetics are very complicated, partly because kids grow and sometimes grow quickly, so prosthetics have to be adjusted or replaced with astonishing regularity. If it doesn't fit right, we would have to get new parts. As Adam was figuring it all out, we started to make plans to return home to North Carolina. We decided to leave after Thanksgiving, but two days before, a part had yet to arrive. Adam hoped we would be able to complete the prosthetic while we were still in town rather than ship the part to us when it came in.

A prosthetic is not like assembling a Lego kit. A lot of

adjustments have to be made when each piece is attached. There's an art to it, making sure the piece is doing what it's supposed to, making it level, watching how Jameson walks, things like that.

But we were planning to leave by midnight the night before Thanksgiving and head to Orlando to celebrate with some friends before heading home. It was a three-hour drive, and I didn't mind driving through the night if it meant Jameson would have his new leg, and it would be working well. Adam was getting ready to leave too; he was planning to spend Thanksgiving with his daughters in Tallahassee. So we all were watching for FedEx shipping updates, just counting down the minutes until this thing was supposed to arrive, right up until the moment it became clear it wasn't going to arrive in time.

We were disappointed and trying to figure out what this would mean for us. But Adam said, "Don't worry. We'll figure it out." He had his assistant one-day ship the part to Tallahassee when it arrived, and then he drove to meet us on the road to Orlando. We met in an Academy Sports parking lot on Black Friday. He brought his tools and got everything perfectly balanced, which was just another of those unique moments when you have to stop and be thankful for the many blessings God gave us along the way. Adam is a generous man who gave up some of his vacation time to drive for hours to a parking lot and help Jameson on his journey.

When Jameson was fitted to his new leg, Adam took his crutches and said, "Now, you're going to walk without these." And he did! He started walking, barely, without his crutches. Just moving in circles around Adam. What a moment that was! Seeing our son take his first steps after losing his leg.

It takes time to get used to wearing a prosthetic. To walk normally takes therapy and hard work. Everyone told us that Jameson would eventually get back to his old capacity, and it was even possible he'd exceed some of his old abilities. Jameson responded excitedly about the process. "The doctor told me my prosthetic will have a spring in it, and I'll be able to jump and dunk!" His joy and effervescent personality really shone through.

We have learned a lot about prosthetics, including how people often have specialized ones for different needs. With the expertise of prosthetist Matthew Phillips and the kindness of the Hanger Clinic, Jameson Jr. was able to receive a custom-designed running leg that we're deeply thankful for. When we finally made it home to North Carolina, we were surprised to see more than two hundred people lining the streets to welcome us with signs. We had a police and fire escort and the neighbors decorated the front of our home with Christmas lights. The news media was there too. But maybe most important for Jameson, all of his friends were there. It was as if everything suddenly clicked back into place for him. What does he love to do with his friends? Ride bikes, skateboard, run, and play baseball and basketball. It was a reminder of all these things he loves in a super healthy way.

MARY

Coming back home to this enormous, over-the-top block party was a huge deal and a big day, with so many wonderful emotions. We were all crying tears of joy. There were so many people! They had posters and bells and they lined the streets for blocks.

They had been planning this for months. They decorated our

house for us with lights and ornaments. There was a gigantic Welcome Home sign in the yard. There was food and hot chocolate, and T-shirts that said "Brave Like Jameson," which so many people were wearing.

The police had come too, and they'd brought a police dog for the kids to meet and all sorts of toys wrapped for Christmas. It was like going to the fair, but it was all a surprise and all a party for *you*.

All of this was so sweet, such a special homecoming. It's funny to think that if we had experienced a regular vacation, our return would have been unremarkable. But now we had been gone for months instead of weeks, and all of our friends wanted to make sure our return was fun and that we knew we were loved. And we did!

Christmas was almost upon us too, so they had strung lights everywhere. The whole neighborhood was packed.

Up until this point, Jameson had been very protective of his leg. He didn't want anyone to see it, not even his brothers or his sister. But all the kids were so curious about his leg, and for the first time, Jameson seemed to believe it was time to give people a good look. This was a very big deal for him.

So his Grandma Cece got a chair for him, and he hopped onto it and unwrapped his leg so they could all see. The scars, the marks, the toothmarks of the shark. Then he asked us to bring him his prosthetic, and he showed them all how to put it on and how it worked.

The whole thing was a big party. Everyone was so kind and generous to each other. It was really beautiful.

And his Cece had been so thoughtful. She hadn't just been part of the big plan for our return; she had also remembered the boys' beds had had shark sheets on them when we left for our trip. She had changed the sheets before we got back.

JAMESON SR.

It was early January when we started talking about playing baseball again for real. The season started at the end of February, so we were less than sixty days out. Jameson told me that he wanted to play, that he wanted to get to the place that he could, and I told him, "I want you to play baseball more than you want to play baseball. And yet there's going to be an additional tryout for you."

"What do you mean?" he asked.

"You have to prove to me that you can do more than be on the team." This season he was actually moving up a league. He'd be the youngest kid in the new league. They threw harder, and a lot of the kids were faster and stronger than what he was used to. "Son, you have to show me that you can do more than throw, catch, hit, and run to first. You have to do those things at a level I can feel confident you can contribute to your team and not get hurt."

It was a sobering conversation. For him, he was just thinking, *I've got this.* That has practically been his motto since the attack. It's his favorite phrase for when anyone has any doubt he can do what he's decided to accomplish. The Bible talks about having faith like a child, and I feel like that's what Jameson truly possesses: he possesses an overwhelming belief that he can accomplish anything.

I wanted to believe that grit and belief would be enough to

get Jameson to his goal, but I knew those were only part of moving forward. This would take a lot of hard work. There would be a lot to do to get him ready for the upcoming season.

Jameson started by telling his physical therapist, Dr. Kate, that he wanted to play baseball. Together with her, they came up with a list of all the things he'd have to be able to do if he wanted to play. "Here are some goals for the end of February."

We started working in the backyard to get ready. Noah, Jameson, and I worked on hitting and throwing. Jameson had to learn to throw again. He had to transfer his weight from his back leg—his prosthetic—to his left foot as he followed through with a throw. It's a whole new way to throw; the balance is all different. And he had to learn how to do it again.

Jameson's swing was already incredible, so it wasn't impacted by the new leg, but it was still something he could work on improving. So, he started focusing on that too. I was really pleased by the hard work he was putting in, and every week he looked better.

There were tryouts, but those were more like skill assessments. Where would be the best place to play this kid? I was working with him more than usual, because I was still concerned that he could hurt himself physically and mentally. Physically, because there are bigger guys in this league and on this team. He was playing second base, and I needed to be sure he could catch the ball, get out of the way of runners, and that he could get out of the way of a pitch in the batter's box. He's got such a pure spirit and people-oriented focus, I didn't want something happening during a game to hurt that spirit.

I decided not to coach because I wanted to be able to ping back and forth between all the kids' teams, and especially keep a close eye on Jameson. Also, the coaches are like me, volunteers.

They're not professional coaches with special training for over-seeing kids with new prosthetics on their team.

MARY

Jameson started playing that season, and in one of the early games, he hit the ball then fell when he started to run. My heart dropped when I saw that. I wanted to rush out there to him, but I managed to hold myself back until he got up and it was clear he was fine. He was okay.

JAMESON SR.

We had told the coaches to play Jameson however they needed to. If they thought Jameson needed someone to run for him, we were not going to complain. We just wanted him to get to play. They said, "Let's let Jameson tell us. If he wants to run, let him run." Jameson didn't even think about it for a second. He said, "Of course I want to run!"

Those first three weeks, the whole team looked a little rough. The team was doing fine, winning a decent number of games, but it's sports with young boys. Kids don't have much self-control. They get angry easily. It's a volatile time. At that age, it's hard for a young pitcher to find the strike zone, so there were lots of strikeouts and lots of walks.

Around week three, something amazing began to happen. Jameson started hitting *everything*. Line drives into the outfield and sometimes past the outfield. And all of a sudden, he started cheering from the dugout, shouting for his team. "We're in this thing together! Let's go, team!" And something about seeing Jameson give his everything really helped his team gel and come together.

We started winning not just some of the games but almost all of them. One of the coaches told me halfway through the season, "Your son is the leader of this team. It's incredible to watch him. We love having him on the team, and my son loves playing with him." It was such a sweet conversation, and it made me thankful for the hard work Jameson had put in so he could play.

Then they made it to the playoffs! They lost in the semifinals by just one run. It was an unbelievable season.

During the first part of the season, one of the new coaches spent time with Jameson and didn't know he had a prosthetic. The coach was giving him a pep talk, trying to encourage him. He was being kind, but also pointing out ways he wanted Jameson to do better. So he said, "You need to hustle so you aren't always the last person on the field."

Jameson didn't give him any excuses. He just said, "Yes, sir."

Then he asked Jameson, "What's your biggest fear?"

Jameson said, "Letting my team down."

"Outside of baseball, I mean," the coach said.

"Well," Jameson said, "it used to be something else. But since the attack, I guess I'd say sharks."

The coach was baffled by this. Jameson pulled his pant leg up and showed him his prosthetic. The coach was completely floored. He told me later, "I had no idea how special this kid is. He never once asked for special treatment. He just wanted to play."

CHAPTER TEN

Back in the Water

JAMESON SR.

Before baseball, Jameson wanted to get back in the water. It was late October, ten weeks after the shark attack. We loaded the family into the boat and headed for the open sea. Jameson had been out of the hospital for two weeks and wasn't cleared to get his leg wet. His doctors said the risk of infection or other problems was too high. They advised, "Let's give it a few more days." We had to tell Jameson there was still more healing to do. We were changing bandages every day, and there was new skin still growing, trying to keep a healthy connection to Jameson's leg.

The weather was getting colder. But Jameson was determined to return to the reef. He didn't want to put it off any longer. He wanted to go back to Looe Key Reef and jump into the water where the bull shark had attacked him. He wanted to swim around, put his snorkel on, and dive right in. He was minus a leg but had twice as much heart as ever. It wasn't that he wanted to prove something or even to prove something to himself. He just wanted to get in the water and show he wasn't allowing fear to change him, to keep him from doing the things he loves to do.

I told him, "You don't just get back in, son. We need to put the proverbial big toe in first." I suggested we go somewhere else, somewhere more familiar that our whole family loved. Somewhere we could relax a little, get in the water for the first time, and not be worried about being in the exact place where the attack occurred. Someplace where we had only seen juvenile sharks, where we knew he could get his feet wet, but only figuratively since the doctors didn't want his leg getting into the ocean at all.

So, I pitched him to go to Marvin Key, one of our favorite places. Marvin Key is on the "bay side" of the Keys, meaning the water is calmer and shallower there. Looe Key Reef—where the attack took place—is on the "ocean side," where it's deeper. You go to Looe Key Reef to snorkel and see gorgeous coral reefs and beautiful fish and to Marvin Key to anchor your boat, set up folding chairs on the sandbar and walk in the shallows. We love both places, but what I liked about starting out at Marvin was we could beach the boat, sit in our chairs and just hang out as a family.

We continued texting Dr. Aaron Berger, hoping he would give us the okay to get in the ocean. We had been sending him pictures and videos of the wounds, hoping he'd say things were good enough for Jameson to get into the water so we could get back to the reef at Looe Key while the weather was still good. We knew we couldn't go out there if the sea was too choppy or conditions were too rough.

So we decided to do our first trip to Marvin Key. Getting

back in a boat for the first time since the attack, I had a lot of mixed feelings. I love being in a boat. I love being on the water. It's such an enjoyable part of my life and something I always look forward to. But I couldn't help remembering the last time we were all together in this same boat almost three months ago. The horrific feelings, the blood, the race for shore. I had been worried that driving by those bridges might give me a suffocating feeling, that just seeing the same stretch of ocean again might be too much. But everyone seemed to be doing okay.

MARY

For a while, I have felt unsettled every time I saw a commercial for a cruise or we watched a movie that had a lot to do with the ocean. It saddened me that an ocean I once loved so much now scared me. Maybe not scared, but uncomfortable. I love the ocean. I love looking at it. I want to retire near the ocean so I can see it, so I can take a short walk and get in it. It's such an amazing creation, so vast and beautiful. But now, whenever I saw a picture of the ocean, I felt *anxious*. Is this how I was going to feel for the rest of my life? I hoped not.

Some things were changing, even as we were going back to the ocean for the first time. It's not like we all just ran down the beach and splashed in the water. Even though we all love the ocean and have been in it thousands of times, we felt differently about the sea now. I wanted to somehow feel safer.

A company called Sharkbanz reached out to us after they heard the news about Jameson's shark attack. One of their products is a band or bracelet that produces an electromagnetic field that messes with a shark's electroreception, its sensory organ. The field is much stronger than anything in nature. It doesn't hurt the shark. It's equivalent to shining an extremely bright light into someone's eyes. We just naturally squint and look away. If a shark gets within three to six feet of you, they sense this field and they hate it. They may come up close to check things out, but when they come within range of Sharkbanz, it deters them and they swim the other way.

Sharkbanz contacted us and said they knew we loved to swim and surf, but now had some worries about getting back into the water. They offered us some free samples. They suggested we needed one or maybe two per family member. I accepted their offer and asked if there was a chance we could get one brace-let per limb. I wanted to be sure every limb was covered, not just every person. I'm sure they thought that was overkill, maybe they thought it was funny, but a couple of days later, a box arrived with twenty-four bracelets for the six of us.

I have to say, the Sharkbanz bracelets made me feel better, even though I knew we had nothing to worry about at Marvin Key. Still, they gave me confidence more than anything.

JAMESON SR.

A big storm had just come through. It was still hurricane season. The weather, however, was beautiful, but it meant we were the only people out on the water. My cousin Jacob came with us, so we had three adults on the boat, which made all of us feel a lit-tle more secure. It was just good to know there were three of us keeping an eye out. We came to Marvin Key and found it was deserted. We were the only people there. On the one hand, that was really cool, but in another way, it was spooky. I'd never seen it so empty.

I beached the boat and dropped anchor, and we all got into the water. Again, it was really shallow. You could lie on your back on the sand and have your face out of the water during low tide. Which is more or less what we did. We brought out our folding chairs, and we read or played in the sand or splashed around in the shallow water in front of us. We all stayed pretty close to the boat. It felt safe, and we could get out of the water quickly, if needed.

As the sun started to set, the tide began to come in. The water was getting deeper. Junior asked, "Will you tie up one of these little rafts and pull me around?" He couldn't walk, and he wasn't supposed to get his leg wet (well, it might have gotten a little wet, but we really tried to keep him out of the water).

I tied some floats together and started pulling him around. He was having a great time. I was having a great time. But, a little way ahead of us we saw what appeared to be a dorsal fin.

My first thought was that it was a bonefish in the mud flats. Bonefish are a trophy fish. Anglers come from all over the world to the Florida Keys to catch one. And if you catch more than one, that's a big deal. The flats are their habitat, and with the tide coming in and Jameson and I being a little farther from the boat, well, it made sense we might come across one. The Lower Keys have the best habitat and conditions for this fish. Lots of people try to get a saltwater grand slam, which is when you catch three trophy fish in one day: a tarpon, a permit, and a bonefish. Bonefish aren't dangerous. They're nothing to worry about.

I told Jameson, "There's a bonefish up there. See the dorsal? I want to check it out. You good with that?" I wanted to make sure he wasn't scared, that he was okay with getting closer to a wild fish, especially after our last experience.

"Yeah, let's go!" he said, and I started pulling him that way.

But this wasn't a bonefish. It was a three-foot shark in very shallow water. As we got closer and I realized this, I start narrating this to Jameson so he wouldn't be afraid. "It's a shark, not a bonefish. A small one."

He said, "Dad, pick me up!"

I didn't know if it was so he could get a better view or so I would be a little closer to him or both. But I immediately scooped him up.

You have to understand, we've been to this place dozens and dozens of times and we've never seen a shark here. Maybe a twelve-inch one, a tiny baby shark. Nothing of this size. Never. I didn't think we'd even see a foot-long one. But, three feet long!?! I couldn't believe it.

When you're in the ocean in the Florida Keys, you can choose whether to go to the back country, which is just another way of saying the Gulf of Mexico side, or you can go to the front, the Atlantic side. Marvin Key is in the back country. Which is to say, we were in the Gulf of Mexico. But we had never seen sharks of this size near Marvin Key. We wouldn't have chosen this as our first outing back to the ocean, if we had seen them there before.

We turned and saw *another* shark!

It was slightly bigger. Maybe three-and-a-half feet.

Mary and the other kids had joined us, so we were all standing together at that point. I said, "You know what, let's head for the boat."

And as we started toward the boat, there was a *third* shark. This one was bigger still. A four-footer. And it was between us and the boat!

MARY

A nauseous feeling came over me as soon as I saw the shark. I was trying not to throw up. I was paralyzed with fear.

JAMESON SR.

The tide was coming in. It was getting deeper. I was already holding Jameson, but I wanted to scoop the other kids out of the water.

MARY

Again, we grew up by the ocean. We have both been around sharks in the past. When I was in fifth grade, there was a field trip the students took every year called Sea Camp, which was staged in the Upper Keys. We got to go to the beach and do interactive things as a class. I remember being so excited because they told us, "You'll get to swim with some sharks!" We'd pick up sea cucumbers and swim with nurse sharks.

I grew up on the water. My dad and my uncle Todd were both surfers. My grandfather had a pharmacy right on the beach. My other grandfather had two boats. My family had a house within walking distance. But now my heart was pounding and I was almost dizzy looking at this shark between us and the boat. I couldn't believe this was happening. I'd never been afraid of sharks before, but now we were trying not to panic, just trying to get back to the boat.

JAMESON SR.

Since Jameson's attack, we've spent a lot of time with some shark experts. Scientists, fishermen, ship captains, and researchers. According to the Florida Museum of Natural History, the number of unprovoked shark attacks have been going straight up since the 1970s, with more attacks every decade. And while they're still not a gigantic risk—you're statistically more likely to be struck by lightning five times than attacked by a shark once—that doesn't change the way you feel when you're standing knee-deep in the ocean with a four-foot shark between you and your boat.

Shark bites and shark attacks are also very different. Shark bites are usually accidents and happen more often than shark attacks. Jameson wasn't only bitten, he had been violently attacked by an ambush predator that was massive and fierce. He was bitten multiple times in the lower leg and the shark violently shook its head, tearing away all of the muscle and flesh, and completely breaking and removing one main bone in the lower leg.

One expert we met with was a man known as Mark the Shark. He does charters out of Miami and is one of a kind. He takes people out to catch sharks and often catches giant ones. One of the things he does afterward is dissect them and take a look at what's in their stomachs. He takes the shark meat to a homeless shelter. He was great to all of us and invited us to his home, where we met his wife and son. Mark was great to Jameson and to us and actually gave us the jaws of a bull shark that was about the same size as the one that attacked Jameson. We've got it on the wall in our office now. Mark said he does about a hundred shark autopsies a year, and in the last year and a half, their stomachs are nearly always empty. They're having trouble finding food. Which means, as Mark said, "We're on the menu." Not exactly comforting words.

MARY

Jameson's attack was even rarer than other attacks at the shoreline, around the beach. Sharks hunt the shoreline because they try to trap their prey in the shallow water, giving them nowhere to escape to. Sharks want to corner their prey in sandy areas where there's no reef or rock to hide under.

Shark attacks are usually near the shore, where swimmers tend to be. But we had been nine miles offshore at one of the most frequented dive sites in Florida, a place where there's never been a documented shark bite, let alone an attack.

JAMESON SR.

Florida has more shark bites than anywhere else in the world, and one of the places I grew up and surfed was the shark bite capital of the world: New Smyrna Beach. It's a popular surfing destination, with tons of surf competitions because the wave action is epic, consistent, and reliable. Some of the best swells on the East Coast are at New Smyrna.

A lot of my best surfing days were there. But New Smyrna is also known for shark sightings—and bitings. I've actually paddled out into the lineup and kicked sharks. I lived in Orlando, so we'd drive out for the day, stop at our favorite surf shop to get wax for our boards, check the local weather report, and study the day's aerial photos to see how many sharks were sharing our space. Some days there were anywhere from eight to ten sharks, so there was an awareness of and a regular reminder of the fact that New Smyrna was popular with both surfers and sharks.

And yes, we saw shark-bite victims sometimes. What was more common was seeing a shark and giving the other surfers hand signals about the shark's presence. We'd point out dorsal fins to each other. But seeing a shark didn't mean we'd get out of the water, and it didn't mean we needed to be afraid. Sure, one day my brothers and I helped *two* shark bite victims get out of the water, but it didn't scare us. I'm not sure how to explain it, other than to say that if you've seen a pretty bad car accident, you're not afraid every time you get behind the wheel. You take precautions. You learn how to be safe. And you trust that your chances of getting hurt are low. In fact, you're much more likely to be hurt in a car accident than by a shark attack.

The day after Jameson's attack, a local dive shop near Looe Key was telling people we must have been chumming the water. They couldn't imagine an attack could have happened on the reef, because it had never happened before. "They must have been fishing and riled the sharks." We were pretty upset that people were saying things like that. We always follow the rules, and we're really careful about personal safety and responsibility in the ocean in general and especially in a marine sanctuary like Looe Key Reef.

But things are changing. There had never been an attack in Looe Key Reef until there was. We've talked to world-class captains, and they told us, "There are more bull sharks out there right now than we've ever seen." They're just everywhere.

Which means, as humans share their space, there are going

to be more encounters. But it might not always mean an attack. One of the craziest experiences I ever had was when I was in the Keys with Grant. It was so murky, we couldn't see our hands in front of our faces. I was swimming in a canal and bumped into something and realized it was a shark. I told Grant about it, and he said, "I bumped it too!" No one was hurt. It was just the three of us—two men and a shark—swimming blindly.

Another time I was at Grecian Rocks near Key Largo, a large coral reef state park. I was with my friend Captain Mick, who used to help with those youth group trips to the Keys. The reef is beautiful, the water is blue and as clear as crystal, and I was in the shallow part of the reef. There were lots of fish everywhere, which usually means there's going to be someone looking for a snack. A Caribbean reef shark chased a fish into the shallows and breached right near me. Again, there was no attack, but when you're sharing the water with an apex predator, you need to be aware and be careful.

But there are more sharks and more shark attacks happening right now than we've ever seen in history. According to experts, there are three reasons for this. One, there are more people in the water than there have ever been. We're sharing more space with sharks. Two, there are fewer fish in the sea than ever before, because of overfishing. And fish are the main food source for sharks. Three—and this is where places like the Looe Key dive shop is right—more and more people are doing unwise things to get good photos or great videos. They'll chum the water to get sharks to come up for pictures.

I saw a video of a guy who had just caught and cleaned a fish and said he was going to wash his hands in the ocean. His buddy told him not to do that because they'd seen some bull sharks around. But the guy said, "It'll just be two seconds." And in those two seconds a shark grabbed him by the arm and pulled him out of the boat (an EMT at the park treated him and sent him to the hospital afterward).

Because we have irresponsible and unwise behavior at the same time, sharks and people are increasingly in each other's space—and the sharks are hungry. They start to associate the sound of boat engines with food, because people are fishing and throwing scraps into the water or chumming. Then someone jumps in for a swim or throws a line in to fish, and the sharks don't know any better. They hear the engine sound and think it's a dinner bell.

For several years I was a certified lifeguard. My family always talks about having a healthy fear of the ocean. Which just means be respectful. Know what's sharing the water with you. Know about the dangers of the ocean. And don't engage in any purposefully risky behaviors. When it comes to sharks, be aware but not afraid.

And now we were at Marvin Key, as the water was getting deeper and the sky was getting darker, and we had just seen three sharks somewhere we'd never seen any before. If we were all adults, that would have been one thing. But we had kids with us. A four-foot shark is longer than most of our kids!

Our boat was beached with the back side in deeper water so we could get it off the sandbar. That back part of the boat was getting pretty deep now. I didn't think there would be more sharks out there, but then again, I didn't think there would be any. I'd already been proven wrong three times on that count.

I was thinking through what I would do if one of those sharks decided to come after us. We were already carrying the kids, and the sharks weren't that big. There weren't a lot of good options.

The best thing would be if the sharks stayed away from us. But what if one didn't? Could I jump on it and crush its skull? What could I possibly do if this thing came toward us?

MARY

I'm so glad to say we managed to get everyone back into the boat without any trouble. But we were taking some deep breaths and laughing and trying to shake off the stress. I was already so nervous about the idea of Jameson going back to Looe Key Reef to swim where he'd been attacked by the shark. *And this was supposed to be our trial run?* So much for taking it slow and dipping a toe in the water. We went straight to a shark encounter on our first time back in the water.

JAMESON SR.

Most people probably would have seen this as a sign. Don't go back to the reef! You should definitely cancel.

I think what's important to remember is that going back to the reef was Jameson Junior's idea. We weren't prodding him. He was leading this. He'd wanted to go back to the reef. And the sharks didn't discourage him at all. If anything, they strengthened his resolve.

Our trip to Marvin Key happened on October 4.

"We have to go back to the reef," Jameson said.

We were planning the return for October 22, a little over two weeks away.

MARY

Our reason for going wasn't just about getting in the water again. We knew Jameson would get in the water again. He's brave and so committed to overcoming this fear.

JAMESON SR.

We went because we wanted to reconnect as a family to this place that we love. We wanted to really enjoy the mystique, the mystery, the allure of the Keys. We always built up our trips there and talked about pirates, superheroes, and things in Key West, like The Shipwreck Museum. Listening to certain Beach Boys songs, eating at our favorite restaurants, we wanted to reattach ourselves to the culture of the Keys.

We wanted to reengage our memories with something other than that horrible day two months earlier. We wanted to get that scream out of our heads, wash the images of that day out and replace them with the many, many other memories we'd had together as a family in that place. We wanted to remember what we loved about being not just in the water but around the culture of the watermen communities down there. We wanted to enliven, reaffirm our connections to the ocean.

We wanted to bring the love and joy of the water back.

MARY

In this new post-shark-attack world, we were given opportunities not only to reconnect with our old life but to build something new too. Jameson was invited to a *ball*. And not just any ball. A really special event, where he would be a guest of honor, complete with wearing a tuxedo, arriving in a limo, and walking the red carpet.

So, before we got back in the ocean again, we were going to need to get ready to dance.

CHAPTER ELEVEN

Walking the Red Carpet

Three months after the shark attack

MARY

There's a nonprofit in South Florida called Little Smiles, and it specializes in finding kids who need a smile and providing that through gifts, events, snacks, and things like that. Their entire mission is to support, encourage, and cheer kids up. They partner with pediatric hospitals, foster homes, medical support clinics, and similar places. It's purely focused on giving kids a few moments of enjoyment in difficult situations. And once a year they do a big fund-raising event called the Little Smiles Stars Ball. This year, Jameson was invited to attend.

We almost didn't go, because we didn't understand exactly what it was or what a big deal it was to be invited. It's a ticketed event, so donors buy tickets to be a part of it, and before the event, the kids who are invited make crafts and things to be sold at an auction. It seemed like a lot, though, and the event was in West Palm Beach. We were about 230 miles away in Key West.

But Karla from the hospital PICU called to talk about it, and she explained it was a big honor. Each facility partnered with Little Smiles nominates one kid to be part of this thing. She said it would be a huge, fun event that really made the kids feel like stars. The hospital staff at each facility discuss who should be nominated for the event, and they had chosen Jameson! It's a really big deal and another example of how much the staff at Nicklaus Children's Hospital are thinking beyond just the immediate treatment of their patients. They're always thinking about the children they work with and how to make their lives outside the hospital better too.

So, we decided it would be okay. We would drive up for the pre-meetings and make it happen and stay in West Palm Beach for the night. At the first event, Jameson was so interested in all the other kids who were there. He wanted to get to know them, hang out with them, and hear their stories.

JAMESON SR.

A lot of the kids have ongoing challenges, whether it's their health or traumas they've experienced. Some of them have processing disorders or come from very difficult family backgrounds. This was another reminder of how much we have to be thankful for. Jameson had been hurt, yes, but he was healthy otherwise and seemed to be returning more and more to his old self with each passing week.

If you're looking through the eyes of a parent, your heart isn't just heavy, it's rebreaking in some ways. We were reminded that our son's life was spared. It was fun to watch Jameson at those meetings, though. His spirit is unbelievable, and he has a way of becoming a ringleader of the kids around him. He has a unique calling, I think, to see and love and embrace the people he's around. And it was amazing to see him doing this with these kids, all of whom were going through something tragic in one way or another.

MARY

We thank God every day, every second of every day, that Jameson's alive. And now he would be surrounded by kids who were having a very hard time. One sweet girl passed away from a terminal disease before the ball, which was sad for everyone.

Jameson kept track of the other kids after the meetings and couldn't wait to see them all again, each one in the midst of a terrible difficulty. Which is the point of Little Smiles, to give them an amazing night they will never forget. There was another girl who was like Jameson, so outgoing and just wanting to move around and make sure everyone was having the best time. She and Jameson really hit it off.

Jameson and the kids and I were all invited, and then Senior's parents and Aunt Linda bought tickets too, so they could go with us. It was such a fun time.

JAMESON SR.

On the day of the event, all of the kids were taken to a salon to get their hair done. Jameson was fitted for a tuxedo, and then they were all picked up in a limousine. The parents and everyone else headed off for the venue while that was happening. When the kids showed up, we were all waiting outside the place on either side of a red carpet. The kids came down the carpet, and we all crowded around them to get their autographs.

It was so fun. There were probably a thousand people trying to get autographs; a radio station was there, and a disc jockey. And all of this went on at the Palm Beach County Convention Center. One of the donors there was a recent winner on *The Bachelor*. And there were models and yacht brokers, all of us standing in line to get the kids' autographs. The police were there in dress uniforms and white gloves, and they held their swords over the red carpet for the kids to walk under.

MARY

It was like the Emmys or the Oscars. As each child came out of the limo, the radio station would announce them, and the whole crowd—everyone dressed in black tie or formal dresses—would cheer. Jameson was second from the last to come out of the limo, and he had the hugest grin on his face.

Lights were flashing everywhere, and there were people on stilts covered in lights. The whole thing was like a giant festival for the kids, where they were the center of all the attention.

They gave Jameson his walker, and he started making his way down the red carpet. Everyone wanted his autograph, and Jameson was making sure to stop and talk to every single person and sign something for them.

JAMESON SR.

As you can imagine, it took a long time to sign all those autographs.

My dad said to Jameson, "You know you don't have to stop for everyone to sign their book."

But Jameson was going to sign every single one. "Well, Sabba," he said, "everyone is special."

MARY

Even at this fun event, we learned new things. We'd never had Jameson in a tuxedo with only one leg. What do you do? Hem the tux? Fold the pant leg? We had no idea. I was asking people at the salon, and they didn't know either. So even at amazing events like this, we were still discovering how far we had to go as we were learning about our new life.

During the dinner, Jameson sat with us and kept turning around and looking at the other kids. He would tell us, "There's so-and-so with his family," and then tell us his story. He cared a lot for the other kids; he wanted to be in their lives.

There was a presentation portion of the evening when each kid would go up front where prerecorded videos of their stories were shown, and then the Little Smiles people would give each kid an individualized bag of gifts. The kids had filled out some long questionnaires about what they loved, what kind of snacks they liked to eat, and other things they liked, and they were given bags packed with those things. They even had staff to help the kids carry it all. At this event, Jameson didn't have his prosthetic, so he was pushing his walker around with one leg. He definitely couldn't carry much in addition to that.

When that part was done, the disc jockey took over and the dance began. Jameson went out there without his walker, hopping around on one leg and having a blast.

JAMESON SR.

Jameson was also helping all the other kids have fun. If someone was shy, he pushed his walker over and grabbed them from their table, telling them to come dance with him. And it was really fun, of course, to see all four of our kids out there on the floor together and having the best time, especially given everything else they'd been through the last couple of months. It was fun, too, to see the guy from *The Bachelor* or the fire chief out there interacting with the kids. But it was better to see Jameson interacting with the other kids. It was so great to see his face all lit up, full of joy, and having fun. We hadn't seen that carefree side of Jameson in a while.

MARY

At the end of the night, they wheeled out this gigantic candy station. It was twelve feet long and four feet wide, and there was candy on all sides and plastic bags, so the kids could take the candy home. They were over there just shoveling candy into their bags. They were so excited.

Nehemiah, our youngest, was too small to reach the goodies, so some adults filled candy bags and handed them down to him.

There was something really beautiful about all of these adults who were trying to give a nice night to these kids who were going through such hard times. I'm really glad we went. It was a night we'll always remember.

JAMESON SR.

We overheard so many people talking about Jameson that night. "Can you believe Jameson? He wants to get back in the water!" "He's so excited to be here." "How brave!"

At one point during the night, Jameson turned to Sabba, and with a giant smile on his face said, "This is the greatest night of my life!"

CHAPTER TWELVE

Return to the Reef

JAMESON SR.

A day or two after the attack, before Jameson said he wanted to go back to the reef, I was talking to my friend Grant, who was on a business trip in San Francisco. Grant owns a boat company, and he's a proficient and talented waterman who is at home on the water. Watermen often live or work in or on the sea. They are action-sports type of guys. Intense divers who push the limits and want to see how far, how long, and how deep they can dive. They know well the waters where they dive because they are talented people who have ocean water in their veins.

We were talking about the attack, and the subject came up that Jameson had been holding a GoPro on a selfie stick when the shark struck him. We hadn't recovered it, and I knew it was lying somewhere on the bottom of Looe Key Reef. I told Grant, "I've gotta get that GoPro somehow, but I don't know how. I'm in Miami, you're in San Francisco, and that GoPro is in the Atlantic somewhere."

Grant thought for a second and said, "I heard that Zack is supposed to be in the Keys right now. Let me see if I can reach him."

Zack is Zack Spurlock. He had a video go viral a while back of when he was spearfishing and a great white shark came over to check on what he was up to. Instead of trying to scare the shark off or jumping back into the boat, Zack took some up close and personal video of the shark as they swam around together.

I didn't know Zack at the time, but I've gotten to know him since. He's one of the greatest watermen I know. Grant thought he was out near Dry Tortugas, a national park about seventy miles west of Key West. He has a commercial license for lobstering and harvests them regularly. He takes a bag and swims to the ocean floor, picks them up, sticks them in the bag until it's full, drops them into the live well on the boat, and goes down again.

In fact, Zack and a friend were near Dry Tortugas—about two hours from Key West—and something went wrong with one of his engines. After bringing his boat back to shore to get worked on, they decided to rent a boat and continue lobstering and fishing in an area not far from Looe Key Reef.

Zack was in the water and had just caught a huge black grouper when he saw he had missed a call from Grant. He called Grant back, who told him, "My friend's son just got attacked by a shark yesterday out at Looe Key Reef." He told Zack the whole story.

"Dude, I'm not far from Looe Key Reef right now," Zack said. "I'm a mile or two outside the SPA." The SPA is the Sanctuary Preservation Area, the area where you can't fish or put down anchors and so on. "Send me the coordinates," he said. "I'm gonna find this kid's GoPro."

I sent him all the maps of where our family had been the day before, described the camera, and told him how deep the water was, things like that, hoping he could find it. Meanwhile, my friend Shane called me. Shane's a divemaster who goes out to

Looe Key Reef frequently to snorkel and dive. He had also been on the lookout for the GoPro, and he knows Looe Key Reef like the back of his hand because he's dived there maybe a thousand times. Nobody knows that reef better than Shane.

"I couldn't find it," he said. "Either someone else saw it and grabbed it, because the water is so clear—in which case we should start asking around, checking the internet, see if someone mentions they have it—or I was just diving the wrong spot or missed it."

I was bummed, but I said, "No problem, Shane. I'm so thankful you looked for it. And yeah, let's start asking around and see if someone's picked it up."

"One other thing," Shane said. "I've never seen the resident sharks as stirred up as they were today. The sharks were following me around while I was diving. This bigger one kept passing me. I was so uncomfortable. I've never felt that way at Looe Key Reef. Never seen them so stirred up."

We filled Zack in on all this and he said, "Thanks for telling me, but no worries, no big deal." Zack treats Looe Key Reef like a swimming pool and goes there with his wife and daughter. He thinks it's one of the safest places to dive. Even if it wasn't, Zack has fought off sharks in the past at other dive locations. He's not afraid of them.

He was committed to giving it a shot. Zack checked in with the locals on the radio and the internet before getting in the water, though.

Someone told him, "Look, we haven't seen a GoPro, but we have to tell you that there was an eight- to ten-foot bull shark out there that we haven't seen around before. We cleared the water."

To cover the most area in the quickest way, Zack decided to be towed around behind the boat, holding a rope. And sure enough, sharks started following. "I was spooked," he said. "I've never seen sharks acting like that. These sharks were looking at us like bait. I don't know what the heck is going on out there."

After searching for a while, he traded places with his buddy, but a short time later, they pulled him out. "Nah, man, I'm out. The sharks are following me the whole time I'm in the water."

Zack called me. "Look, we've combed what I think is the right area. The sharks are acting weird. And there's this large bull shark hanging around. We didn't find the camera, and I'm not sure we're going to."

I told him I understood and thanked him for trying.

Then Zack decided to get in the water one more time. And that, of course, was when he found it. He called me from the reef and described it. I couldn't believe it.

"Yes! That's it, bro!"

It was amazing, just amazing that he actually found it.

Zack headed in to get his computer. He loaded the GoPro footage and then we FaceTimed so we could watch it together.

"This is some of the most beautiful turtle footage I've ever seen," he said. "Unreal."

We had almost forgotten how beautiful the day had been and how Jameson was having a deep connection with creation in the moments before the attack.

As we came closer to the actual moment, I told him, "You've got to mute it. I can't hear that again."

I wasn't sure I wanted to see it either.

Zack asked me that exact question, and I told him I didn't know, I wasn't sure.

So, Zack watched it. "I don't see it," he said. "I don't see the shark."

It was my brother, Joshua, who spotted it. Zack drove to Miami to give us the camera and meet Jameson. He's just an incredible guy, and Joshua is a pretty incredible waterman and surfer himself. Joshua started going through the footage frame by frame. And he found it!

"Look, there's the shark in the shadows, way out on the edge of visibility."

And then, "There's the tail . . . the fin."

Then a cloud of blood.

And then a shark's tooth appears on the screen before drifting out of view.

Later, we showed the footage to Neil Hammerschlag. He is a marine ecologist, a research associate professor at the University of Miami, and the director of the Shark Research and Conservation Program. Neil looked at the evidence and said, "Yes, that's a bull shark. And not just *a* bull shark. It's a massive one." He also told us that the shark that was lurking around the reef the next few days was likely the culprit. "After a shark gets a kill, they often stay in the same area to keep hunting."

The same area, of course, that we were going back to.

MARY

Jameson Jr. had told Senior about wanting to go back to Looe Key Reef. He had called his dad while he was under the sheets when I was out of the room to make sure I didn't hear him. When Senior told me about it, I remember exactly what I told him: "You're all nuts! I'll be on the boat with a shotgun!"

At the same time, I knew if they were going back, that's where

I would be too. I wasn't angry that Jameson wanted to go back, but I had a lot of other emotions. Mostly I was shocked. At the time it was like, "Really?! It's so early!" At the same time, I was happy he wanted to get back into the water. I didn't want him to be afraid. And I was sad at the thought that our relationship to the water might change.

Deep down I knew we would overcome all these challenges, even though it felt like they were overwhelming us right then. I knew there was a story unfolding in our lives, but I couldn't change the fact that I also felt hurt by the loss we were experiencing, felt joy that Jameson was alive, and felt terrified. All of these emotions, often at the same time.

JAMESON SR.

We were all feeling that mishmash of emotions. But that didn't change what had to be done from day to day, and it didn't change that we still had goals and things we were working toward, including getting back to the reef. Other people wanted to go with us too. We wanted to do it in community, all together.

So ten weeks after the attack, despite our exciting encounter at Marvin Key a few weeks before, Jameson was still committed to going back to the reef. He wanted to overcome his fear and get back into the same water. We made a big production out of it.

All six of us from my family were present, plus my brother from North Carolina and my brother from Maui and their older children. The rest of their families had been in town for several weeks while we were in the hospital, and had gone home a couple of weeks before. So getting everyone back together was just too much. The sisters-in-law and some of the younger cousins weren't able to make it back again with such a quick turnaround. But accompanying us were both our parents; Mary's sister and brother-in-law, Danielle and Anthony; Zack and his daughter and his dad; the guys from Fish and Wildlife, Lieutenant Bulger and John Hettel; Mike and Tammy Nolan, who owned the Sugar Shack; Sarah and Mike and their two girls; Marc and Holly and their children; nurse Jennifer; paramedic David, and a videographer. There were about forty of us altogether.

We had five boats, and the plan was to go to Looe Key Reef and get into the water together. But the weather wasn't cooperating. It was during an unpredictable weather season when getting out to the reef had to be seriously evaluated day by day, even hour by hour. Ultimately, we realized we weren't going to be able to get out there.

But we could get close. We could get into the same waters, just not at the exact place we'd been. So we did that.

We all met up at a home on the water. It felt like a reunion. There was so much excitement, but we were there for a specific reason: to support Jameson. He'd been looking forward to this since four days after the attack. All the boats went out, like a parade. We played music, smiled a lot, and prayed a lot. There were big emotions all around. When we reached a safe place, we roped the boats together. The FWC guys were already out there, waiting for us.

We made a playlist for the big day, and we listened to a few songs and even sang along, including "Here I Am to Worship," the song Jameson sang on the trip from this place to Dolphin Marina. And we prayed together. We shared a few words, and we talked about what we were there to do. It was like a church service on the water, completely made up of people who loved Jameson and us and were there to support and cheer Jameson on.

I asked Jameson, "How are you feeling?"

"I'm scared, Dad."

No surprise. I think we might have all been feeling a little scared. At the very least, we were a little nervous.

"You don't have to jump in," I said. "We had an awesome ride out here. We sang. We remembered what happened. That's enough. You don't have to do anything."

"No, Dad. I want to jump in!"

I was so proud and moved in that moment. Jameson acknowledged his fear and still jumped in. The fear didn't go away; it was still present. That's something we talk about in our family, that the absence of fear isn't what we're expecting, just a reminder that God is with us in uniquely difficult moments. We can be afraid and also trust God at the same time.

The common thought is *I'll trust God and all the fear will go away,* but it doesn't always work like that. Sometimes we're still afraid, but we choose to jump in despite that.

All the men got into the water first. We made a circle for Jameson and invited him to jump into the center. He was standing on one leg, looking down at the center of our circle. Then he took a deep breath, bent his knee...and jumped!

Once he was in, it was as if all his love for the ocean came slamming back into his body. We couldn't get him out of the water. He was laughing and swimming. It was like his old self was back. Slowly, one by one, everyone else started jumping in the water too. All of us got over our fears and nervousness and what-ifs. Jameson was having such a good time, it was hard for it not to be contagious. We all acknowledged our fear and followed his lead.

MARY

I was extremely nervous that day. I had anxiety and didn't want to get back in the water. Just being on a boat was extremely nerve-racking. And not just because of the sharks I imagined. I was worried about all these things I had never worried about before. What if the boat sinks? What if the boat sinks and we're stranded? I've never once had thoughts like those before. But now, after the unthinkable had already happened to us, it seemed like any and every terrible thing was possible.

I didn't want to let fear control me, but at the same time, everything was frightening. I was going along for the ride. I wasn't about to let my children out of my sight on the water. But I wasn't planning to get into the ocean. I was prepared to get in but planning not to. I had my swimsuit on, I'd brought a towel, but I wasn't thinking either would get any use.

Then Noah told me that he was afraid to get in. He didn't have to, of course. None of us had to. And we weren't going to force anyone—even Jameson!—to get in.

Noah said, "If you go, I'll go," and that decided it for me.

He, true to his word, took my hand, and we jumped in with everyone else.

When I first got in, I tried to keep an eye on my kids every second. It was a lot less scary with so many of us in the water, but at first, all I could think about was that I had to protect our children. Then I saw that Jameson was good. And not only good, but laughing, with a giant smile on his face. It was a very joyous occasion, especially when we had all these people who knew us and loved us, alongside us in the water.

It was a very happy occasion, and all the people who came

down to visit saw that it was a big deal. And dinner that night was a big deal.

We had to pull Jameson out. He didn't want to come out of the water, but we had planned a big dinner with all the people who had come down for the return to the reef. We had been there for two hours, and we had to practically throw a net over Jameson to get him out.

JAMESON SR.

It was at the one-year anniversary when we actually made it back to Looe Key Reef. One year since the attack.

We had tried a few other times to get out there, and every time the weather didn't cooperate. One day we were headed out—just my family and my mom and dad—and we called Lieutenant Dodd Bulger with Fish and Wildlife, but they couldn't meet us because a plane had just gone down in the Atlantic. They were on a rescue mission! We thought about going out ourselves, but the weather was just too harsh. My parents had to leave the next day, so we were all pretty bummed they couldn't make it with us for the anniversary trip.

When we finally made it to Looe Key Reef, there were only a few of us, and Lieutenant Bulger and his crew went out there ahead of us. There were a total of four boats. The six of us were in one boat with some of our friends. Lieutenant Bulger and his FWC team were in their boat. Anneliese and Jennifer and her husband, Jason, were in the third boat. More friends were in the fourth.

As we headed out to the reef, Lieutenant Bulger called and said, "We've seen a couple of big bull sharks out here. Wanted you to know before you come out."

We were already on the way out, so I said, "Wow, that's crazy. Thanks for letting us know, but we'll come out and see how we feel about it."

Once we got there, we saw the water wasn't as clear as it had been a year ago. Visibility was very low. And it didn't have quite the same celebratory feeling as months ago when forty people jumped into the water. I asked Jameson how he was doing, and he said he was pretty scared, but he wanted to get in the water.

"When you can see farther in the water, it just feels better. But yeah, Dad, I want to get in."

We did see some sharks when we got in the water. Four of them at different times. That definitely made it more frightening, more intense. But Jameson still jumped in.

MARY

All those feelings I had months before—the worry, the fear— all those feelings were gone. I was worried I'd be anxious again when I got to the water, but I wasn't. Even getting on the boat felt fine. Even swimming. The time that had passed helped, I think. And the fact that we'd done it before. I just enjoyed the day.

JAMESON SR.

There were so many times that things didn't work out and we had to change our plans. I think sometimes we get a little friction and say, "We have to abandon our plans! I guess we shouldn't do this." But we really believe if there's something we want to do or we're called to do, then we have to keep going, we have to push through. We should be cautious, we should be wise, we should be measured. But we have to be bold.

Looking back at the year, there were so many little bits of the story we didn't know at first. So many other people who had seen something, been a part of something that we hadn't known or noticed in the moment.

MARY

For instance, that very first day of the attack, Jameson Sr. asked Noah to take Eliana and Nehemiah to the bow while he was driving back to Dolphin Marina. We learned later that when Jameson had driven our boat over to Anneliese's boat to tell them about the shark attack, Anneliese and her sister-in-law had heard Noah and the kids singing "Jesus Loves Me." Later, at the dock, when Jennifer went to check on the kids, she and Anneliese and a couple of Jennifer's relatives huddled with the kids and prayed and sang "Jesus Loves Me." What a sweet moment. Even during that horrible time, the kids knew to turn to God, remembered that He is in control, and that no matter what was happening, God loved them.

So many people were amazed by Junior and his story and wanted to keep checking in on him. The EMT who jumped into the boat to get Jameson to the ambulance, David, said he knew Jameson was special. Even in the moment when we thought he might still lose his life, Jameson was asking David about the cross necklace he was wearing and asking about his relationship to God. David told us more than once, "Jameson's calling in the world is bigger than he knows. He touched my life. I'm never going to forget him. I've never been this impacted by a patient in my entire career." He visited us in the hospital many times and invited us out for a tour of the firehouse. He also came and joined us for our return to the reef.

JAMESON SR.

We also talked with Kyle, a guy who had been on Todd's boat, the boat that had taken Jameson to shore. It was a month or so after the attack, and they were all at dinner together and called to check in on Jameson. I told them Jameson was doing great, all things considered.

Kyle then told me that when they were running the boat in toward Dolphin Marina, and Jennifer was taking care of Jameson's leg, Jameson kept saying something that they couldn't make out. So Kyle leaned in close, and Jameson said, "My leg hurts."

Kyle said, "We're almost there and we'll get you some pain meds!"

Then Jameson said, "It's okay. I see the light."

Kyle said his heart climbed into his throat and he said to himself, "Oh, kid, don't look at the light!" Jameson was white as a sheet, and Kyle thought for sure he was on his way out.

Then Jameson said, "It's okay. I see Jesus."

Kyle got choked up as he was saying this. It took him a minute, and then he said, "I can't explain this. There's no good explanation for this, but when he said, 'I can see Jesus,' all of his color came back to his whole body."

I was overcome with emotion and understood what this moment was about. There was a good explanation, a great explanation for what had happened there, but it just wasn't a medical explanation. The only explanation was, when you see the Lord, He changes everything.

JAMESON SR. AND MARY

The day of the attack, we had no idea what was coming. We had returned to the beauty of the familiar waters of Looe Key Reef as a family. The ocean is big and wide, and we don't always know what's in it. Our experience had been over and over that the ocean was powerful, not that it's safe, but something to be revered and respected, something healing and something beautiful. To have it become a place of violence and harm and tragedy wasn't what we were expecting. The shark attack really caught us off guard.

We know that you have had or are going to have huge, unexpected moments of tragedy. Who doesn't these days? Whether it's the loss of a loved one, a global pandemic, an accident, an attack, the loss of a job, an emergency at work, a broken relationship.

We want to tell you that what Jameson has gone through since the attack, what we have all gone through, has reminded us of certain things. One, we are so thankful for what we have even in the midst of loss. And we are so thankful for the many ways that God reminds us in the midst of these things that He is at work and doing something amazing. And Jameson has reminded us over and over that we can be in pain or scared or angry or healing but also be kind and hopeful and even cheerful and friendly and loving as we navigate really horrific things.

We believe God positions His people perfectly. The night before the attack, Anneliese had put a new marine-grade first aid kit on her boat, complete with ACE bandages and special ice packs. Her best friend Jennifer, a nurse assistant had been on the boat with her. They'd found their buoy in Looe Key Reef two minutes before we'd arrived. Another boat, a faster boat, was moored nearby and had a captain ready to take us to shore. There were paramedics, doctors, helicopter pilots, friends and family, and so many strangers who went out of their way to rescue and keep our son safe. God positioned all of these people like precious angels. These folks *were* angels. They were all God's agents for our family.

In this journey there have been people learning to hope and be courageous from Jameson's story. If you have a shark in your life, a challenge, a wound, a storm, and you need help to overcome that fear, we encourage you to hold on to your loved ones and trust that God will hold on to you. Look up and look out and see God and the people around you.

If you're afraid, the God we know brings comfort and courage. If you're in need, our God describes Himself as a loving father. He loves you and wants to be a part of your life.

The future is like the ocean. It's big and wide and we don't know what's in it. But we know there are beautiful things and moments of healing and fun and community and probably some dangers and difficult times to come too.

But that doesn't mean we need to be afraid of getting into the water.

Even though the shark attack is behind us, we know there are a lot more challenges coming our way. More surgeries, more emotional fallout, more processing of that one afternoon and everything that's happened since. And who knows what other disruptions await us? But this is a way that all of us as human beings are the same. We can't prepare for the unexpected any more than we can see the future. But we can build our lives in the present so that when the unexpected comes, we can be ready to respond, rebuild, and realign ourselves to our new reality. One of the ways we do that is through building family, faith, and communities that are strong and can stand any storm that comes our way. Remember, the world isn't falling apart; it's falling into place just like the Lord said it will in scripture

Our family and others started saying let's be "Brave like Jameson." Sometimes that's the reminder we need. Our son Jameson could do these incredible things because of his courage and the love and support of his Savior and his family. We can do amazing things too and so can you.

Let's take a deep breath and grab each other's hand and dive in!

Epilogue

As we write these words, it's almost exactly two years since the shark attack, and in many ways we're still recovering. On the anniversary of the attack, Jameson is going in for another surgery because his bones are still growing... He'll have to keep getting his bones shortened occasionally until he stops growing, and he's having a lot of nerve pain too. He's going to be in the PICU for at least a week, and we'll all stay nearby for a couple of weeks until the doctors say we're good to go home. He'll continue to need revisions to his muscles, tissues, and bone until he stops growing.

In typical Jameson style, he's looking forward to the surgery because he wants to see all of his "hospital friends," the doctors and nurses and security guards that he befriended during our time in and out of the hospital two years ago.

In some ways we still have those moments where we see Jameson acting like he was born to this life... the way he runs around and flings himself over the couch, and how he doesn't usually let his leg slow him down much.

But there are frustrations too. His leg hurts more right now than it did five months ago, and he's going to need new prosthetics and more doctors' appointments in the years to come.

In the midst of all that, we're reminded nearly every day of the fact that he's still with us is a miracle. We are so thankful to God for keeping him alive and to the many people who moved fast and worked hard to keep him that way.

We're still learning from Jameson.

We put off this surgery because his baseball team made it to the national tournament in their division, which meant a trip to Cooperstown Dreams Park and the home of the Baseball Hall of Fame. Even though he was in pretty major pain, Jameson didn't want to pass up the chance to go to nationals with his team, and asked if he could postpone his surgery.

And he had an amazing time. The boys all stay in dorm-style housing during the tournament, so he was running around with all of his friends the whole time, and of course it's this historical baseball location and just a really exciting and fun event.

Like everywhere we go, everyone knew Jameson. There aren't a lot of kids with a prosthetic leg playing in this league (just Jameson so far as we know), and he has that way of showing care and interest in everyone around him.

During one of the games, Jameson was up to bat, and the pitcher on the other team pitched a fastball that was just a little off, and because of his leg, Jameson couldn't get out of the way fast enough. The ball hit him in the helmet and then the chest, and he went down hard. He was crying. The *pitcher* was crying. Everyone was upset and it took a minute to shake it off and get going again in the game.

Jameson ended up making a spectacular catch a few minutes later in the game, the kind of moment where you couldn't believe this was a guy with a prosthetic: he was running backward close to the back wall, and he reached up and caught the ball. He fell, and the whole crowd was leaning forward to see how he was.

Then Jameson held up his glove and showed that he had the ball. Everyone went wild cheering in the stands. It was a key moment in the game.

But the really amazing thing, the really sweet thing, was that later in the day, Jameson and this young pitcher crossed paths and you might expect some of that bluster and competitive spirit that sometimes comes between rival teams. Instead, it turns out that the pitcher's whole team had been looking for Jameson. Once they found him, they hurried over and circled around him, and the pitcher asked Jameson, "Are you okay?" and Jameson told him he was fine and asked the same question back, "Are you okay?" and in that moment they both made a new friend.

These are the kinds of lessons we come across every day. We love this idea that if we could just be paying attention to each other, recognizing how these unexpected crises come along and looking out for each other's needs, maybe we all would be better cared for and have more friends.

We are thankful for God's presence in our lives throughout all of these difficult moments, and we're thankful for the many friends He has put in our path along the way. We are looking forward to many more stories of God's provision in the years to come!

A final thought Jameson Jr. wants to leave you with: "I want to tell everybody who's going through something hard...*you've got this*! Face your fears, and I promise you, it will change your life."

Join the movement by
following our continued
story and sharing your story at
BraveLikeJameson.com.

Spiritual Reflections

When writing this book, we wanted to make sure that people from any faith or background could read Jameson's amazing story. While we were up front about our faith, we didn't want anyone to be turned off by long sections of spiritual thoughts. On the following pages are a few spiritual topics we've been thinking a lot about for those who are interested in reading more. Thank you so much for reading about Jameson and our family.

Stewarding Your Story

MARY

Jameson Jr. has been invited to tell his story to so many people. His baseball team. So many media outlets. All of his friends want to know what happened. Not too long ago we heard him embellish the story a little. Who wouldn't be tempted to do that? It wasn't a major change, just something to make it a little more dramatic, a little more exciting. But it wasn't anything he had said before. It wasn't something he had said the many times we told the story to each other in the hospital or afterward. We pulled him aside and told him the story he has is already perfect and it's true. "Let's protect the story God has given you," we said. "There's no need to make it bigger or different than what really happened. It's already so dramatic, beautiful, and powerful."

Jameson's story is already amazing. I keep remembering how we were so far outside our abilities when Jameson was hurt. If it had been just us out there, Jameson almost certainly wouldn't have survived. But God provided a faster boat and people who were ready to speed him to shore. God provided a nurse at the reef, ready to jump into the water and get to work. God kept us from beaching on the reef when Senior sped toward the other boats. God provided amazing paramedics and a helicopter and doctors. And God was there, too, keeping our son alive, holding our boy in His arms until we could get him safely to Nicklaus Children's Hospital.

And we have Jameson, who has been such a sweet and beautiful and brave presence in the midst of it. We're amazed by the courage he's shown and by how kind and generous he has been to people along the way. Just that piece of it is unbelievable to most people. They've never had a kid ask them if they'd like to pray with him before he goes into surgery. They've never had a kid ask to give gifts to all the other children in the hospital. People see his love and faith and his deep care for other people, and they ask themselves, "What is happening here? Who is this child?" It's as rare and remarkable as a shark attack. People are as amazed by Jameson as they are by his story.

We are so, so proud of him and deeply thankful for him.

JAMESON SR.

As Jameson's story spread, churches started inviting us to come share about it, especially in South Florida. When we were still in Florida, waiting for more parts for Jameson's first prosthetic—children's prosthetic parts are hard to come by; they're not as common as adult parts—we were invited to come to Nest Church, Miami.

The night before we spoke at Nest, Eliana—who was six at the time—was sharing her heart with us about all that had happened in the last several weeks, and we all felt that God was very close to us in this conversation. I could tell it was her moment to put her faith, hope, and trust in Jesus as her Savior.

I asked Jameson and Noah to share the gospel message, how we are sinners in need of a Savior and how Jesus is the only one mighty enough to save us from our sin. Then we led our darling daughter to profess her saving faith in Jesus.

What a powerful and precious night that was. The whole family rejoiced that Jesus saves even in the midst of difficult and disruptive times. This made the next morning at Nest Church extra special. At the end of the service, the senior pastor, Rigo Figueredo, prayed over our family. It was a beautiful moment as he shared many words about Jameson Jr. and what he could see God doing through him in the future. I won't share all those things here. But if you reach out to us, we would love to tell you more!

MARY

We didn't choose this story, but God chose us to steward this story. We believe this story isn't just for us, but that God wants to show His love for other people through us and through our story too. We don't want to hide it, change it, or be ashamed to tell it just like it happened.

JAMESON SR.

I flew back to South Florida to speak at a church, the same church I spoke at the week before the shark attack. It was just me, not Mary or the kids. I walked through the terminal toward baggage claim, and I saw a guy with a big beard, a Cuban guy, standing at a travel kiosk. He was selling travel items, like phone chargers and neck pillows.

And the Lord clearly said to me, "Go talk to that guy."

Now, this doesn't happen all the time. I'm not saying it was a voice I could hear, but a really strong communication from God that I couldn't ignore.

I was in a hurry, and the person picking me up was waiting for me to go to lunch with him. I was hustling and needed to get to my ride.

I'd been thinking about this idea of how we'd been entrusted with this story, that we are stewards of this story, and I felt God just asked me, "How can I trust you with this story? How can you be a steward of it if you won't talk to the people I tell you to talk to?"

I felt weird. I didn't want to do it. I said to God, "You want me to tell this guy our story? Are you sure about that?"

I walked up to the man, and the whole thing felt so awkward. I said, "I know this might seem a little peculiar to you, and it is for me too. My name is Jameson and God told me to tell you a little bit of our story."

He told me his name was Eduardo.

I started to tell him our story: "God saved my son's life a year ago. He was attacked by a rogue bull shark, and without a miracle he would have died."

I started laying it all out, how all these unimaginable things happened that saved Jameson's life and so many of the things that are in this book.

There we were, on a Saturday afternoon in Fort Lauderdale, and I was telling a stranger our shark attack story as quickly as I could. I thought he might decide I was a little crazy and wonder why I was talking to him.

But that's not at all what happened. He put his head down and started to cry. I didn't know what was going on, but I started to cry too. A minute went by and he was still crying. He hadn't even looked up. I put my hand on his shoulder to let him know I was still there. I could tell our story had really spoken to him, but I wasn't sure how.

Finally, he looked up and said, "Just this morning I prayed and I said, 'God, if you're real, you have to show me.'"

Jameson's story, our story, had assured Eduardo that God is real and God was listening to him. That's what we keep reminding ourselves: The stories God gives us may not be the stories we would choose for ourselves, but when they're handed to us, we become God's stewards for those stories. And this story we've been given is a story of real hope, not hype, and a story about light when we're worried there's only darkness.

Eduardo and I exchanged phone numbers and took a picture together. God spoke to him through our story, and God spoke to me through gifting us with that conversation.

As I headed out from the airport, I realized, when God calls us to go public with a story, we can't keep it private. Even when it seems strange or weird, we don't have to be afraid to take our story from personal and private to public.

And what's really beautiful is that we don't have to embellish or exaggerate our story, or worry about how or when to share it, or even how to get an audience. We can trust that the story God has given us is enough, and trust that God will open doors for us to tell it. The apostle John wrote, "That which was from the beginning, which we have heard, which we have seen with our eyes, which we have looked at and our hands have touched—

this we proclaim concerning the Word of life" (I John 1:1 NIV). Our job is—just like the earliest followers of Jesus—to say what we have seen with our eyes and heard with our ears. We've seen God moving in our lives, and that is story enough for anyone. Our responsibility is to be good stewards of this story.

Grief and Comfort When You Don't Know What to Say

JAMESON SR.

From the very beginning, even in the first hours after the shark attack, something that became very evident to us is that people didn't know what to say. Whenever we talked to people at the hospital, texted people to tell them what was going on, over and over, they would say, "I'm so sorry this happened to you." It's the easiest thing to say. Maybe the only thing to say. It was such a bizarre situation that it was hard to know how to respond. People tried to apologize for there being a rogue shark, which is funny if you think about it. At the same time, people felt awkward saying nothing. They love you and want to communicate something appropriate and necessary, and the default is to apologize for the situation.

Don't get me wrong. I'm not complaining. Our friends and family have been so kind, so generous, so caring throughout this time. Even strangers. Jameson wouldn't be here if it weren't for the many kindhearted and helpful strangers that day on the reef. But we kept hearing over and over, "I'm sorry this happened to you."

I have several Jewish friends, and one of the things I've learned from them is the idea of sitting shiva. The idea is that when you lose someone close to you—a first-degree family member—a whole week is set aside for the family to grieve. Friends and family come to the house, feed and care for the grieving, and sit with them. The idea is to be present, not to have the right words to say. By being together, we are reminded of connections that come with community during a time of loss.

We don't have a tradition like that in the Christian faith, and sometimes that results in people searching for something to say and sometimes coming up with something meaningless or even hurtful. Yet those words have really beautiful and well-meaning intention. In my time as a pastor, I had to learn about this, how to say things that are helpful or meaningful in difficult times.

What's the best way to grieve with people? It's really difficult, so I do my best to give grace when people say silly things or try to create meaning in ways that aren't that helpful. For instance, I've heard people say to parents who have lost a child, "Heaven needed another angel." That's not comforting for the parents. It's someone trying to find sufficient words for an impossible situation when sometimes there are no words.

On the other hand, while we might not have a tradition like our Jewish friends who spend a week to sit shiva, our friends and family did something much like it. They came to the hospital, made sure we had what we needed, paid our bills, sent us gift cards, had food delivered to the hospital, and prayed for us. We can do things to care for people even when we're not sure what to say.

MARY

It didn't bother me when people said awkward things in response to what happened to Jameson. I understood where they were coming from and what they meant. We've all lost someone and tried giving comfort that seems inadequate.

I just love the person. I do my best to be encouraging or bring them a meal. There were many people doing that for us and with us after the attack, and they're still doing it today.

I keep reminding myself that God is in control. I have faith that He knows what is happening. We have struggles every day. All of us do. In the midst of the chaos and craziness, there's this underlying picture that God is in charge. We lean on Him, have faith, and pray that we're doing what He wants, that we're walking in His will, that we're following Him on the right path.

When I'm talking to someone who's grieving, any kind of grief, what I'm asking myself is how can I show understanding? How can I bring peace to them? How can I make sure I'm open to God's guidance? When I can, I share my own experiences. Not only my life experiences, but also my experience of faith.

"Have faith" is one of my life mottos. It's like a tagline on my life. I've even had it in my email signature. It's a way that I remind myself who is in charge and who I should be looking to for help. It's a reminder that He is stronger and more powerful than anything I'm facing in life.

Keeping an "Up and Out" Focus During Hard Times

JAMESON SR.

When I was frequently asked, "What happened?" I realized the question made me look down and inward. Questions like, "Why did this happen?" and "How did this happen?" were pushing me toward negativity. It was as if I was saying, "Look at me! This terrible thing happened!" I understood that people needed to say, "We're sorry this happened to you," and I needed to be okay with how people expressed themselves.

Four weeks after the attack, I was praying—I keep a journal and I remember this so clearly—and I heard the Lord say, "This is not what has happened to you, but what I am going to do through you. Look up and look out." This wasn't an audible voice. I didn't have a vision or hear anything with my ears. But I knew immediately these words were straight from God.

"This is not what has happened to you, but what I am going to do through you. Look up and look out."

What I understood "look up and look out" to mean in that moment was that we needed to be prepared. We needed to be ready for whatever God was going to do next. And I thought immediately of Psalm 121:1: "I lift up my eyes to the mountains—where does my help come from?" (NIV). I needed that reminder to look to God, the source of our help. In that moment I felt things shift. Instead of asking why did this happen to me or what's wrong with the world, I was able to connect with the Lord. It moved me away from these isolating thoughts and toward a deeper connection with God.

When I am looking up and out, there's a compelling call from the Lord. Genesis 50:20 says that what was intended for evil, God intended for good. Sometimes a terrible thing in our lives that we think could cause us harm is instead used by God to do something good. Not to say the shark was evil or had bad intentions... It was just doing the thing that came natural to it. But we could look at this harmful thing happening in our life and say, "God can take even this terrible thing and bring something good out of it."

It reminded me of what we were doing just before the attack, the whole reason we were in Florida to begin with. I had been going there for work, for our ministry, and for the RISE campaign. Through RISE, which stands for a Rich Investment in a Soul's Eternity, we were going to expand our ministry to thirty major cities and help build churches that are really deeply committed to one another and their communities, because we wholeheartedly believed that would make a difference in the lives of people who come across it. We know this work is of eternal consequence; therefore our great conviction is that it will make a difference in the lives of those we get to minister to. We had been working in this direction for a long time, and I realized there was a danger that this unprecedented and horrific moment for our family could upend this whole ministry expansion we were working on—if we let it.

After all, there was nothing I could have done to stop the shark. No way I could have known that would happen. No way to prepare for it. But God knew. God could have intervened. But He didn't. Does this mean—this was my fear—the Lord had abandoned me and abandoned Jameson and abandoned our family? This whole event was directly attacking the ministry and vision that we as a family had accepted and were passionate about. We believe that the evil one does not want us to accelerate in this vision.

"This is not what has happened to you, but what I am going to do through you. Look up and look out." Those couple of sentences from God woke me up out of all those fears.

There were two ways to look at this attack. One was to keep focusing on that moment when I was holding Jameson in my arms on that boat and crying out "No!" The other was remembering handing Jameson's bleeding body over to the people in the other boat and hearing God say, "I'll hold him until you get here."

I saw Jameson embracing life after the attack. He wasn't focused on what had happened to him but on what he was going to do next. He was going to get a new leg. He was going to play baseball again. He was going to go back to the reef. And along the way, he kept impacting people's lives. Asking about them, checking on them, encouraging them, and sharing his own vision for where he was going. He cheered them on and kept his eyes on the prize that was at the finish line, not on the loss of his leg. He had every intention to keep running.

And I realized, God could use this story in our lives to bless a lot of people, just as Jameson was doing. We could all be a part of that, because his story is our story too.

MARY

When Jameson shared what he'd discovered in his prayer time, I thought, well, that's brilliant, the idea that this is not happening *to* us, but what God is doing *through* us. It reminded me of when we finally got Junior into that ambulance, and there was a moment when it clicked in my head, and I thought, *He's going to be okay.* I went running back to find Jen. I didn't want to lose touch with her. I needed to get her number, because she was part of what was happening. "You're part of his story." That's what I told her. And I'm sure she thought I was crazy, but in that moment, I was so certain Jameson was going to be fine.

Jameson Sr. said this was our story too, and that really resonated with me. That's also what I thought.

JAMESON SR.

We've been talking about this with the kids at a level they can understand, trying to talk about having a sense of purpose, not just about this event, but in our lives in general. We want them to know that God is working in the world. We talk about how God is looking for people who want to be a part of what He's doing in the world.

God says, "The eyes of the LORD run to and fro throughout the whole earth, to give strong support to those whose heart is blameless toward him" (2 Chron. 16:9 ESV). God is watching. God is looking for people who have set themselves apart to do His work, because He has a story He wants to tell. He's looking for a faithful cast whom He can tell that story through. The Lord is using the story of saving Jameson to tell a story about God's grace and saving power.

MARY

Not when it first happened, but much later—even in the last few months—I've realized that God is asking me to let go of the guilt I feel about all of this. That's hard when horrible things happen in the world and to your family. You're thinking, "I'm their mom! I should have done something different. I should have protected my children."

I feel this deep guilt in the core of me. Replaying that moment over and over, wishing I hadn't let Jameson swim. Wishing I'd been in the water with him. Wishing I had been there or could have gotten to him faster. But none of those things would have helped. And that's the core thing: I've been wrestling with how helpless I've felt. I did my best and it wasn't enough to protect him at that moment.

But what God has been bringing me toward is this idea that we co-parent with God. These children—mine, yours, all of them—they're a gift from God. He's gifted them to us and allowed us to care for them. I wasn't in the water and I couldn't have gotten to Jameson any faster than I did. But God was with him, and God never left him for a second!

I have to lay aside that guilt, because if I'm to blame for what happened, then how much more would God be to blame? I don't believe that. I don't believe God does harm to kids like that. So I need to be able to let my kids go adventuring in the forest sometimes. Be wise, be smart, but also know I cannot control everything. Because obviously I can't. None of us can.

Things could have been so much worse, and I keep thinking, God held more harm back from Jameson. I continue to hear Jameson's screams, yes, but the thing I keep seeing is that moment when I saw Jameson jutting so far out of the water, like someone was holding him in a giant hand. Holding him out of the water until we got to him.

Our Family Secret

JAMESON SR.

As followers of Jesus, we have this secret. In our family we call it the family secret. It's Romans 8:28: "And we know that in all things God works for the good of those who love him, who have been called according to his purpose" (NIV). We believe that God takes everything that happens—good things and terrible things—and reshapes them, reworks them, transforms them to be something good for the people who are working with God.

As a dad, there are times, lots of times, when I look at myself and say I'm doing this wrong. I'm struggling. I feel over my head, not really sure what to do. I get tired sometimes or worn out or distracted or angry. I make mistakes like anyone else. I think any dad who is looking at himself honestly, anyone who's paying attention, thinks this sometimes. But also I look at my kids and what amazing people they are. I look at Mary and the amazing person she is in my life and in my children's lives. And I think, wow, we're doing a lot of things right.

As for those things that aren't the best, the things we aren't doing quite right or the things we can't control—bad actions by other people, wars, crime, disease, shark attacks, or natural disasters—we have to trust that God means it when He says He's with us in those moments. When He says He works all things together for good, when we trust Him, when we let go, when we know He has called us according to His purposes, we must believe that God really and truly means it. And that means we can give the Lord our full devotion. We don't have to spend a lot of time dwelling on difficult situations or circumstances or the past. We can trust He's making something good and inviting us to tell a better story together with Him.

So now it's time for us as a family to move past what has happened *to* us and move toward what God is going to do *through* us. God has given us this amazing story and invited us to tell it, and that's what we're going to do now!

MARY

One thing we've done to try to keep us focused on the good things to come from all of this is that we gave keys to each member of the family on the one-year anniversary. These are physical reminders to keep our focus on God and our vision as a family to bring good things into the lives of other people. We bought each family member a key charm that has a word or phrase on it to connect us back to the story we're telling, to keep our attention turned up and out.

JAMESON SR.

The Florida Keys (we like the idea of connecting back to the Keys) used to be inaccessible unless you had a boat. Then the Flagler East Coast Railroad was built, and later the Overseas Highway made access to the Keys easier, which is how we get there today. The physical reality of the Keys is that they were locked off until we built roads to reach them.

We've talked a lot about keys as a family. Keys are about authority in one sense: A key indicates the key holder has authority to access what is kept safe by the lock that the key opens. Keys give us access. We don't give someone a key to our house unless they've earned our trust, built a relationship with us where we know they are someone who can be trusted to come into our family spaces.

So the keys we gave to our family members represent the keys God has given us. We were invited into an event. We saw something on the day of the attack that was an invitation to see God more clearly. God showed up and did something amazing that day.

Each of our keys is inscribed with something different.

Jameson Jr.'s key has the date on it: *August 13, 2022*. It's a reminder of the miracle that happened that day on the reef. Not the attack but everything after it. God orchestrated a response team of people who kept Jameson alive.

My key has the word *With* inscribed on it. It's a reminder that God is with us. He hasn't abandoned us. He was in the water with Jameson and with me. He was in the boat with Mary and the kids. In the moments when we were separated from each other, God was with Mary and Jameson in the boat, in the ambulance, in the helicopter, in the hospital, and God was with me and the kids in our boat, in the Sugar Shack, and on the drive to Miami. We are not abandoned. We are never alone.

Noah's key is inscribed with the word *Belief*. Belief turns the key of faith. It's belief that ushers us into a relationship with God that gives us access to all the gifts, all the perks of knowing God.

It's the first step toward becoming a part of the family secret: God working everything for the good of those who believe.

Nehemiah's key is inscribed with the word *Faith*, because faith is opened up by belief. Faith is the key that turns the lock in the door of fear. We don't have to be afraid if we trust in the Lord. We don't have to be afraid if we have faith that God is doing something good even in the midst of our darkest days.

Eliana's key is inscribed with the word *Hope*, because hope is always looking into the future, believing that something better is coming. Hope is like faith in that it's about trusting God, but it's specifically about the days and months and years to come, the things that are ahead of us. Hope says, even in the midst of the difficult things of today, I know this won't last forever and something better is coming. The best is yet to come!

I don't think anyone would be surprised to know that Mary's key is inscribed with the word *Love*, because love is the key that keeps us all connected to one another, regardless of the circumstances. Love keeps us connected to God. Love keeps us connected to our family. Love keeps us connected to the people we meet along the way. So much of Jameson's story is about love: the love he showed to people every step of the way, the love we have for him, the love our family and friends showed to us, the love even of strangers, and of course, the deep, deep love of Jesus for Jameson Jr. and for us all. Mary has shown such love to everyone around us and to Jameson and to our family throughout this time.

We told our kids to grab hold of all of these keys: *faith, hope, love,* and *belief* that the One who is with us is greater than anyone or anything in the world. Let's turn the key in the lock of the door of fear and together we can walk through that door and find what's on the other side: freedom!

God's spirit is with us, and that means nothing can stand in our way. Not a shark. Not a giant. Nobody. Nothing. Let's unlock our door and see what God has on the other side of this door for us. We believe as we look up and out, with a clear vision of God and a clear vision of our mission in the world, that God can work through us and we can tell the story He wants told.

Acknowledgments

First, we would like to thank our Lord Jesus Christ for saving our son's life.

We want to thank *everyone* who prayed for us and gave to us in some way—financial support, food, child care, gifts, and your time!

Please forgive us if we do not mention by name each and every one of you who has been a part of our story from the beginning. We truly hold you very close to our hearts and will forever be grateful!

We would like to take this opportunity to thank *all* the first responders that were there on August 13, 2022, and stood ready to help save our son's life: David Milan, Samantha Seco, Alex Amezaga, Jason Pearson, Steve Hudson, Lieutenant Dodd Bulger and John Hettel of the Florida Fish and Wildlife Conservation Commission (FWC), and the helicopter crew that picked us up at St. Peter the Fisherman Catholic Church on Big Pine Key. These we know by name, but many we did not get to meet personally but are so grateful for the role they played in this life-saving event.

To all those out on the water who dropped everything to come to our rescue: Crew on *The Invincible*—Todd, Lisa, and

Tyler Grooms, Kyle and Kristin Kelly; Jennifer (CNA) and Caddy Cadwell and family, and Anneliese Dietrick and family

To *all* the wonderful staff of Nicklaus Children's Hospital (NCH), Dr. Aaron Berger, Dr. Monica Payares-Lizano, Dr. Maggie Wright, Dr. Chad Perlyn, Lizzie Diaz, Karla Filosa and the entire Child Life staff; Amanda Clifton, Yesha Patel, and Samantha Crespo with NCH Physical Therapy; Mia Kalisch; Mark Amador, Mr. Wonderful, and the many, many other doctors, nurses, and staff that we now call family.

Good Morning America, the *Today* Show, all the local news outlets, SharkBanz, Captain Tim Carlile, Captain Jim Sharpe, Donna Hart with 104.1 US1 Radio (the Florida Keys radio station), Positive Behavior Supports Corporation, Lynn Bell and the Square Grouper, Little Smiles, GiveSendGo, Deerfield Beach Island Water Sports and the Gathering Grounds, Catholic Charities, Grant and Trey with Pompano Beach Galuppi's Restaurant, Oakland Park Wings Plus, Bob Babbitt and Challenged Athletes Foundation (CAF), the Florida Marlins Major League Baseball team, the Tampa Bay Rays Major League Baseball team, MLB Network, and Cooperstown Dreams Park, New York.

To our family and friends: Bob and Celia Reeder; Michael and Donna Marchione; Justin, Jennifer, Jack, Josie, Esther, and Sarah Reeder; Joshua, Jacki, Jemma, Jonah, Justus, and Jaylah Reeder; Anthony, Danielle, Luca, Sonny, Micah, and Lino Valletta; Mary Marchione; Dana Marchione; Rod and Linda Sidway; Todd and JoEllen Marchione; Jessica Reeder; Jacob Reeder; Michael and Tammy Nolan; the Mike and Sarah Hajduk family; the Marc and Holly Douglas family; Jackie Anduiza; Dr. Rob and Jen Pacienza family with Coral Ridge Presbyterian Church Fort Lauderdale; Arturo and Catelyn Cervantes; Rebakah Mitchell; Bob and Cathy Meehan; Richard, Claudia, Lenny,

Lorrie, and Jesse Axelrod; Myriam Carrion-Quijano; the Cynthia and Jeremy Cobb family; Jason and Lauren Scruggs Kennedy; the Bill and Jennifer Agganis family; the Chris and Melissa Sosa family; the Zack and Jamie Spurlock family; Shane Farmer; the Tim and Tahra Cessna family; Grant Brooks; Mike Coots; the Adam and Bethany Hamilton Dirks family; the Becky and Noah Hamilton family; the Timmy and Kyah Hamilton family; Kelly Slater; Frank and Michelle Briganti; Shannon Meigh; Jessica Bradley; Dr. Neil Hammerschlag; Dr. Steve Whitaker and Brian Rose of the First Academy Orlando; Carol Lamey; Hadassah Johnston; Jimmy Miller and family; Mark the Shark; Jessie Lanzi; Parker Byrd; the Jeff and Mitzi Byrd family; the David and Roxanne Avakian family; Jim Quick; Dr. Chris Chen and family; Dr. Daniel Chan and family; the J. J. and Tiffany Johnson family; the Brian and Tami Mamo family; Casey Fitzsimons with Stewards of the Game; Pastor Stephen and Bonnie Geiger with Grace Point Church, Fort Lauderdale; Pastor Rich Wilkerson family and Vous Church, Miami; Pastor Rigo Figueredo family and Nest Church, Miami; Pastor Jon Andrickson family and Turning Point Church, Miami; Caleb Hogan; Ed and Linda Rice; Jim and Darla Rice; Captain Jay Lingham; Captain Michael "Mick" Nealy; the Andy and Katie Cook family; Laura Montwaid; Adam Finnieston with Prosthetic Orthotic Designs; Dr. Josh and Becca Buck with re_Health Studio; Matthew Phillips and Hanger Clinic; Matt Glover, Alex Ford, and the team at OrthoFit; Dr. Loeffler and Dr. Gaston with OrthoCarolina; Atrium Health; Melissa Dupree and Kate Huseby with Novant Health Rehabilitation Center, Huntersville.

Those who made our neighborhood "Welcome Home" such a success; Huntersville police and fire departments; all the neighbors and friends who lined the streets with "Welcome home" signs; those who planned the event, decorated our home for Christmas, and provided food: the Richard and Abena Kwarteng family; the Andy and Jessica Schmidt family; the James and Brandi Fonseca family; the Alan and Ellen Morgan family; the Jose and Laura Garcia family; the Gary and Debbie Evering family; the Brandon and Lisa Smith family; the Josh and Lisa Edwards family; the Guy and Marnie Horton family; Susan Lippert; and the Chris and Jillian Hansen family. Thank you for making it the most memorable day for our family!

A special thank-you to Worthy Books and Hachette for believing in our story and publishing our book, along with Wes Yoder, our literary agent, Matt Mikalatos, our ghost writer, and Jenny Baumgartner, our executive editor.

About the Authors

Jameson Reeder is the founder and president of Urban Village, a nonprofit movement designed as a missional arm of the church to reach beyond the walls of the building in major cities of the western world. Jameson speaks about this ministry in churches around the United States.

Mary Catherine Reeder is a Board Certified Assistant Behavior Analyst who coaches parents and other caregivers of children with autism and related disabilities. She helps children with autism to live, work, and play in their communities as independently as possible.

Jameson and Mary live in North Carolina and share in the joy of raising their five children: Jameson Jr., Noah, Eliana, Nehemiah, and Malachi.